AF835022

Introduction to Modern Randomization-Based Design and Analysis for Causal Inference

Design of experiments is, in essence, a disciplined way to learn about cause and effect. Modern experiments can involve a few to millions of units and hundreds or thousands of covariates. These settings demand tools that are flexible, transparent, and faithful to the underlying design in order to reach reliable conclusions about which interventions work and which ones do not. This book provides a modern, accessible, and computationally supported introduction to experimental design grounded firmly in randomization and the formulation of ideas and methods in terms of potential outcomes. Instead of prescribing a model for each design, we begin with the treatment assignment mechanism and link it directly to the observed outcomes through the potential outcomes framework. This formulation illuminates how changing the design changes the analysis, and it naturally distinguishes finite-population inference from super-population modeling. The book also incorporates new developments at the interface of causal inference and experimental design, many stemming from the authors' recent collaborative research efforts.

- Strengthens the link between design and analysis, enabling students to see immediately how the structure of an experiment shapes the exact tools used to analyze it.
- Teaches foundational concepts without assuming linear-model assumptions.
- Equips readers with the tools needed to analyze non-standard and complex experiments, whose randomization mechanisms fall outside the scope of traditional textbooks.
- Support students with limited programming experience by providing algorithms and code throughout the book, enabling them to implement randomization-based methods easily and efficiently.

This book is a textbook for one/two semester course on introductory experimental design.

CHAPMAN & HALL/CRC
Texts in Statistical Science Series

Joseph K. Blitzstein, *Harvard University, USA*
Julian J. Faraway, *University of Bath, UK*
Martin Tanner, *Northwestern University, USA*
Jim Zidek, *University of British Columbia, Canada*

Recently Published Titles

Linear Models with R, Third Edition
Julian J. Faraway

Statistics in Survey Sampling
Jae Kwang Kim

Time Series
A Data Analysis Approach Using R
Robert H. Shumway and David S. Stoffer

Bayesian Statistical Methods
With Applications to Machine Learning
By Brian J. Reich, Sujit K. Ghosh

Experimental Design for Data Science and Engineering
By V. Roshan Joseph

Survival Analysis
Principles and Applications in Clinical Trials and Beyond
Song Yang

Foundations of Bayesian Statistics for Data Scientists
With R and Python
Alan Agresti, Maria Kateri, Ranjini Grove, and Antonietta Mira

A Course in Regression and Smoothing Methods
Zhiqiang Tan

A Course in Large-sample and High-dimensional Theory
Zhiqiang Tan

Markov Chain Monte Carlo
Stochastic Simulation for Bayesian Inference
Dani Gamerman, Hedibert Lopes, and Flávio Bambirra Gonçalves

Introduction to Modern Randomization-Based Design and Analysis for Causal Inference
Tirthankar Dasgupta and Donald B. Rubin

For more information about this series, please visit: https://www.routledge.com/Chapman--HallCRC-Texts-in-Statistical-Science/book-series/CHTEXSTASCI

Introduction to Modern Randomization-Based Design and Analysis for Causal Inference

Tirthankar Dasgupta and Donald B. Rubin

CRC Press
Taylor & Francis Group
Boca Raton London New York

CRC Press is an imprint of the
Taylor & Francis Group, an **informa** business
A CHAPMAN & HALL BOOK

Front Cover Image: Tirthankar Dasgupta and Donald B. Rubin

First edition published 2027
by CRC Press
2385 NW Executive Center Drive, Suite 320, Boca Raton FL 33431

and by CRC Press
4 Park Square, Milton Park, Abingdon, Oxon, OX14 4RN

CRC Press is an imprint of Taylor & Francis Group, LLC

© 2027 Tirthankar Dasgupta and Donald B. Rubin

Reasonable efforts have been made to publish reliable data and information, but the author and publisher cannot assume responsibility for the validity of all materials or the consequences of their use. The authors and publishers have attempted to trace the copyright holders of all material reproduced in this publication and apologize to copyright holders if permission to publish in this form has not been obtained. If any copyright material has not been acknowledged please write and let us know so we may rectify in any future reprint.

Except as permitted under U.S. Copyright Law, no part of this book may be reprinted, reproduced, transmitted, or utilized in any form by any electronic, mechanical, or other means, now known or hereafter invented, including photocopying, microfilming, and recording, or in any information storage or retrieval system, without written permission from the publishers.

For permission to photocopy or use material electronically from this work, access www.copyright.com or contact the Copyright Clearance Center, Inc. (CCC), 222 Rosewood Drive, Danvers, MA 01923, 978-750-8400. For works that are not available on CCC please contact mpkbookspermissions@tandf.co.uk

For Product Safety Concerns and Information please contact our EU representative GPSR@taylorandfrancis.com. Taylor & Francis Verlag GmbH, Kaufingerstraße 24, 80331 München, Germany.

Trademark notice: Product or corporate names may be trademarks or registered trademarks and are used only for identification and explanation without intent to infringe.

ISBN: 978-0-367-50098-6 (hbk)
ISBN: 978-1-041-30818-8 (pbk)
ISBN: 978-1-003-04879-4 (ebk)

DOI: 10.1201/9781003048794

Typeset in CMR10 font
by KnowledgeWorks Global Ltd.

Publisher's Note: This book has been prepared from camera-ready copy provided by the author.

To my parents Late Malay Kumar Dasgupta and Mira Dasgupta who made me who I am today; and to my wife and best friend Koyeli Dasgupta.

Tirthankar Dasgupta

To the families and friends of my youth, my younger days, and my later years, especially to M.A. and our Little Man.

Donald B. Rubin

Contents

Preface

Design of experiments is, in essence, a disciplined way to learn about cause and effect. By deliberately intervening on input variables and observing the consequences on outcome variables, experimenters can illuminate the mechanisms that govern scientific, engineering, biomedical, and social systems. In today's world, this methodology touches nearly every domain—from 3D printing and nanotechnology to stem-cell biology, education research, business analytics, and the rapidly evolving landscape of online platforms and social networks.

For nearly a century, random assignment has been the cornerstone of rigorous experimentation. It is the primary device that allows investigators to disentangle the effects of controlled interventions from the myriad uncontrolled forces that also shape outcomes. Although the use of randomized experiments, in some form, is ancient, randomization-based analysis and a catalog of randomized designs are remarkably recent. Sir R.A. Fisher—the architect of modern statistical thinking—insisted that "the actual and physical conduct of an experiment must govern the statistical procedure of its interpretation." Randomization-based analysis honors this principle directly. By grounding inference in the randomization distribution of a test statistic, it links the conclusions drawn from data to the very mechanism that produced those data. This connection is not only conceptually clean but also remarkably intuitive, especially for students encountering experimental design for the first time.

Despite its conceptual appeal, randomization-based inference did not dominate practice for much of the twentieth century. Monte Carlo methods were computationally burdensome, and enumerating randomization distributions was therefore often infeasible. Linear-model-based analyses—powerful in many physical, engineering, and industrial (PEI) settings—therefore became the dominant pedagogical and practical tools.

The widespread availability of modern computing has profoundly changed the landscape. With today's virtually unlimited computational power and the explosion of large-scale randomized experiments—ranging from multinational collaborations in scientific studies to the massive online A/B tests conducted continually by technology companies—randomization-based approaches to analysis have become not only practical but essential for reaching reliable conclusions. Modern experiments routinely involve millions of units and hundreds or thousands of covariates. These settings demand tools that are flexible, transparent, and faithful to the underlying design in order to reach reliable conclusions about which interventions work and which ones do not.

This textbook emerges from that need. Our goal is to provide a modern, accessible, and computationally supported introduction to experimental design grounded firmly in randomization and the formulation of ideas and methods in terms of potential outcomes. We aim to:

1. Strengthen the link between design and analysis, enabling students to see immediately how the structure of an experiment shapes the exact tools used to analyze it.
2. Teaches foundational concepts in a way that meets the needs of students from both PEI and SBB disciplines, without assuming linear-model assumptions that may not hold in practice.
3. Equip readers with the tools needed to analyze non-standard and complex experiments, whose randomization mechanisms fall outside the scope of traditional textbooks.
4. Clarify the scope of inference, helping students understand not only what an experiment shows but to whom its conclusions apply.
5. Supports students with limited programming experience by providing algorithms and code throughout the book, enabling them to implement randomization-based methods easily and efficiently.

How This Book Differs

Although courses on experimental design are ubiquitous, most existing textbooks are heavily reliant on linear-model theory. Texts often introduce increasingly complex linear models: from randomized blocks and Latin squares to split-plot designs. The algebraic forms associated with such models can appear to be disconnected from the way experiments are actually conducted. This dis-connect becomes even more pronounced for categorical (e.g., binary) responses or when the population of interest is finite.

Our approach is fundamentally different. Instead of prescribing a model for each design, we begin with the treatment assignment mechanism and link it directly to the observed outcomes through the potential outcomes framework. This formulation illuminates how changing the design changes the analysis, and it naturally distinguishes finite-population inference from super-population modeling. Although such ideas are central to modern causal inference (e.g., Imbens and Rubin, 2015), they appear only peripherally—if at all—in most standard design texts, with the notable partial exception of Kempthorne (1952) [46] and Hinkelmann and Kempthorne (2007) [41].

The book also incorporates new developments at the interface of causal inference and experimental design, many stemming from the authors' recent collaborative research efforts. These contemporary perspectives illustrate how classical principles continue to apply and evolve in response to modern scientific challenges.

We hope that this book serves not only as an introduction to the design and analysis of experiments, but also as an invitation to think deeply about causality, randomness, and the logic we exploit to learn from data. In an era of unprecedented computational capability and unprecedented mountains of accessible data, randomization-based thinking provides a rigorous, intuitive, and unifying foundation for scientific discovery.

Acknowledgements

We are deeply indebted to many collaborators, colleagues, and students from whom we have learned much about the topics discussed in this book over the years. These include Guido Imbens, C. F. Jeff Wu, Rahul Mukerjee, Kosuke Imai, Peng Ding, Valeria Espinosa, Jiannan Lu, Avi Feller, Joseph Lee, Arman Sabbaghi, Colin Fogarty, Natesh Pillai, Edo Airoldi, Luke Miratrix, Sam Pimentel, Kari Lock Morgan, Jonathan Hennessy, Matteo Bonvini, Cassandra Wolos Pat-tanayak, Xinran Li, Nicole Pashley, Marie Abéle Bind, Zach Branson, Ying Hung, Arun Ravichandran, Xiaokang Luo, Minge Xie, Regina Liu, Lukian Wernyj, Janine Yanes, Willem Weigand, Arka Sanki and Ayush Baran Sen.

We are also grateful for stimulating discussions over the years with Xiao Li Meng, Carl Morris, Joe Blitzstein, David Harrington, John Stufken, Rosemary Bailey, Angela Dean, Robert Mee, V. Roshan Joseph, David Woods, Samuel Kou, Frederick Phoa.

Portions of the research results described in this book were carried out with support from the National Science Foundation under Grant numbers DMS 1107004 and SES 2217522. Whereas these grants supported the underlying research, they did not directly fund the preparation of this book. Any opinions, findings, and conclusions or recommendations expressed here are those of the authors and do not necessarily reflect the views of the National Science Foundation.

Symbols

Symbol Description

Symbol	Description
N	Total number of experimental units
$Y_i(j)$	Potential outcome of experimental unit i when exposed to level j of a treatment factor.
y_i	Observed outcome for experimental unit i
Y_i^{obs}	Observed outcome for experimental unit i (alternative notation)
Y_i^{mis}	Missing outcome for experimental unit i
W_i	Random treatment assigned to experimental unit i
w_i	Observed treatment assignment for experimental unit i
$\mathbf{y}$	$N \times 1$ vector of observed outcomes, also denoted by $\mathbf{Y}^{\text{obs}}$ in Chapter 8
$\mathbf{Y}^{\text{mis}}$	$N \times 1$ vector of missing outcomes for all units
$\mathbf{W}$	$N \times 1$ treatment assignment vector (random)
$\mathbf{w}$	$N \times 1$ vector of observed treatment assignments
τ_i	Treatment effect for unit i
$\mathbb{E}(X)$	Expectation of a random variable X
$\text{Var}(X)$	Variance of a random variable X
$\text{Cov}(\mathbf{X})$	Covariance matrix of a random vector $\mathbf{X}$
$\mathbf{X}$	Matrix of covariates, typically of dimension $N \times k$
k	Number of covariates associated with study units
D_{E}	Euclidean distance between two vectors
D_{M}	Mahalanobis distance between two vectors
J	Number of treatments being studied
N_j	Number of units allocated to treatment j
B	Number of blocks for experimental units
N_{bj}	Number of units in block b allocated to treatment j
$\Pr(A)$	Probability of an event A
$\text{pr}(\cdot)$	Proportion in Chapter 2, Probability function (mass function or density function) in Chapters 8 and 9

About the Authors

Tirthankar Dasgupta is a Professor of Statistics at Rutgers University, New Jersey. Prior to joining Rutgers University, he served as a faculty member at Harvard University. His primary research interests include experimental design and causal inference. He is a fellow of the American Statistical Association. He has published about 50 peer-reviewed research articles and has served on the editorial boards of several leading journals of statistics including the Journal of the American Statistical Association, Journal of the Royal Statistical Society (Series B) and Statistical Science.

Donald B. Rubin is an Emeritus Professor of Statistics at Harvard University, and is currently affiliated with Temple University. He is most well known for the Rubin causal model, a set of methods designed for causal inference with observational data, and for his methods for dealing with missing data. Professor Rubin is a fellow/member/honorary member of the Woodrow Wilson Society, Guggenheim Memorial Foundation, Alexander von Humboldt Foundation, American Statistical Association, Institute of Mathematical Statistics, International Statistical Institute, American Association for the Advancement of Science, American Academy of Arts and Sciences, European Association of Methodology, the British Academy, and the US National Academy of Sciences. He has authored or co-authored about 450 publications (including 10 books), has four joint patents, and is one of the most highly cited authors in the world, with nearly 450,000 citations.

1

Understanding Experimental Design: Fundamental Concepts

In recent years, there has been an explosion of research on "big data", much of which has focused on methods for finding interesting patterns in data and studying correlations and associations among variables of interest. However, as recently well stated by Shiffrin [86], "Finding patterns of correlations in the data is a necessary first step, but not in itself the goal: one must explain and use the information. Human explanation is almost always couched in terms of causal forces; that is the way we try to understand the infinite complexities of the world we inhabit."

The distinct difference between association and causality has been emphasized by many scholars and researchers [44] for many years. It is well understood that causality can only be accurately assessed by utilizing known or assumed knowledge about how the data were collected, and consequently is far more difficult to establish than association. The importance of properly *designing* randomized experiments to draw objective causal inference is now widely recognized [83]. The design of experiments is a systematic approach to explore causal relationships by deliberately introducing interventions that manipulate input variables. In today's scientific world, such experiments find application in engineering (e.g., manufacturing), biomedicine (e.g., stem-cell technology), physical sciences (e.g., materials science), social sciences (e.g., education), law (e.g., access to justice), business (e.g., product marketing), and computer sciences (e.g., social networks). Examples of such experiments will be briefly discussed in the last section of this chapter.

Even until a few years ago, well-designed studies covering large populations were hard to conduct physically, often leading to the paucity of data. The ubiquitous availability of the internet in every corner of the world and the advent of social media have dramatically changed, even reversed, the situation in the past few years. For instance, now researchers conducting a randomized experiment in New York can collaborate with researchers from all across the world to obtain data from similar experiments, sometimes in minutes. Many such studies have information on hundreds, or even thousands, of background measurements (often called covariates because they vary with outcome variables of interest) from each experimental unit (e.g., each person), generating huge sets of data that can be examined before designing the experiment. Some designed large email experiments are being regularly conducted

DOI: 10.1201/9781003048794-1

by social and behavioral scientists. Also, all technological companies, such as Meta, LinkedIn, Microsoft and Google, have been conducting enormously large randomized experiments online that are commonly referred to as "A/B testing" because they compare two treatment interventions labeled A and B (e.g., Jin and Ba (2023) [45]). Many such experiments typically involve millions of subjects, again with a plethora of covariate information available on each subject.

Parallel to the conduct of such new generation experiments, a world with almost unlimited computational power is allowing experimenters and researchers to use the powerful, intuitive and flexible tool of randomization-based analysis to draw causal inference from experimental data – something that was not possible even a few decades earlier because of past computational limitations. Techniques for the design and analysis of experiments, developed almost a hundred years ago, are therefore resurging across all scientific disciplines in our search to discover what works. This chapter will introduce the key concepts of experimental design, introduce common statistical terminology and provide examples of experiments conducted in various scientific fields.

1.1 Experiments: Basic terms and definitions

A *designed experiment* is one in which one or more manipulable inputs, or *treatment factors* (e.g., type of medicine), are deliberately varied, and different versions are applied to groups of subjects or objects called *experimental units* (e.g., patients) to investigate the treatments' (individual and combined) *causal effects* on outcome variables of interest (e.g., mortality). Each of these experimental units is assigned to a specific *level*, or *level combinations*, of treatment factors using an *assignment mechanism* that is determined by the experimenter. The key difference between a randomized experiment (the formal focus of this book) and an observational study is the fact that in a randomized experiment, the assignment mechanism is randomized and determined by the experimenter, and thus known to the data analyst, whereas in an observational study, the assignment mechanism is at least partially unknown.

Once the experiment is conducted, the outcomes are recorded for the units. The observed outcomes are then analyzed to draw inferences about the causal effects of the treatment factors. Here, the analysis is determined, at least to some extent, by the assignment mechanism, and the analysis is therefore called "randomization-based". In this section, we introduce the common terminology used in traditional experimental design literature in statistics.

1.1.1 Factors, levels, treatments

As mentioned earlier, the inputs or interventions that are deliberately modified to assess their effects on a population of experimental units (e.g., persons) are known as treatment factors. In some studies involving exactly two levels of a treatment factor, the active level (e.g., administering a medicine) is referred to as the *treatment*, whereas the passive level (e.g., administering a placebo) is referred to as the *control* level or simply control. In the rest of this book, we will typically refer to treatment factors, whose levels are modifiable, as "factors" in contrast to typically unmodifiable covariates such as the sex or age of a person.

Treatment factors can often be usefully classified as *qualitative* or *quantitative*. A qualitative factor is one for which the levels are pre-specified. For example, in an experiment conducted to compare three new drugs with respect to their ability to cure a population of patients suffering from a specific medical condition, the factor "type of drug" is qualitative, perhaps with three levels.

A quantitative factor is typically a more continuous quantity, and the experimenter often must determine the levels that are to be used in the experiment. The choice of levels is often guided by prior knowledge or the specific scientific question addressed by the experiment. For example, if the experimenter is interested in exploring a nonlinear, possibly quadratic, relationship between a continuous treatment factor (e.g., temperature in an engineering process) and the outcome of the experiment (e.g., defects in the manufactured products), then choosing at least three treatment levels is an obvious strategy, whereas choosing two levels is not as helpful. In *screening experiments*, which consider several factors, the objective is to identify a subset of factors that are *active* in the sense of increasing or decreasing the outcomes being studied, and then choosing only two levels of each factor can be a reasonable initial strategy.

1.1.2 Experimental units

Experimental units are the objects that are exposed to receive the treatments or levels of factors, and each object yields measured values of the outcome variables of the experiment. Defining the experimental units is crucial to any experiment. In many situations (particularly in industrial experiments), the experimental units may not be as obvious as in the case when the units are people at a particular point of time. The choice of experimental units is closely tied to the scope of inference that can be learned from the experiment. One natural question that an experimenter may have is the extent to which the conclusions drawn from the experiment can be logically extended to situations beyond those in the experiment.

Such a logical extension may be possible when the experimental units of the study can be considered a representative sample drawn from some larger population of units. However, the units in most real-life experiments are rarely randomly chosen samples from a larger population. For example, it is never possible to obtain a representative sample of units from the future, to whom current treatments will be applied. Hence it is important to clearly understand the scope of inferences to be drawn.

An inference that pertains only to the experimental units in the study is typically called a *finite-population inference*, whereas the inference for a larger population is referred to as a *super-population inference*. The scope of this book will by and large be formally restricted to finite-population inference. However, some key concepts necessary to extend such an inference to super-populations will be introduced in the later chapters. Extending inference to larger populations is sometimes informal, whereas a formal extension involves statistical techniques that compare background characteristics of the finite population in the experiment to those characteristics in some target super-population.

1.1.3 Response variables – outcomes

A measured outcome of the experiment is typically referred to a *response*. The correct response may not be obvious in an experiment, and defining the one or several appropriate response is often an important aspect of designing an experiment. For example, in a manufacturing experiment, a binary response variable indicating whether each manufactured product is defective or not may be of interest and may be easy to record. However, a continuous response variable that measures the quality of a product and provides a threshold for the product being defective or not may be more informative, although more expensive to record and store. In a clinical trial, survival of a patient after a specific period of time may be an obvious response variable. However, critical health variables associated with the disease being studied (e.g., results of a blood test) and capable of predicting survival may be more useful, although additional efforts may be required to collect such information (e.g., conducting such blood tests). A response can be binary (e.g., good or bad), nominal (e.g., one of several categories), ordinal (e.g., ranging from poor to excellent) or essentially continuous (e.g., diastolic blood pressure).

1.2 Objects of inference: Potential outcomes and estimands

Before conducting experiments, it is important for the experimenters to have a clear idea about what quantities they are aiming to estimate from the experiment. Such quantities of interest are called *estimands* and are functions of fundamental quantities called *potential outcomes*, which are introduced now.

For simplicity, at this point consider a two-level treatment where 0 and 1 denote two levels of the treatment factor, with 0 representing the control level and 1 the active treatment level. Also, assume for now that we have only one experimental unit. The experimental unit has two potential outcomes $Y_{(0)}$ and $Y(1)$ when exposed to the control and treatment respectively, out of which only one can be observed because they are both defined at the same point of time. A comparison of these two potential outcomes is the causal effect of the treatment for this specific unit, which is well-defined but never is directly observable.

Because a single unit can generate only one observed outcome, no causal inference can be drawn solely from the one observed value. We thus require multiple experimental units. Throughout this book, unless stated otherwise, we will denote the number of units by N. Each of these N experimental units (e.g., individuals participating in the study) can be exposed to either level of the treatment factor. For $i = 1, \dots, N$, two potential outcomes associated with the ith unit will be denoted by $Y_i(0)$ and $Y_i(1)$. The basic object of inference—often referred to as the science—thus is a $N \times 2$ matrix where the rows represent the N units and the columns represent potential outcomes under each possible treatment exposure. Such a representation of the science is adequate under the stable-unit-treatment-value-assumption (SUTVA) as defined by Rubin (1980) [81], which means (1) that the potential outcome of unit i depends only on the treatment combination it is assigned, and NOT on the assignments of the other units, and (2) that there are no hidden versions of treatments not represented by the two levels of the factor.

Often, designing a good experiment involves ensuring the plausibility of SUTVA, so that $\{Y_i(0), Y_i(1)\}$, $i = 1, \dots, N$, represent all values that can be observed for the N units in the experiment. Although cases where SUTVA is violated are beyond the scope of this textbook, readers may benefit from thinking about situations where such violations may occur. See Exercise 1 of this chapter, for example.

The causal effect of Treatment 1 versus Treatment 0 for the ith unit is the comparison of the corresponding potential outcomes for that unit: $Y_i(1)$ versus $Y_i(0)$ (e.g., their difference or their ratio). The "fundamental problem facing inference for causal effects" [79, Rubin(1978)] is that only one of the potential outcomes can ever be observed for each unit. Therefore, unit-level causal effects cannot be known and must be inferred. We will see that this framework permits prediction of unit-level causal effects from either the Neymanian [65, Neyman(1923)] perspective or from the Bayesian perspective [79, Rubin(1978)], although such unit-level estimations are generally imprecise relative to the estimation of population or subpopulation causal effects.

Consider the following hypothetical example from Chapter 1 of [44, Imbens and Rubin (2015)] involving four units representing medical patients, and two possible medical procedures labeled 0 (Drug) and 1 (Surgery). The response is years of post-treatment survival. Assuming SUTVA, the hypothetical potential outcomes of each patient under the two procedures are displayed in Table 1.1.

TABLE 1.1
Potential outcomes (Years of survival) for four patients under drug and surgery

Experimental unit	Potential outcomes $Y_i(0)$	$Y_i(1)$	Causal effect $Y_i(1) - Y_i(0)$
1	1	7	6
2	6	5	−1
3	1	5	4
4	8	7	−1
Average	4	6	2

The table also shows the four unit-level causal effects $Y_i(1)-Y_i(0)$, $i = 1, \ldots, 4$ and the average causal effect $\overline{Y}(1) - \overline{Y}(0) = 6 - 4 = 2$.

Although average causal effects $\overline{Y}(1) - \overline{Y}(0)$ are common estimands in many fields of application, other summaries of unit-level causal effects, e.g. difference of medians of potential outcomes corresponding to treatments 0 and 1, may also be of interest in specific scientific studies. As has been emphasized by Rubin (1974, 2010) on several occasions (e.g., [75, 84]), there is no reason to focus solely on average causal effects, although this quantity is especially easy to estimate using standard statistical tools in randomized experiments under simple assumptions.

Neyman [65, Neyman (1923)] was apparently the first to introduce the formal notation for potential outcomes for randomization-based inference in randomized experiments. The notion was subsequently used by several authors including Kempthorne (1952) [46] and Cox (1958) [19] for causal inference from randomized experiments. The concept was formalized and extended by Rubin (1974, 1975, 1977, 1978) [75, 76, 78, 79] for other forms of causal inference from randomized experiments *and* observational studies, and an exposition of this transition appears in Rubin (1980) [81].

The definition of the potential outcomes, and consequently that of The Science, depends on the number of factors and their levels. For example, suppose an experimenter wishes to investigate the causal effects of two factors, labeled F_1 and F_2, and each factor has two levels: the current (control) levels denoted by 0 and the new (active) ones denoted by 1. Then there are four treatment combinations for (F_1, F_2): $(0,0), (0,1), (1,0)$ and $(1,1)$. Thus, the ith unit has four potential outcomes $Y_i(0,0)$, $Y_i(0,1)$, $Y_i(1,0)$ and $Y_i(1,1)$ for $i = 1, \ldots, N$, only one of which can be observed for each unit. In such a situation, unit-level causal estimands of interest often include *main effects*, which are defined as average effects of factors, the *conditional main effect* of one factor given a fixed level of the other factor, and also the *interaction effect* between the two factors. Roughly speaking, an interaction effect between two factors measures how much the conditional effect of one factor changes when the level of the other factor is varied.

1.3 Assignment mechanism and the three fundamental principles of experimentation

The assignment mechanism is what the experimenter does to enable inference about values for The Science (and in particular, the estimands of interest). Consider a factor with two levels. Critically, every unit can be assigned to *one and only one* level of the factor. Then a possible way to conduct the experiment is to assign one of the levels to each of the N experimental units by tossing a fair coin, for example, to control if a head shows up, and to treatment if a tail appears.

Whereas the simple treatment assignment mechanism just described is a valid assignment mechanism, it may not be a wise strategy to adopt. Under this mechanism, with a small positive probability, all the N units will be assigned to either the active treatment or control, making the experiment useless by itself for estimating the causal effect of the treatment versus control. An alternative strategy is to adopt the following assignment mechanism: at random, assign a pre-determined number of units, say N_0 to control, and the remaining $N - N_0 = N_1$ units to treatment. The total number of randomizations under this mechanism is $\binom{N}{N_0} = N!/(N_1!N_0!)$. See Appendix 1 if not familiar with this "combinatorial" notation. Such a treatment assignment is called a *completely randomized treatment assignment.*

If $N_0 = N_1 = N/2$, the design is said to be *balanced.* Balanced designs have some attractive properties that will be discussed in subsequent chapters.

Randomization is one of the three fundamental principles of experimentation as stated by R. A. Fisher [33], arguably the father of modern statistics, because it (i) prevents the treatment-control comparison from being confounded (inseparable) from effects of unobserved covariates or latent variables, (ii) reduces the possible influence of subjective bias in treatment assignments, and most important, (iii) ensures validity of statistical inference, in senses formally expounded later here.

To illustrate the critical role of the assignment mechanism, and randomization in particular, consider again the simple hypothetical example in Table 1.1 discussed in Chapter 1 of Imbens and Rubin (2015) [44]. From the table, it is clear that on average, surgery (1) is better than drug (0) by two years' life expectancy. In other words, the average causal effect of surgery versus drug is two years for these four individuals. As in Imbens and Rubin (2015) [44], consider the situation where the doctor, through expertise or magic, knows enough about these potential outcomes, and thus assigns each patient to the treatment that is more beneficial to that patient. In this scenario, Patients 1 and 3 will receive surgery, and Patients 2 and 4 will receive the drug treatment. The observed outcomes for patients receiving surgery will be (7,5) with an average survival of 6 years, whereas those for patients receiving the drug will be (6,8), with an average survival of 7 years. Consequently, the average *observed* outcome with surgery will be one year less than the average observed outcome with the drug treatment, leading to the erroneous conclusion that on average, the drug treatment is superior to surgery. As seen in Table 1.1, if surgery and drug were uniformly applied to a population like these four patients, the average survival for patients receiving surgery would be two years greater than those receiving the drug treatment.

This example suggests that drawing causal conclusions simply by looking at the *observed* outcomes under different treatments while ignoring the assignment mechanism is clearly *not* the correct approach. In order to draw valid causal inferences, we must consider *why* some units received one treatment rather than another.

Why is a randomized treatment assignment a better strategy for inference? Suppose in the above example, two out of the four patients are randomly chosen to receive surgery and the remaining two receive the drug treatment. The total number of randomizations is $4!/(2!2!) = 6$. Table 1.2 shows all of these possible assignments.

Such an assignment essentially partitions the set of patients numbered $\{1, 2, 3, 4\}$ into two subsets. Note that assignment number 4 in Table 1.2 is the perfect doctor's assignment described earlier. Thus, it is important to realize that randomization may generate an assignment that leads to an incorrect point estimate of the average causal effect, and in reality, only one of six assignments in Table 1.2 will be observed. Therefore, it is important to make an assessment of the uncertainty associated with the observed point estimate. The standard measure of uncertainty is called the *sampling variance* of the point estimator, which measures how much the point estimator varies across the possible assignments.

TABLE 1.2
All possible assignments under a completely randomized design

Assignment	Units assigned to		Observed outcomes for		Observed
no.	drug	surgery	drug	surgery	average difference
1	(1,2)	(3,4)	(1,6)	(5,7)	2.5
2	(3,4)	(1,2)	(1,8)	(7,5)	1.5
3	(1,3)	(2,4)	(1,1)	(5,7)	5.0
4	(2,4)	(1,3)	(6,8)	(7,5)	−1.0
5	(1,4)	(2,3)	(1,8)	(5,5)	0.5
6	(2,3)	(1,4)	(6,1)	(7,7)	3.5
Average					2.0

A completely randomized assignment implies that each of the six assignments in Table 1.2 are *equally likely*, in the sense that they will occur with the same probability, 1/6. Denoting the observed average outcome under treatment by $\overline{y}(1)$ and that under control by $\overline{y}(0)$, the estimated average treatment effect $\overline{y}(1) - \overline{y}(0)$ is a random variable taking values 2.5,1.5,5.0, −1.0, 0.5, 3.5 (each with a probability of 1/6), as shown in the last column of Table 1.2.

Consequently, the expected value of the estimated average treatment effect with respect to its *randomization distribution* is $(2.5 + 1.5 + 5.0 - 1.0 + 0.5 + 3.5)/6 = 2.0$, which is identical to the true value of the average causal effect. Thus, because of this average correctness, when using a completely randomized assignment, the estimated average treatment effect is called an *unbiased estimator* of the true average treatment effect.

As will be explained in later chapters, in addition to justifying a point estimator of the average causal effect as *unbiased*, randomization will also provide a theoretical justification for an interval estimator that will contain the true effect with a guaranteed probability over the randomization distribution, at least in experiments with a large number of units.

Now suppose, N_m out of N experimental units in a medical study are males and the rest $N_f = N - N_m$, are females. Will a completely randomized treatment assignment still be appropriate? Perhaps not! Assuming N_m and N_f to be even integers, a more reasonable strategy might be to assign half of the male participants to control and half of the female participants to control, and the other halves to treatment. The total number of randomizations under such an assignment mechanism is $\binom{N_m}{N_m/2} \times \binom{N_f}{N_f/2}$, and the strategy of doing distinct randomizations for the two groups of participants is referred to as *blocking*, a second fundamental principle of experimentation.

More generally, the experimental units are divided into relatively homogeneous groups or strata called *blocks*, and units within each block are then randomly assigned to the different treatments. In the words of Fisher (1942) [33]: "Block what you can and randomize what you cannot." When the number of experimental units N is even, an extreme case of blocking may be achieved by creating $N/2$ pairs (blocks) of "similar" (e.g., of the same sex and roughly the same age) individuals, and randomly assigning one unit of each pair to treatment and the other to control. An experimental design with such a treatment assignment mechanism is called a *matched-pair design* or a *paired-comparisons design* and is a special case of the blocked design. However, a notable feature of many modern social, behavioral and biomedical experiments is the availability of multiple pre-measured covariates prior to designing the experiments. How to implement blocking is not obvious with a large number of covariates. We discuss this issue further with an example shortly.

The variation among observed outcomes of units exposed to the same treatment can be attributed to differences among units, and also to manipulable factors that are not controlled during the experiment (e.g., factors like humidity in an industrial experiment that vary over time). This variation is typically referred to as *residual variation* or more commonly (but less correctly in our views) as *"experimental error"*. To estimate the residual variation, it is important to consider assigning multiple units to the same treatment condition - a strategy called *replication*.

Replication is Fisher's third fundamental principle of experimentation, and is natural in many social, behavioral and biomedical experiments. However, time and cost constraints in industrial experiments sometimes compel engineers to conduct unreplicated experiments, the formal valid analysis of which can be more challenging.

1.4 Focus of this book: ONE principle, MANY designs

The focus of this book is on (a) designing *good* randomized experiments and (b) analyzing the resulting data using randomization as a "reasoned basis for inference" [33, Fisher (1942)]. The random assignment of interventions to the objects being studied (i.e., the units) has served for nearly a century as the foundation for experimentation and as the primary tool to distinguish effects of the controlled interventions from those of unknown and uncontrolled variables. Quoting Sir R.A. Fisher, "the vital principle has often been overlooked that the actual and physical conduct of an experiment must govern the statistical procedure of its interpretation". Using randomization-based analysis of experimental designs is the *most natural way to make the analysis of experimental data match the manner in which the experiment is conducted*, and is the most intuitive way of implementing this connection. Thus (a) and (b), mentioned at the beginning of this paragraph, will be closely connected through this principle of randomization-based inference.

1.4.1 Designing a good experiment

Although the "design of an experiment" can include several tasks, such as identification of factors and their levels, determining the number of experimental units and selecting them, etc., in this book we will restrict the definition of the design of an experiment to identification of the most appropriate treatment assignment mechanism, defined in Section 1.3 as the (probabilistic) law that governs the assignment of treatments to experimental units. Thus, unless specified otherwise, the factors, levels, experimental units will be assumed known and fixed. For example, consider a clinical trial in which a new drug will be tested against a placebo on N human beings. The "design" in this case will be restricted to the assignment of these N units to the two treatments. This notion will be formalized in subsequent chapters.

Even with this restricted focus, designing a good experiment is not a straightforward task. The inputs that typically guide the design of an experiment include, but are not limited to: (a) scientific goals of the experiment, (b) number of factors and their levels, (c) available covariate information for each experimental unit, (d) resource constraints affecting the number of available experimental units, (e) practical considerations (e.g., ethics) leading to randomization restrictions, and (f) domain knowledge, that can help identify the most important covariates or the most important anticipated interactions between treatment factors. The treatment assignment mechanism, that determines which experimental units will receive which treatment combinations, will be the primary output of the design phase of the experiment. Failure to utilize inputs (a)-(f) wisely when designing the experiment will generally lead to an inferior experiment.

1.4.2 A unified principle for analysis

Once the design is determined, implemented, and responses are observed, the next step is to analyze the experimental data to infer (learn) the causal effects of the treatment factors. The focus of this book is to present a *unified* and *intuitive* randomization-based approach to perform this analysis, using the output of the design process, i.e., the treatment assignment mechanism, as a crucial input to the analysis process. The randomization-based analysis consists of the following steps:

Procedure 1.1 ***Steps in randomization-based inference:***

Step 1: Hypothesize a model, often referred to as the "null" model, used to anticipate and explain the results of the experiment.

Step 2: Identify a suitable statistic (a function of the observed outcomes, also called the estimator) that estimates the causal effect, the estimand of interest.

Step 3: Calculate the value of this statistic from the observed outcomes. This value is called a point estimate of the causal effect of interest.

Step 4: See whether the data (as summarized by the statistic) is consistent with the model in Step 1, in the sense that the observed value of the statistic could have been plausibly generated if the hypothesized null model were true.

Step 5: If step 4 is answered "yes", then the null model should not be rejected as unacceptable with respect to the statistic (note the awkward double negative in this sentence), which could be replaced by "may be considered as plausible".

Step 6: If not, then it's time to search for a better working model, and not live with the current null model.

To illustrate the above steps, again consider the example discussed in Sections 1.2 and 1.3 involving four units representing the patients, and two possible medical procedures labeled 0 (Drug) and 1 (Surgery). In Step 1, we can hypothesize a model that assumes that the choice of the medical procedure has *no* effect on the outcome of *any* of the four patients. In Step 2, the estimated average treatment effect $\overline{y}(1) - \overline{y}(0)$ can be considered a suitable statistic that estimates the average causal effect $\overline{Y}(1) - \overline{Y}(0)$. Assuming that the observed assignment is the one in the fourth row of Table 1.2 that assigns drug to unit (2,4) and surgery to units (1,3), the estimated treatment effect in Step 2 i $(7+5)/2 - (6+8)/2 = -1$. Step 4 is more complicated, because in reality w will observe only the fourth row of Table 1.2, and consequently, we observe the data shown in Table 1.3 in which "?" denotes the missing observations One has to obtain the randomization distribution of the statistic under th hypothesis in Step 1, and determine whether the observed value is unusua or extreme with respect to that distribution. Details of this procedure will b described in the following chapters.

TABLE 1.3
Observed outcomes (Years of survival) for four patients under drug anc surgery

Experimental unit	Potential outcomes $Y_i(0)$	$Y_i(1)$	Causal effect $Y_i(1) - Y_i(0)$
1	?	7	?
2	6	?	?
3	?	5	?
4	8	?	?
Average	7	6	–1

1.4.3 Three versions of randomization inference

In this book we will discuss three versions of randomization-based inference - Fisherian, Neymanian and Bayesian. All of these versions utilize the six steps outlined in Procedure 1.1, differing primarily with respect to Step 1 and Step 4. Among the three versions, Fisherian inference is most intuitive and will be the major focus of the book. Neymanian inference is more complex because it mostly relies on large sample approximations. However, it has the following three advantages:

(1) It can help Fisherian inference by providing insights and inputs regarding "suitability" of the test statistic to be chosen in step 2 (e.g., unbiasedness as an estimator of a causal estimand).

(2) Issues related to "optimality", such as determining the best possible distribution of the total number of experimental units to the different treatments can be easily addressed by using Neymanian inference.

(3) The approach helps extend the scope of inference to a super population if the finite population units in the experiment can be assumed to constitute a probabilistic sample from such a population.

The Bayesian approach is the most flexible because it includes the Fisherian approach as a special case, permits potential outcomes to be random variables, and allows extension of inference to a real or super-population.

However, it requires background in Bayesian inference, and thus will be discussed in a separate chapter after introducing the Fisherian and Neymanian perspectives.

1.4.4 Illustration: a New York school experiment

A few years ago, the New York Department of Education (NYDE) considered conducting an experiment with five different "incentive programs" to be introduced to high schools "which desperately need performance improvement." The supplemental programs being considered included a quality review, a periodic assessment, inquiry teams, a school-wide performance bonus program and an online resource program. The goal of the experiment was to examine the causal effects of these interventions (and their possible interactions) on the performances of the schools as measured by scores (numbers between 1 and 100) in the schools' progress reports, issued annually near the start of each school year. Prior to designing the experiment, the NYDE also collected data on several covariates that were suspected to be correlated with the school performance: previous year's performance score, total number of students, five different race variables (proportion of white, black, Asian, Native American and Latino students), proportion of female students, enrollment rate and poverty rate in neighborhood. Further information about this experiment can be found in Dasgupta et al. (2015) [20] and Branson et al. (2016) [12].

One factor at two levels, completely randomized design

To illustrate utilization of available information when designing the experiment, consider first a simplified version of this experiment with 80 schools (the geographically separated experimental units) and a single factor: a performance-based bonus scheme for teachers at the school, with two levels: 0 (not implemented) and 1 (implemented). In the absence of any covariate information, the simplest design to study the causal effect of this factor would be to conduct a completely randomized experiment, i.e., implementing the bonus scheme in N_1 schools chosen at random and not implementing it in the remaining $N_0 = 80 - N_1$ schools. A related question is the choice of N_0 and N_1. Intuitively, in the absence of additional information, a *balanced* design with $N_1 = N_0 = 40$ seems natural. The data collected from this experiment can be analyzed by first assuming a null hypothesis of no effect, and then using Procedure 1.1 to check if the observed data contradicts that hypothesis. The difference between average (or median) performance scores of schools exposed to levels 0 and 1 of the factor (or its scaled version) is a natural test statistic.

Randomized block designs

Now consider the availability of covariate information for each school. Assume the simplest case of one covariate, the total number of students. When education experts believe this covariate to be associated with the performance scores of the schools (fewer students per teacher may imply more attention per student), then it should be accounted for in the design. A straightforward way is to construct two blocks defined by schools with total number of students below and above a certain threshold score (e.g., 1000). The schools within each block can then be randomly assigned to treatment and control groups using a completely randomized assignment. Such a design is a randomized block design, discussed in Chapter 3. Similar to the case of completely randomized design, one can draw inference about the causal effect of the treatment using the approach in Procedure 1.1. In this case a suitable statistic may be the average estimated treatment effect within blocks, perhaps weighted by the respective block sizes. Such statistics will be discussed in detail in Chapter 3.

It should be noted that the choice of a randomized block design over a completely randomized design is motivated by the need to prevent *confounding*, i.e., the inseparabaility of the treatment effect from the effect of a covariate. When two effects are inseparable or indistinguishable, they are said to be *confounded.* Blocking is one of the easiest ways of preventing such confounding. However, as we shall see later, blocking may always not be as obvious as it seems in this example.

Matched pair designs

One may argue that treating the total number of students as a dichotomous covariate may be an over-simplification, and the performance of schools with covariate values on the same side of the dichotomy can be vastly different. In that case one can take recourse to an extreme case of blocking by pairing the 80 schools based on their similarity, so that each pair represents "near-identical twins" with respect to size. One school in each pair would then randomly be assigned to treatment and the other to control. This is an example of a randomized block design with 40 blocks (half the number of units), and is commonly called a *matched-pairs* design. The randomization distribution of a suitable statistic will be based on the 2^{40} possible assignments and will be used to infer the causal effect of the treatment following the steps in procedure 1.1. A matched pair design is a special case of a randomized block design where the number of blocks is half the number of experimental units, or, in other words, each block consists of only two units. Matched pair designs are discussed at some length in Chapter 3.

The case of multiple continuous covariates: defining unacceptable assignments (randomized allocations)

The design and the analysis issues become more complicated when we have many covariates, especially if each of them can take many values. This was actually the case for the school experiment described above [12]. In this case, how to create blocks or even matched pairs is not intuitively obvious. How should one create a reasonable design that provides some protection against obtaining treatment and control groups that are unbalanced with respect to one or more covariates?

Note that as we try to protect our design more and more from potential confounding effects of covariates, the total number of treatment assignments reduce from those of a completely randomized design to those of a matched-pair design. The first experiment had $(80)!(40!)^2 \approx 10^{23}$ possible assignments, whereas the second had $\{(40)!(20!)^2\}^2 \approx 2 \times 10^{22}$ (assuming equal block sizes and balanced allocation within each block), and the third has $2^{40} \approx 10^{12}$. Thus, *improving a design* by preventing confounding of treatment and covariate effects essentially entails identifying several possible assignments in a *basic design* (e.g., the completely randomized design) that are deemed unacceptable and eliminating them from consideration. The "acceptability" of a particular treatment assignment depends on the type and perceived prognostic nature of the covariates. Thus, blocked or a matched pair designs can be viewed as designs that disallow certain unacceptable randomized allocations to be used.

This design idea can be generalized to include any precise definition of "undesirable or unacceptable" allocations. This idea and some recently developed methodologies can be found in Chapter 4. Using randomization as the only basis for inference permits analysis of the experimental data following Procedure 1.1.

Single factor to multiple factors

The education experiment actually involved five treatment factors (the five incentive programs), with each factor having two levels : 1 (application) or 0 (non-application). Application of a program means implementing that program whereas non-application means not implementing it. The total number of treatment combinations is $2^5 = 32$. Here each treatment is a combination of the binary factor levels and can be represented by a binary vector of length five, ranging from 00000 (none of the five programs applied) to 11111 (all five programs applied). This type of experiment is called a 2^K factorial experiment, with $K = 5$. Here, each school is an experimental unit, and the response is the annual school performance score. The design and randomization-based analysis of 2^K factorial experiments with $K = 2$ are discussed in Chapter 6, and this discussion is extended to the general case of $K > 2$ in Chapter 7.

If a *full factorial* experiment is conducted, which means all the $2^5 = 32$ treatment combinations are assigned to the N schools involved, then the *Science* is a $N \times 32$ matrix of potential outcomes $\{Y_i(00000), \ldots, Y_i(11111),\ i = 1, \ldots, N\}$. In each row, only one of the 32 potential outcomes will be observed and the other 31 will be missing.

In this case, some care will be required to define the estimands of interest. The most common estimands are average main effects of each of the factors, and average interaction effects between different factors. These estimands will be formally defined in later chapters. Depending on the amount of covariate information available, one can adopt a completely randomized design, a randomized block design or a design with any pre-defined acceptable set of randomizations. The analysis will again be based on the basic idea outlined in Procedure 1.1, whose details change as the specific design changes.

Fractional factorial experiments

a 2^5 factorial experiment has 32 treatment combinations. Clearly, conducting an experiment with 32 treatment combinations involves substantial effort and resources. Running an experiment with a fraction, say half or one-third, of these 32 treatment combinations would be considerably easier from administrative and resource perspectives, e.g., requiring fewer experimental units. Such experiments are called *fractional factorial designs.* However, because "there is no free lunch", the price to be paid for eliminating some treatment combinations from the experiment is that, estimates of certain factorial effects will be indistinguishable from certain others. The key ideas associated with the design and analysis of fractional factorial experiments are discussed in Section 8.1 of Chapter 8.

Complex and restricted randomizations

Consider again a version of the school experiment with two factors – a mid-year quality review by a team of experts, and a bonus scheme for teachers. Each factor has two levels denoted by 1 (application) and 0 (non-application). A completely randomized assignment of the 40 schools to the four treatment combinations $00, 01, 10, 11$ is likely to disperse the schools assigned to level 1 of the first factor (i.e., schools to undergo review) all over the state. However, such a design may be eliminated from the consideration because of travel cost and time to implement. A more practical alternative would be to partition these 40 schools by geographic proximity into four groups called *whole-plots*, where the terminology comes from the agricultural origins of the design. Two of these whole-plots would then be randomly assigned to level 0 and the other two to level 1 of the first factor. The teacher bonus scheme can then be applied to half of the schools chosen randomly within each whole plot. A design with such a randomization scheme is called a *split-plot design.* Formal definitions can be found in Kirk (1982) [48], Cochran and Cox (1957) [17], Box, Hunter and Hunter (2005) [10], Wu and Hamada (2009) [93] and Montgomery (2005) [60]. Although the analysis of such a design is relatively complicated, in principle it can still be conducted using Procedure 1.1. The design and analysis of these experiments are discussed in Section 8.2 of Chapter 8.

1.4.5 Leveraging today's computational power

As a final comment in this introductory chapter, we emphasize that the basic inferential approach adopted in this book is based on the premise that the reader has access to essentially unlimited computing power, so that the design and analysis procedures being recommended rely on adequate machines for their implementation. This was only slightly apparent in this chapter, but will be more apparent in later chapters in both design (where we advocate the consideration of many randomized allocations, before eliminating all but one) and analysis where we rely on possibly complicated analyses in each of many randomized allocations under a variety of null hypotheses.

1.5 Glossary

Experimental units or units: Objects that are exposed to receive the treatments, yielding measured values of the outcome variables of interest.

Finite-population inference: An inference that pertains only to the experimental units in the study.

Response or outcome variable: The measured outcome of an experiment.

Estimands: Quantities of interest that are functions of potential outcomes.

Estimators: Quantities that are functions of observed outcomes and used to draw inference about the estimands.

Assignment mechanism: What the experimenter does to enable inference about values for the estimands.

1.6 Problems

1.1 The owner of a Chinese restaurant wants to improve the taste of fried rice they serve in their lunch boxes (free with a lunch entree) to improve customer satisfaction. He believes that the key ingredient that is crucial to the taste is the brand of soy-sauce, so he wants to find out whether the improvement he is seeking can be achieved by using a more expensive brand of sauce rather than the existing one. If the evidence in favor of superiority of the new brand is not strong enough, then he would rather stick to the existing brand and think about other means to improve the quality. His plan of action is to prepare two large portions of fried rice, one using the existing brand of soy-sauce, and the other using the new brand. Fifty individuals participating in the study will be fed one of these two portions of fried rice.

(i) What is the experimental factor here? Which is its treatment level and which is its control level?

(ii) Who or what constitutes the experimental units?

(iii) What do you think should be an appropriate response or outcome variable and how can it be measured?

(iv) Discuss whether SUTVA is reasonable in this experiment, and discuss the choice of units (e.g., people eating at the same table or each person individually).

(v) What can the experimenter do to avoid the complications arising from possible violations of SUTVA?

1.2 The experimenter in Problem 1 realizes that simply making two large batches of rice for the experiment will not reflect the variation in quality of the fried rice from one batch to another. He therefore asks his chef to prepare ten small portions of fried rice: five with the existing soy-sauce and five with the new soy-sauce. Each portion will be served to one and only one individual.

(i) Does the definition of experimental unit change under the above strategy? If so, clearly describe the experimental units.

(ii) If the five portions with control soy sauce are cooked first, followed by five portions with treatment soy sauce, do you anticipate any problems with such a strategy? Explain your answer. What additional precautions can you take to avoid possible problems?

1.3 Suppose, in the context of the fried rice experiment described in Problems 1 and 2, the owner of the restaurant feels that soy-sauce is not the only determining factor and the following other factors need to be considered when designing the experiment: (i) brand of rice (existing versus new), (ii) the amount of soya sauce per portion (currently 0.5) and (iii) the rice-to-water ratio (currently 1:2).

(a) How many factors are under consideration? Which among these are qualitative and which are quantitative?

(b) For which of these factors are the levels already determined? For each of the other factors, suggest how many levels you should be choosing. Give reasons for your answer and give examples of the levels that can be chosen.

(c) Based on your answer to part (b), write down notation representing potential outcomes assuming that there is no interference among units of the population

1.4 Consider an experiment with one factor with two levels (treatment and control). Suppose 51 units are available for the experiment. In an attempt to maximize balance between the groups, the experimenter proposes two methodologies:

(a) Randomly assign 26 units to treatment and the rest to control.

(b) Choose one unit at random and select treatment by tossing a fair coin, for the remaining 50, randomly select 25 for treatment and 25 for control.

How many possible randomizations are there for each scheme? What is the unit level probability of being assigned to treatment in each case?

1.5 Suppose you want to conduct a randomized experiment to compare two keyboards, A and B, with respect to their typing efficiencies. Your resources consist of (i) a one-page 400-word manuscript and (ii) twenty individuals who are willing to help you implement the experiment by taking part in typing the manuscript. The response variable (Y) is the time taken to type the manuscript. Assume that each individual is willing to spend as much time as it takes to type the manuscript but only once.

(a) What are the experimental units in this experiment?

(b) Write down a natural *estimand* for this problem in terms of potential outcomes. Is this the only possible estimand? If not, provide some alternatives.

(c) Explain how you would assign the 20 individuals to one of the keyboards if:

(i) All the 20 individuals are males of roughly same age with no formal training in typing.

(ii) The 20 individuals comprise 10 males and 10 females. None of the individuals have any formal training in typing.

(iii) The 20 individuals comprise 10 males and 10 females. Although none of them has a formal training in typing, you ask each one of them to take a simple typing pre-test and you record the data (see Table 1.4) on their typing speed measured as words per minute (WPM).

TABLE 1.4
Pre-test typing speeds of 20 individuals

Individual	1	2	3	4	5	6	7	8	9	10
Sex	F	M	F	F	F	M	M	F	M	M
WPM	30	34	21	33	22	22	30	22	20	20
Individual	**11**	**12**	**13**	**14**	**15**	**16**	**17**	**18**	**19**	**20**
Sex	M	M	F	F	M	F	M	F	M	F
WPM	32	30	23	30	22	22	21	17	20	30

(d) Write down how many randomizations are possible in cases (i)-(iii) described above.

2

Completely Randomized Designs with a Single Treatment Factor at Two Levels

2.1 Introduction to completely randomized experiments (CRD)

We first discuss what are called completely randomized experimental designs, and in particular, the simple case with a single experimental factor at two levels. When one of these two levels can be considered active and the other passive, then the former level is typically called the *treatment* and the latter the *control*. We will use a simple agricultural example from Box et al. [10] to illustrate this design. In this example, two fertilizers A (standard) and B (new) are compared with respect to their ability to increase yield (measured as weight in pounds) of tomato plants on 11 adjacent plots of land. In this study, the 11 plots of land constitute the experimental units, and there is one qualitative factor (fertilizer) at two levels (A and B).

Before designing the experiment, the estimands, or causal quantities of interest should be defined and understood, using the potential outcomes framework introduced in Chapter 1. The most commonly used estimand is the difference of average potential outcomes (yields) under fertilizers A and B. The choice of the average, rather than some other measure of central tendency, such as the median, is common, not because it is necessarily more scientifically relevant, but because the associated traditional mathematical results are simpler to derive. Another more formal way to describe the scientific goal of the experiment is to state a hypothesis regarding the differential causal effects of the fertilizer on the yields in the 11 plots. In the absence of any prior information about the fertility of the 11 plots, an obvious way to design the experiment is to use a completely randomized treatment assignment mechanism. Such a mechanism entails: (i) fixing the number of plots to be assigned to each fertilizers in advance (e.g., 5 to A and 6 to B respectively), and then (ii) randomly choosing one assignment among all possible assignments that satisfy (i).

This chapter is organized as follows: In the next section, we define the potential outcomes and the estimands of interest. Then we define the assignment mechanism, observed outcomes, estimators of the estimands and their

DOI: 10.1201/9781003048794-2

statistical properties, using two methods of finite-population inference: Fisherian and Neymanian, both of which rely on the random assignment of treatments to units, but in different ways.

2.2 Examples of two-armed CRD

1. In one part of the experiment reported by [85], 235 undergraduate students from introductory psychology classes at a large mid-southern U.S. public university received either the treatment (vocabulary training) or control (math training) through a completely randomized treatment assignment. The number of students assigned to treatment and control were 116 and 119, respectively. The outcome we focus on is the vocabulary test scores after the experiment. We will refer to this experiment and its hypothetical versions (considered in the future chapters) as the "partial Shadish experiment" throughout this book.

2. The LEAP trial was a randomized, open-label, randomzied trial of infants conducted in the UK in 2009 [29]. To be eligible for enrollment, infants had to be least 4 months and less than 11 months of age and had to have severe eczema, egg allergy, or both. Possible participants were subjected to skin-prick test for peanut allergy. If they tested negative they were randomly assigned to one group (treatment) in which dietary peanut would be consumed or a group (control) in which its consumption would be avoided. The primary outcome was whether the participants developed peanut allergy at 60 months of age, which was determined by means of an oral food challenge where passing the food challenge meant the participant had not developed peanut allergy. Among the 542 infants in the group with a negative result on the initial skin-prick test, 530 could be evaluated for the primary outcome and were included in the analysis. We will refer to this experiment as the "LEAP trial" throughout this book.

2.3 Potential outcomes and estimands

Consider a finite population of N experimental units, each of which can be exposed to either of two treatments labeled 0 and 1. In the tomato experiment $N = 11$, and the units are the plots. Let Y denote the response of interest (e.g., yield of tomatoes on a plot) and let $Y_i(j)$ denote the potential outcome corresponding to unit i when assigned to treatment $j = 0, 1$. Note that under SUTVA, this notation is adequate for any treatment assignment vector with N-components used to represent all values that can be observed. Table 2.1

TABLE 2.1
Potential outcomes for the experimental units and causal effects

Experimental unit	Treatment		Causal effects
	0	1	
1	$Y_1(0)$	$Y_1(1)$	$\tau_1 = Y_1(1) - Y_1(0)$
2	$Y_2(0)$	$Y_2(1)$	$\tau_2 = Y_2(1) - Y_2(0)$
...	...	...	...
N	$Y_N(0)$	$Y_N(1)$	$\tau_N = Y_N(1) - Y_N(0)$
Average	$\overline{Y}(0)$	$\overline{Y}(1)$	$\overline{Y}(1) - \overline{Y}(1)$
Median	$\underset{i}{\mathrm{Med}}[Y_i(0)]$	$\underset{i}{\mathrm{Med}}[Y_i(1)]$	$\underset{i}{\mathrm{Med}}(\tau_i)$
Variance	$\frac{\sum_{i=1}^N (Y_i(0)-\overline{Y}(0))^2}{N-1}$	$\frac{\sum_{i=1}^N (Y_i(1)-\overline{Y}(1))^2}{N-1}$	$\frac{\sum_{i=1}^N (\tau_i-\overline{\tau})^2}{N-1}$
Range	$\max_i[Y_i(0)] - \min_i[Y_i(1)]$	$\max_i[Y_i(0)] - \min_i[Y_i(1)]$	$\max_i[\tau_i] - \min_i[\tau_i]$

shows notation for the potential outcomes for the N experimental units. Note that only one of the two potential outcomes from each experimental unit (i.e., in each row of Table 2.1) can be *actually observed*. Let $\mathbf{Y}(j)$ denote the N-component column vector of all potential outcomes corresponding to treatment j, i.e.,$\mathbf{Y}(j) = (Y_1(j), \ldots, Y_N(j))^{\mathrm{T}}$. The $2N$ potential outcomes can be represented in an $N \times 2$ matrix $\mathbf{Y}$ shown in Table 2.1. It is important to remember that in this context $\mathbf{Y}$ is considered fixed.

The unit-level treatment difference for unit i (which is the common definition of the unit-level treatment effect) is denoted by

$$\tau_i = Y_i(1) - Y_i(0). \tag{2.1}$$

To define the population-level average treatment effect, we first define the average of all potential outcomes for treatment $j = 0, 1$:

$$\overline{Y}(j) = \frac{1}{N}\sum_{i=1}^{N} Y_i(j). \tag{2.2}$$

The population-level average treatment effect is then defined as

$$\overline{\tau} = \frac{1}{N}\sum_{i=1}^{N} \tau_i = \overline{Y}(1) - \overline{Y}(0), \tag{2.3}$$

The quantity $\overline{\tau}$ is the causal estimand of interest in many studies. The fact that $\overline{\tau}$ can be expressed as a linear combination of the unit-level treatment effects, τ_i, leads to easy evaluations of the properties of estimators, discussed later. However, there can also be other estimands of interest, for example, treatment medians and their differences, expressed in the final rows of Table 2.1. Note that $\underset{i}{\mathrm{Med}}(\tau_i)$ is generally not equal to $\underset{i}{\mathrm{Med}}[Y_i(1)] - \underset{i}{\mathrm{Med}}[Y_i(0)]$, unlike the average differences, which is one reason for the popularity of the average. The last two rows of Table 2.1 contain measures of dispersion of individual columns of potential outcomes and also that of unit level causal effects. Inferential properties of standard estimators of the population level causal effect τ depend on the variances shown in the second last row of Table 2.1, as will be seen later in Section 2.6. Solving part (a) of Problem 2 at the end of Chapter 2 should give readers a clear understanding of the causal effects and estimands described above.

2.4 The completely randomized assignment mechanism

For $i = 1, \dots, N$, define assignment variables W_i such that

$$W_i = \begin{cases} 1 & \text{if the } i\text{th unit is assigned to treatment } 1 \\ 0 & \text{if the } i\text{th unit is assigned to treatment } 0. \end{cases}$$

Let $\mathbf{W} = (W_1, \dots, W_N)^{\mathrm{T}}$ denote the $N \times 1$ vector of treatment assignments.

A very simple probabilistic assignment mechanism *independently* assigns each unit to treatment 1 with a fixed probability $0 < p_1 < 1$ and to treatment 0 with a fixed probability $0 < p_0 = 1 - p_1 < 1$. Such an assignment mechanism is called *Bernoulli assignment mechanism,* because in this case the assignment variables $W_1, \dots, W_N$ are independently and identically distributed Bernoulli variables with the probability of taking value one as p_1. Specifically, if each assignment is decided by the toss of a fair coin, then $p_0 = p_1 = 1/2$, and there are 2^N possible ways in which N experimental units can be randomly allocated to the two treatments. The joint distribution of the assignment vector $\mathbf{W}$ can thus be written as

$$\Pr(\mathbf{W} = \mathbf{w}) = \frac{1}{2^N},$$

where $\mathbf{w}$ is any realization of $\mathbf{W}$ and thus can be any of the 2^N binary vectors from $(0, 0, \dots, 0)$ to $(1, 1, \dots, 1)$.

Clearly, a Bernoulli treatment assignment mechanism can lead to all N units being assigned to only one of the two treatments, in which case the experimenter would have no observed outcomes from the other treatment and consequently would not be able to draw any empirically informed causal inference at all from the experiment. For example, a Bernoulli assignment mechanism could have assigned all the eleven plots in the tomato experiment to the old fertilizer, providing no outcomes for the new one. To avoid such *worst* possible results, we must impose restrictions on the assignment mechanism by *apriori* deciding the number of experimental units to be assigned to each treatment. For example, we can consider a *balanced* experiment in which an equal number of units are assigned to the two treatments.

In the tomato experiment example with two levels of the treatment factor a balanced experiment cannot be implemented using all units because of the odd number of experimental units. Typically for a completely randomized treatment assignment mechanism, it is decided apriori that N_j experimental units will be randomly assigned to treatment $j \in \{0, 1\}$, where $N_0 + N_1 = N$. Then, there are $\binom{N}{N_1} = N!/(N_0!N_1!)$ possible ways in which N experimental units can be randomly allocated to the two treatments, which implies that

$$\Pr(\mathbf{W} = \mathbf{w}) = \begin{cases} \frac{1}{N!/(N_0!N_1!)} & \text{if } \sum_i w_i = N_1 \\ 0 & \text{otherwise.} \end{cases}, \tag{2.4}$$

Like a Bernoulli assignment mechanism, in this case the treatment assignment variables, W_i, are still marginally Bernoulli random variables with

$$Pr(W_i = 1) = N_1/N,$$

although $W_1, \ldots, W_N$ are not independent. From (2.4) the following results can easily be established for $i = 1, \ldots, N$ (Problem 4):

$$\begin{aligned} \mathbb{E}(W_i) &= \frac{N_1}{N}, & (2.5) \\ \text{Var}(W_i) &= E(W_i^2) - [E(W_i)]^2 = \frac{N_1 N_0}{N^2}, & (2.6) \\ \text{Cov}(W_i, W_{\tilde{i}}) &= E(W_i W_{\tilde{i}}) - E(W_i)E(W_{\tilde{i}}) = -\frac{N_1 N_0}{N^2(N-1)}, \quad i \neq \tilde{i}. & (2.7) \end{aligned}$$

Suppose that it was pre-decided that five out of the eleven plots would receive fertilizer A, and the remaining six fertilizer B. Adoption of an assignment mechanism described in this section for the tomato experiment with $N_0 = 5$ and $N_1 = 6$, yields $11!/(6!5!) = 462$ possible randomizations. As seen from Table 2.2, the observed assignment vector for this experiment is $\mathbf{w} = (0, 0, 1, 1, 0, 1, 1, 1, 0, 0, 1)^{\mathrm{T}}$.

2.4.1 Observed and missing potential outcomes in a completely randomized design

Let y_i denote the observed outcome for unit i. Clearly, y_i is *one* of the two potential outcomes $Y_i(0), Y_i(1)$, depending on the random variable W_i, i.e., the treatment that the ith unit receives. Thus, we can write for unit i's observed outcome,

$$y_i = Y_i(W_i) = W_i Y_i(1) + (1 - W_i) Y_i(0), \tag{2.8}$$

where

$$\sum_{i=1}^{N} W_i = N_1, \tag{2.9}$$

which gives the total number of units assigned to treatment 1. We denote the N-component column vector of *observed* outcomes by $\mathbf{y} = (y_1, \ldots, y_N)^{\mathrm{T}}$.

TABLE 2.2
Table of observed outcomes for the tomato example

Unit	Observed assignment $\mathbf{w}$	Potential outcome under level 0 (Fertilizer A)	1 (Fertilizer B)
1	0	29.2	?
2	0	11.4	?
3	1	?	26.6
4	1	?	23.7
5	0	25.3	?
6	1	?	28.5
7	1	?	14.2
8	1	?	17.9
9	0	16.5	?
10	0	21.1	?
11	1	?	24.3
Average		$\overline{y}(0) = 20.70$	$\overline{y}(1) = 22.53$
Variance		$s_0^2 = 49.43$	$s_1^2 = 29.51$

For *each* experimental unit, only one of the two potential outcomes is observed and the other is missing. That is, it is not possible to observe both potential outcomes $Y_i(0)$ *and* $Y_i(1)$ for *any* $i = 1, \ldots, N$. The fundamental problem of causal inference is thus to infer the missing values from the observed data. Table 2.2 shows the observed assignment vector $\mathbf{w}$ and the observed and missing outcomes (the latter denoted by the symbol ?). The vector of observed outcomes $\mathbf{y}$ is $(29.2, 11.4, 26.6, 23.7, 25.3, 28.5, 14.2, 17.9, 16.5, 21.1, 24.3)^{\mathrm{T}}$ and the vector of observed assignment variables is $\mathbf{w}$ is $(0, 0, 1, 1, 0, 1, 1, 1, 0, 0, 1)^{\mathrm{T}}$. Thus the pair $(\mathbf{w}, \mathbf{y})$ constitutes the observed data from the experiment.

2.4.2 Inferring causal effects from observed data

The inference about the causal effects of interest will be based on the randomized assignment mechanism, specifically following Procedure 1.1. Recall that Step 1 of this procedure entails specifying a hypothesis regarding the model that generates the observed data. In statistical inference, such a hypothesis is typically referred to as the "null hypothesis" and is assessed using the observed data.

We first consider assessment of null hypotheses that involve unit-level causal effects, e.g., all unit-level caual effects τ_i defined in (2.1) are equal to a known constant. Such a null hypothesis is referred to as the "sharp null" hypothesis, and can be assessed using the Fisher randomization test. This topic will be discussed in detail in Section 2.5. Next, we will consider assessment of null hypotheses associated with the population-level average causal effect, i.e., the average causal effect $\overline{\tau}$ defined in (2.3) is equal to a known constant.

Such a null hypothesis is referred to as the "Neymanian null" hypothesis or a "weak null" hypothesis, because, if the strong one (sharp null) holds, so does the weak one, but not necessarily the reverse. Assessment of such hypothesis will be discussed in Section 2.6.

2.5 Randomization test for the sharp null hypothesis of no treatment effect

The null hypothesis of a constant treatment difference of τ^* for *all* units is called a *sharp null hypothesis* (Rubin [81]) and can be formally expressed as

$$H_{0\mathrm{F}}^{\tau^*} : Y_i(1) - Y_i(0) = \tau^*, \quad i = 1, \ldots, N, \tag{2.10}$$

where the subscript "0F" can be read as short-hand for "null Fisher".

A special case of the hypothesis (2.10) is the sharp null hypothesis of no treatment effect for *any* individual, that is $\tau^* = 0$, and can be expressed as

$$H_{0\mathrm{F}} : Y_i(0) = Y_i(1), \quad i = 1, \ldots, N. \tag{2.11}$$

The plausibility of the sharp null hypothesis (2.11) can be assessed by using a randomization test whose idea was introduced in Chapter 1. Our version of the randomization test is inspired by Fisher's idea, but is an enhanced version based on later development, e.g., Rubin(1980)[81]. A suitable test statistic T is chosen, and its observed value T^{obs} is compared with its sampling distribution under the null hypothesis induced by the randomization. This distribution will henceforth be referred to as the "null randomization distribution" or simply the "randomization distribution". If the observed value T^{obs} is found to be "extreme" or unusual with respect to this null randomization distribution, then we have evidence against the sharp null hypothesis H_{0F}.

We now lay out a formal procedure for testing a sharp null hypothesis of no treatment effect, followed by a detailed discussion and demonstration of each step using the data from the tomato experiment shown in Table 2.2.

Procedure 2.1 ***Steps in randomization test:***

Step 1: Choose a suitable test statistic T and define what values (large or small or both) of T will be considered "extreme" in terms of departure from the null hypothesis.

Step 2: Calculate the observed value of the test statistic. Denote this observed value by T^{obs}.

Step 3: Compute the randomization distribution of the test statistic under the sharp null hypothesis, H_0. This involves two sub-steps: (a) imputing *the missing potential outcomes according to the sharp null hypothesis and (b) enumerating all possible assignment vectors, and hence all possible values of T.*

Step 4: Compute the p-value as the proportion of values of T in the randomization distribution that are as or more "extreme" (e.g., $\geq$ or $\leq$ or both) than the observed value. The smaller the p-value, the greater is the degree of belief that the null hypothesis is not true, because the observed value of T is unlikely to be generated under H_0.

Note that steps 1 and 2 in Procedure 2.1 are essentially the same as steps 2 and 3 of Procedure 1.1 when Step 1 of Procedure 1.1 specifies a sharp null hypothesis. Further, steps 3 and 4 in Procedure 2.1 describe the details associated with step 4 of Procedure 1.1.

2.5.1 Choice of test statistic and definition of extremeness

One of the greatest advantages of the randomization test lies in the fact that it can be performed with any test statistic [13]. A reasonable test statistic to assess the sharp null hypothesis $H_{0F}^{\tau^*}$ defined in (2.10) of a constant treatment effect of τ^* is $\widehat{\overline{\tau}} - \tau^*$, where $\widehat{\overline{\tau}}$ is a point estimator of $\overline{\tau}$, the primary estimand of interest identified earlier in this chapter. The value of this test statistic calculated from observed data should be close to zero if $H_{0F}^{\tau^*}$ holds. A natural point estimator of $\overline{\tau}$ is the difference between the observed average responses for the two treatment groups, formally defined as

$$\widehat{\overline{\tau}} = \overline{y}(1) - \overline{y}(0), \tag{2.12}$$

where

$$\overline{y}(1) = \frac{1}{N_1} \sum_{i:W_i=1} Y_i(1) = \frac{1}{N_1} \sum_{i=1}^{N} W_i Y_i(1), \tag{2.13}$$

$$\overline{y}(0) = \frac{1}{N_0} \sum_{i:W_i=0} Y_i(0) = \frac{1}{N_0} \sum_{i=1}^{N} (1 - W_i) Y_i(0), \tag{2.14}$$

are, respectively, the averages of all observed outcomes corresponding to treatments 1 and 0.

It is straightforward to see (Problem 5) that for $j \in \{0, 1\}$, $\overline{y}(j)$ defined by (2.14) and (2.13) is an *unbiased* estimator of $\overline{Y}(j)$ in the sense $\mathbb{E}(\overline{y}(j)) = \overline{Y}(j)$, where the expectation $\mathbb{E}(\cdot)$ of a random variable is defined with respect to its randomization distribution, i.e., distribution of $\mathbf{W}$. Consequently $\widehat{\overline{\tau}}$ defined by (2.12) is an unbiased estimator of $\overline{\tau}$. From Table 2.2, an unbiased point estimate of $\overline{\tau}$ in the tomato experiment can be obtained as $\widehat{\overline{\tau}} = \overline{y}(1) - \overline{y}(0) = 22.53 - 20.70 = 1.83$. Recall that unbiasedness of the estimator $\widehat{\overline{\tau}} = \overline{y}(1) - \overline{y}(0)$ was illustrated in the toy example in Section 1.3 by enumerating all possible values of $\widehat{\overline{\tau}}$ in Table 1.2.

It is also a common practice to scale the test statistic $\widehat{\overline{\tau}} - \tau^*$ by some measure of variability. One popular choice is

$$T_{\text{avg-unegvar}} = \frac{\widehat{\overline{\tau}} - \tau^*}{\sqrt{s_0^2/N_0 + s_1^2/N_1}}, \tag{2.15}$$

where s_0^2 and s_1^2 represent the sample variances of the observations in the treatment groups 0 and 1 respectively. Formally,

$$s_0^2 = \frac{1}{N_0 - 1}\sum_{i=1}^{N}(1 - W_i)\left(Y_i(0) - \overline{y}(0)\right)^2, \tag{2.16}$$

$$s_1^2 = \frac{1}{N_1 - 1}\sum_{i=1}^{N} W_i\left(Y_i(1) - \overline{y}(1)\right)^2. \tag{2.17}$$

The square of the denominator of $T_{\text{avg-unegvar}}$, $s_0^2/N_0 + s_1^2/N_1$, which will be discussed in the following section on Neymanian inference, is an estimator of the sampling variance of $\widehat{\overline{\tau}}$ with a positive bias unless all unit-level treatment effects are equal.

Another possible choice is the conventional t-statistic used to test the null hypothesis of equal means with assumed *equal variances* in the two groups:

$$T_{\text{avg-eqvar}} = \frac{\widehat{\overline{\tau}} - \tau^*}{s_{\text{pooled}}\sqrt{1/N_0 + 1/N_1}}, \tag{2.18}$$

where

$$s_{\text{pooled}}^2 = \frac{(N_0 - 1)s_0^2 + (N_1 - 1)s_1^2}{N_0 + N_1 - 2}$$

is the *pooled variance estimator* and s_0^2 and s_1^2 are defined by (2.16) and (2.17) respectively.

One can consider the unscaled test statistic $\widehat{\overline{\tau}}$ or its scaled versions $T_{\text{avg-unegvar}}$ defined in (2.15), or $T_{\text{t-stat-eqvar}}$ defined in (2.18) with $\tau^* = 0$ as a reasonable choice to test H_{0F}, the sharp null hypothesis of no treatment effect. Whereas these are popular choices for test statistics, there exist various other choices. These include (but are not limited to)

1. The absolute value of the difference in observed median outcomes for treatment and control:

$$T_{\text{median}} = \underset{W_i=1}{\text{Med}}[y_i] - \underset{W_i=0}{\text{Med}}[y_i], \tag{2.19}$$

2. The rank statistic

$$T_{\text{rank}} = \frac{\sum_{i:W_i=1} R_i}{N_0} - \frac{\sum_{i:W_i=0} R_i}{N_1}, \tag{2.20}$$

where R_i denotes the normalized rank of the observed outcome of the ith unit. Here normalized means subtracting $(N+1)/2$ from the original ranks to make the average of normalized ranks zero.

There are other possible test statistics and the reader is referred to chapter 5 of Imbens and Rubin (2015) [44] for further details regarding choice of test statistic. The caveats of using multiple test statistics are explained in Section 2.5.4.

The definition of extremeness of the test statistic formally depends on the type of alternative hypothesis against which the null hypothesis is tested. For example, if one is interested in testing H_{0F} against the one-sided sharp alternative $H_{1F} : Y_i(1) - Y_i(0) = \tilde{\tau}$, where $\tilde{\tau} > 0$, large values of $T_{\text{t-stat-uneqvar}}$ or $T_{\text{t-stat-eqvar}}$ defined in (2.18) will be considered extreme, as they indicate departures from the sharp null. On the other hand, if the alternative is of the form $H_{1F} : Y_i(1) - Y_i(0) = \tilde{\tau}$, where $\tilde{\tau} < 0$, then smaller values of $T_{\text{t-stat-uneqvar}}$ or $T_{\text{t-stat-eqvar}}$ are considered extreme. If the alternative is $H_{1F} : Y_i(1) - Y_i(0) = \tilde{\tau}$, with $\tilde{\tau} \neq 0$, then large values of absolute values of each statistic will be considered extreme.

2.5.2 Computing the randomization distribution of the test statistic under the sharp null and calculating the resulting p-value

To obtain the randomization distribution of the test statistic, first enumerate all possible treatment assignments under the actual assignment mechanism (if the number of such assignments is very large, a random sample of them can be considered). The assumption that the sharp null hypothesis H_{0F} is true permits us to complete the table of all potential outcomes (including those that are missing for the actual assignment), which led to the observed data, using only the observed data. This means imputing the missing potential outcome in each row using the observed outcome. Next, draw a new assignment vector $\mathbf{w}_{\text{new}}$ from the distribution of $\mathbf{W}$ until all possible assignment vectors are exhausted. That is, for each assignment, calculate the value of the test statistic that would have been observed under that assignment and H_{0F}.

The proportion of the computed values of the test statistic (with respect to the total number of possible randomizations) that are as extreme or more extreme than the actual observed test statistic is the p value (i.e., the significance level) of the test statistic under the null hypothesis. The definition of extremeness should be consistent with that identified in Step 1 of Procedure 2.1.

The smaller the p-value, the greater is the degree of belief that the null hypothesis is contradicted by the observed data, because the one observed result, even when combined with all more extreme results, would be a rare event. The p value for testing H_{0F} against the one-sided alternative $H_{1F} : Y_i(1) - Y_i(0) = \tilde{\tau} > 0$ is the proportion of values of the test statistic that are greater than or equal than the observed value. Similarly, the p-value for testing H_{0F} against the one-sided alternative $H_{1F} : Y_i(1) - Y_i(0) = \tilde{\tau} < 0$ is the proportion of values of the test statistic that are smaller than or equal than the observed value. Using the notation $\text{pr}(\cdot)$ for proportion, these two p-values can be formally expressed as $\text{pr}(T \geq T^{\text{obs}})$ and $\text{pr}(T \leq T^{\text{obs}})$ respectively, wher T^{obs} denotes the observed value of the test statistic T. Finally, the p-value fo testing H_{0F} against the two-sided alternative $H_{1F} : Y_i(1) - Y_i(0) = \tilde{\tau} \neq 0$ i the proportion of values of T that are greater than or equal than the absolut of its observed value, i.e., $\text{pr}(|T| \geq |T^{\text{obs}}|)$.

Note that if the total number of randomizations is large (e.g. 235!/(119!116!) in the partial Shadish experiment), it may be awkward to compute the value of the test statistic for all of them. Instead, a commo practice is to obtain a random sample of draws from the distribution of $\mathbf{W}$ instead of a complete enumeration of all realizations, and estimate the p-valu from such a sample of draws. For example, the estimated two-sided p-valu from m draws of $\mathbf{W}$ is

$$\frac{1}{m}\sum_{i=1}^{m} \mathbb{I}\left(|T_i| \geq |T^{\text{obs}}|\right),$$

where T_i denotes the computed value of the test statistic obtained using the ith draw, and $\mathbb{I}(A)$ denotes the indicator function taking value 1 if event A occurs and 0 otherwise.

2.5.3 Randomization test in the tomato experiment

From the observed data in the tomato experiment shown in Table 2.2, $\widehat{\tau}$, the difference of the observed averages for treatment and control is 1.83. The sample variances s_0^2 and s_1^2 are 49.43 and 29.51 respectively. Consequently, the pooled variance s^2_{pooled} is 38.36, and the observed value of the statistic $T_{\text{t-stat-eqvar}}$ defined by (2.18) is $1.83/\sqrt{38.36(1/5+1/6)} = 0.49$, substituting $\tau^* = 0$.

To obtain the randomization distribution of the test statistic $T_{\text{avg-eqvar}}$ under the sharp null hypothesis, we use the fact that the sharp null hypothesis H_{0F} given by (2.11) of no treatment effect permits us to impute the missing values in Table 2.2 as shown in Tables 2.3 and 2.4, where the hypothetical observed outcomes are circled. The hypothetical assignment vectors $\mathbf{w}_{\text{new}}$ in Tables 2.3 and 2.4 are $(0,0,0,0,0,1,1,1,1,1,1)$ (left) and $(1,1,1,1,1,1,0,0,0,0,0)$ respectively, and the computed values of the test statistic are 0.75 and 1.42 respectively. Recall that under completely randomized treatment assignment, we have $11!/(5!6!) = 462$ possible treatment assignments. The histogram constructed from 462 such values is shown in Figure 2.1, with the vertical line representing the observed value of T, 0.49.

Finally, the p value for testing H_{0F} (against the one-sided alternative H_{1F} : $Y_i(1) - Y_i(0) > 0$) is the proportion of values of $T_{\text{avg-eqvar}}$ that are greater than or equal than the actual observed value 0.49, and is calculated as 0.30, which suggests essentially no evidence against the sharp null hypothesis.

The randomization test for the completely randomized design with data from the tomato experiment against a one-sided (greater than type) alternative $Y_i(1) - Y_i(0) = \tau > \tau^* \quad i = 1, \ldots, N$ using the four test statistics defined

TABLE 2.3
Computing the test statistic with the randomization (0,0,0,0,0,1,1,1,1,1,1)

Unit	Hypothetical $\boldsymbol{w}$	Control (0)	Treatment (1)
1	0	29.2	29.2
2	0	11.4	11.4
3	0	26.6	26.6
4	0	23.7	23.7
5	0	25.3	25.3
6	1	28.5	28.5
7	1	14.2	14.2
8	1	17.9	17.9
9	1	16.5	16.5
10	1	21.1	21.1
11	1	24.3	24.3
Mean		$\overline{y}(0) = 23.24$	$\overline{y}(1) = 20.42$

by (2.15)–(2.20) is shown in Figure 2.1. The R-codes used to perform these tests and to generate these figures are explained in detail in Section 2.5.5.

Note that in spite of the fact that the ongoing discussion in this section was devoted to testing $H_{0F} : \tau_i = 0$ for $i = 1, \ldots, N$, it is straightforward to generalize the same procedure to test the sharp null hypothesis $H_{0F}^{\tau^*} : \tau_i = \tau^*$ for any fixed τ^* and $i = 1, \ldots, N$. In that case, only the imputation of missing potential outcomes described in Step 3 of the randomization test procedure will change. Under the sharp null $H_{0F} : \tau_i = \tau^*$, the missing potential outcome for unit i will be imputed by $y_i - \tau^*$ if $w_i = 0$ and by $y_i + \tau^*$ if $w_i = 1$. Every other step will remain exactly the same.

2.5.4 Caveat about use of multiple test statistics

The validity of the randomization test procedure described and illustrated in this section depends on deciding apriori (i.e., before seeing the data) a specific pair of a null hypothesis and a test statistic. The corresponding p-values are

TABLE 2.4
Computing the test statistic with the randomization (1,1,1,1,1,1,0,0,0,0,0)

Unit	Hypothetical w	Control (0)	Treatment (1)
1	1	29.2	(29.2)
2	1	11.4	(11.4)
3	1	26.6	(26.6)
4	1	23.7	(23.7)
5	1	25.3	(25.3)
6	1	28.5	(28.5)
7	0	(14.2)	14.2
8	0	(17.9)	17.9
9	0	(16.5)	16.5
10	0	(21.1)	21.1
11	0	(24.3)	24.3
Mean		$\overline{y}(0) = 18.80$	$\overline{y}(1) = 24.12$

valid for each such individual pair, but the p-values for two different pairs of null hypothesis and test statistic are not independent. For example, consider two potential test statistics, $T_1 = T_{\text{avg-eqvar}}$ defined in (2.18) and $T_2 = T_{\text{median}}$ defined in (2.19) with realized values T_1^{obs} and T_2^{obs}. The analyst may consider both options taking into consideration two different alternative hypotheses of interest. Suppose the distribution of the unit-level causal effects expected to be skewed, and not symmetric. Under the first alternative hypothesis considered by the analyst, the mean of the outcome distribution may shift (suggesting a test statistic based on the difference of means, e.g., $T_{\text{avg-eqvar}}$), whereas under another alternative hypothesis, the median may change (suggesting the test statistic T_{median}).

Under the sharp null hypothesis $H_{0F}^{\tau^*} : \tau_i = \tau^*$ for $i = 1, \ldots, N$, one can calculate a p-value with each of the tests, e.g., $p_1 = \text{pr}(T_1 \geq T_1^{\text{obs}})$ and $p_2 = \text{pr}(T_2 \geq T_2^{\text{obs}})$. These p-values are valid for each test separately, but combining these two p-values into a single p-value by using their minimum or taking the average does not lead to a valid procedure. For example, the experimenter

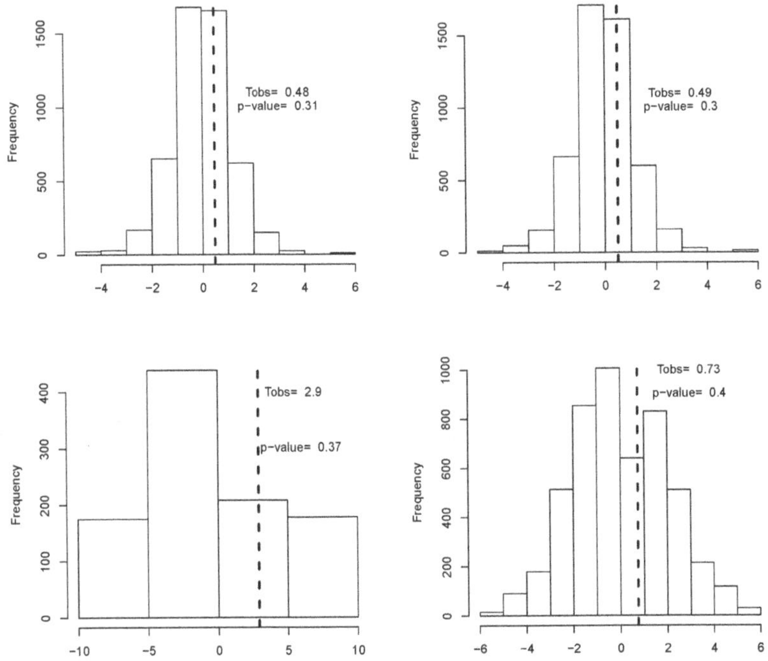

FIGURE 2.1
Fisher randomization test for the tomato experiment to assess the sharp null hypothesis of zero treatment effect against a greater-than type one-sided alternative using four different test statistics: (clockwise from upper left: $T_{\text{t-stat-unegvar}}$, $T_{\text{t-stat-eqvar}}$, T_{median} and T_{rank} defined in (2.15)–(2.20) respectively)

may feel that the inability to find a "significant difference" between the two fertilizers in the tomato experiment is a waste of time and resources, and consequently be tempted to use the p-value that is smaller, but that will not lead to a valid test procedure.

As explained in chapter 5 of Imbens and Rubin (2015) [44], the simplest way to obtain a valid p-value with multiple test statistics is to combine the two (or more) test statistics into a single test statistic.This can be done in two ways: defining a new test statistic that is (i) a function of the two original test statistics, or (ii) a function of the two p-values. Both are functions of the observed data $(\mathbf{y}, \mathbf{W})$ and hence are valid test statistics.

The need for using two (or more) different test statistics may also arise in situations when the experimenter has two outcome variables for each unit. In the tomato experiment, in addition to yield of tomatoes measured by weight, the average height of tomato plants that grow on each plot may be a secondary outcome of interest.

2.5.5 Implementing randomization test in R: modules and functions

We now discuss how the steps in randomization tests described in Procedure 2.1 can easily be implemented using R, illustrating the codes and functions using the tomato experiment. We will assume that the observed outcome data are reported in the form $(\mathbf{w}, \mathbf{y})$, where $\mathbf{y}$ is the $N \times 1$ vector of observed outcomes and $\mathbf{w}$ is the $N \times 1$ vector of observed binary assignments to the units. For example, in the tomato experiment, the vector of observed outcomes $\mathbf{y}$ is $(29.2, 11.4, 26.6, 23.7, 25.3, 28.5, 14.2, 17.9, 16.5, 21.1, 24.3)^{\mathrm{T}}$ and the vector of observed assignment variables is $\mathbf{w}$ is $(0, 0, 1, 1, 0, 1, 1, 1, 0, 0, 1)^{\mathrm{T}}$. The analyst can visualize such data as an $N \times 2$ matrix of observed and missing potential outcomes. To convert the observed outcomes from two separate $N \times 1$ vectors $\mathbf{w}$ and $\mathbf{y}$ to an $N \times 2$ potential outcomes matrix like the one in Table 2.2 with observed and missing outcomes, we can use the function Nbytwo() shown below. The function takes two $N \times 1$ vectors $\mathbf{y}$ and $\mathbf{w}$ as arguments and generates an $N \times 2$ matrix of observed and missing potential outcomes, with the missing outcomes denoted by "NA".

```
-----------------------------------------------------------------
# Function to convert data into an N times 2 matrix of potential
 outcomes

Nbytwo <- function(y,w) {
  N = length(y)
  Ypot = matrix(numeric(N*2), ncol=2)
  for (i in 1:N){
    if (w[i]==1)    {
      Ypot[i,1] = NA
      Ypot[i,2] = y[i]
                   }
      else
                   {
      Ypot[i,1] = y[i]
      Ypot[i,2] = NA
                   }
                       }
return(Ypot)
 }
-----------------------------------------------------------------
```

Application of this function to the tomato experiment is shown below:

```
-----------------------------------------------------------------
# Applying the Nbytwo function to the tomato experiment data:
```

```
ytomato=c(29.2,11.4,26.6,23.7,25.3,28.5,14.2,17.9,16.5,21.1,24.3)
wtomato=c(0,0,1,1,0,1,1,1,0,0,1)
> Nbytwo(ytomato,wtomato)
      [,1] [,2]
 [1,] 29.2   NA
 [2,] 11.4   NA
 [3,]   NA 26.6
 [4,]   NA 23.7
 [5,] 25.3   NA
 [6,]   NA 28.5
 [7,]   NA 14.2
 [8,]   NA 17.9
 [9,] 16.5   NA
[10,] 21.1   NA
[11,]   NA 24.3
-------------------------------------------------------------
```

The first computational task (step 2) in Procedure 2.1 is to compute the observed value of the test statistic chosen in step 1 from the observed data. It is easy to write an R function that takes $\mathbf{y}, \mathbf{w}, \tau^*$ as input and computes the value of the test statistic. For example, to compute the statistic $T_{\text{t-stat-uneqvar}}$, we can use the following function, in which the default τ^* is set to zero.

```
-------------------------------------------------------------
TUNEQUALCOMPUTE = function(y,w, taustar=0) {
# Computes t statistic with unequal variances
N = length(y)
N1 = sum(w)
NO = N - N1
tauhat = mean(y[w==1]) - mean(y[w==0])
s0sq = var(y[w==0])
s1sq = var(y[w==1])
Tobs = (tauhat - taustar) / sqrt(s0sq/NO+s1sq/N1)
return(Tobs)
}


## Application to tomato experiment

> TUNEQUALCOMPUTE(ytomato, wtomato)
[1] 0.4765076
-------------------------------------------------------------
```

A similar function can be written to compute the statistic $T_{\text{t-stat-eqvar}}$:

```
-------------------------------------------------------------
TEQUALCOMPUTE = function(y,w, taustar=0) {
  # Computes t statistic with unequal variances
  N = length(y)
  N1 = sum(w)
```

```
  N0 = N - N1
  tauhat = mean(y[w==1]) - mean(y[w==0])
  s0sq = var(y[w==0])
  s1sq = var(y[w==1])
  spooled = sqrt(((N1-1)*s1sq + (N0-1)*s0sq)/(N0+N1-2))
  Tobs = (tauhat - taustar) / (spooled*sqrt(1/N0+1/N1))
  return(Tobs)
}


## Application to tomato experiment

> TEQUALCOMPUTE(ytomato, wtomato)
[1] 0.4888444
-------------------------------------------------------------
```

The next task in Procedure 2.1 is step 3(a), i.e., imputing the missing potential outcomes under the sharp null hypothesis $Y_i(1) - Y_i(0) = \tau^*, \quad i = 1, \ldots, N$ defined in (2.10). The missing potential outcome for unit i will be imputed by $y_i - \tau^*$ if $w_i = 0$ and by $y_i + \tau^*$ if $w_i = 1$. The function imputeundersharpnull takes an $N \times 2$ matrix of potential outcomes with one outcome in each row missing, and imputes the missing outcome under the above sharp null. The default value of τ^* is set to zero, i.e., if the user does not specify τ^*, then the missing outcomes will be imputed under the sharp null of no treatment effect.

```
-------------------------------------------------------------
# Function to impute missing potential outcomes under the sharp null

imputeundersharpnull = function(Y,taustar=0){
  N = dim(Y)[1]
   for (i in 1:N) {
    if (is.na(Y[i,1])){
            Y[i,1] = Y[i,2]-taustar
                     }
    else
                     {
            Y[i,2] = Y[i,1]+taustar
                     }
                 }
   return(Y)
}
-------------------------------------------------------------
```

Below, we show the application of this function with default $\tau^* = 0$ and $\tau^* = 2$.

```
-------------------------------------------------------------
# Application to tomato experiment

Ypot_tomato <- Nbytwo(ytomato,wtomato)
```

```
# Impute with default taustar=0
> imputeundersharpnull(Ypot_tomato)
      [,1] [,2]
 [1,] 29.2 29.2
 [2,] 11.4 11.4
 [3,] 26.6 26.6
 [4,] 23.7 23.7
 [5,] 25.3 25.3
 [6,] 28.5 28.5
 [7,] 14.2 14.2
 [8,] 17.9 17.9
 [9,] 16.5 16.5
[10,] 21.1 21.1
[11,] 24.3 24.3

# Impute with taustar=2
imputeundersharpnull(Ypot_tomato, tau=2)
> imputeundersharpnull(Ypot_tomato, tau=2)
      [,1] [,2]
 [1,] 29.2 31.2
 [2,] 11.4 13.4
 [3,] 24.6 26.6
 [4,] 21.7 23.7
 [5,] 25.3 27.3
 [6,] 26.5 28.5
 [7,] 12.2 14.2
 [8,] 15.9 17.9
 [9,] 16.5 18.5
[10,] 21.1 23.1
[11,] 22.3 24.3
---------------------------------------------------------------
```

The next task is to enumerate all possible assignment vectors $\mathbf{w}$ (or generate m random draws of them) and compute the value of the test statistic for each $\mathbf{w}$. Whereas enumerating all possible assignment vectors leads to an exact calculation of the p-value, even for an experiment with $N = 11$ as the tomato experiment, this can be computationally expensive. Consider a balanced completely randomized experiment with $N = 4$. Below, we show generation of the six possible permutations of the assignment vector $\mathbf{w} = (0, 0, 1, 1)$ using the permn() function in the combinat package of R:

```
---------------------------------------------------------------
w <- c(0,0,1,1)
library(combinat)
> unique(permn(w))
[[1]]
```

```
[1] 0 0 1 1

[[2]]
[1] 0 1 0 1

[[3]]
[1] 1 0 0 1

[[4]]
[1] 1 0 1 0

[[5]]
[1] 0 1 1 0

[[6]]
[1] 1 1 0 0
-----------------------------------------------------------
```

A computationally faster, but less precise approach is to generate m random permutations of $\mathbf{w}$ using the sample() function in R, where m is large. For example, the following code generates 10000 random permutations of $\mathbf{w} = (0, 0, 1, 1)$ and summarizes the proportion of times each of the six unique permutations occur. These proportions are seen to be close to $1/6 = 0.167$, the exact probability of each permutation.

```
-----------------------------------------------------------
set.seed(999)
m=10000
w = c(0,0,1,1)
wrep = NULL
for (i in 1:m){
  wrep = rbind(wrep, sample(w))
}
library(tidyverse)
colnames(wrep)=c("w1","w2","w3","w4")
wrep %>% as.data.frame() %>%
group_by(w1,w2,w3,w4) %>% summarize(p = n()/m)

# A tibble: 6 × 5
# Groups:   w1, w2, w3 [6]
     w1    w2    w3    w4     p
  <dbl> <dbl> <dbl> <dbl> <dbl>
1     0     0     1     1 0.172
2     0     1     0     1 0.169
3     0     1     1     0 0.163
4     1     0     0     1 0.162
5     1     0     1     0 0.168
6     1     1     0     0 0.166
-----------------------------------------------------------
```

We now create a function that performs the randomization test to assess the sharp null hypothesis $Y_i(1) - Y_i(0) = \tau^*, \quad i = 1, \dots, N$, using the test statistic $T_{\text{t-stat-uneqvar}}$ with τ^* set to zero. The function takes five inputs - **y**, **w**, τ^*, ITER, the number of draws m made from the distribution of **W**, and the type of alternative hypothesis (with two-sided as the default). It uses all the functions defined above and returns one of the following three p-values based on the user's choice of alternative (less, greater, both), pval_plus for the one-sided alternative $Y_i(1) - Y_i(0) = \tau > \tau^* \;\; i = 1, \dots, N$, pval_minus for the one-sided alternative $Y_i(1) - Y_i(0) = \tau < \tau^*; \;\; i = 1, \dots, N$, and pval_twosided for the two-sided alternative $Y_i(1) - Y_i(0) = \tau \neq \tau^* \;\; i = 1, \dots, N$. Application of the function is demonstrated with the data from the tomato experiment, to test the sharp null hypothesis of no treatment effect with $m = 10000$ draws from the distribution of **W** (noting that the number of unique realizations of **w** is 462). The function generates a histogram that is displayed in the top left panel of Figure 2.1.

```
-------------------------------------------------------------
FISHER_TEST = function(y,w, taustar, ITER, alternative="both") {
# Tests sharp null H0:tau_i=taustar against tau_i>taustar
N = length(y) # Number of units
Ypot = Nbytwo(y,w) # Convert data into Nx2 potential outcomes table
Tobs = TUNEQUALCOMPUTE(y,w) # Calculate observed value of statistic
Yimp = imputeundersharpnull(Ypot, taustar)
# Impute missing potential outcomes

Tarray = NULL
for (i in 1:ITER) {
wnew = sample(w) # this command randomly permutes the sequence x
ynew = numeric(N)
for (i in 1:N){
ynew[i]=Yimp[i,wnew[i]+1]
}
Tnew =TUNEQUALCOMPUTE(ynew,wnew)
Tarray = c(Tarray, Tnew)
}

hist(Tarray, main=NULL, xlab=NULL)
abline(v=Tobs, lty=2, lwd=3.0)

# Calculate p-value for greater than alternative
pval_plus = sum(Tarray >= Tobs)/ITER

# Calculate p-value for smaller than alternative
pval_minus = sum(Tarray <= Tobs)/ITER

# Calculate p-value for two-sided alternative
pval_twosided = sum(abs(Tarray) >= abs(Tobs))/ITER
```

```
if (alternative=="both"){
  return(pval_twosided)
}
if (alternative=="greater") {
  return(pval_plus)
}
if (alternative=="less") {
  return(pval_minus)
}

}

> FISHER_TEST(ytomato, wtomato, taustar=0, ITER=10000,
 alternative="greater")
[1] 0.3167
```

2.5.6 Inverting tests of sharp null hypotheses to obtain Fiducial intervals

Although the randomization test is a flexible tool for causal inference, sometimes finding an interval that is likely to contain the true value of a constant treatment effect is desired. Formally, we seek an interval $[\tau_L, \tau_U]$ where τ_L and τ_U are functions of the observed outcomes, that contains the true value of the average treatment effect $\overline{\tau}$ (assumed constant across units under the Fisherian framework) with a pre-assigned probability of at least $1-\alpha$. Such an interval is called a $100(1-\alpha)\%$ interval for $\overline{\tau}$. Using the connection between hypotheses testing and interval estimation, one can invert a randomization test to obtain an interval estimator for the true additive effect τ.

This can be done by considering a set of sharp null hypotheses $H_{0F}^{\tau^*}$: $Y_i(1) - Y_i(0) = \tau^*$ by varying τ^*. Each such hypothesis can be tested against a one-sided alternative using the procedure described earlier in this section, and the p-value can be computed as a function of τ^*. Let $p^+(\tau^*)$ denote the p-value function obtained from a sequence of hypotheses $H_{0F}^{\tau^*} : Y_i(1)-Y_i(0) = \tau^*$ against the alternative $Y_i(1)-Y_i(0) > \tau^*$ for $i = 1, \ldots, N$. Similarly, let $p^-(\tau^*)$ denote the p-value function obtained from a sequence of hypotheses $H_{0F}^{\tau^*}$: $Y_i(1) - Y_i(0) = \tau^*$ against the alternative $Y_i(1) - Y_i(0) < \tau^*$ for $i = 1, \ldots, N$. The lower limit τ_L of a $100(1-\alpha)\%$ interval for the true value of τ can be obtained by solving $p^+(\tau^*) = \alpha/2$ for τ^*, whereas the upper limit τ_U can be obtained by solving $p^-(\tau^*) = \alpha/2$. See Luo et al. (2021) [55] for a detailed description and theoretical properties of this procedure.

To illustrate this procedure for the tomato experiment, we assess the sharp null hypothesis $H_{0F}^{\tau^*}$: $Y_i(1) - Y_i(0) = \tau^*$ against the one-sided alternative $H_{1F} : Y_i(1) - Y_i(0) > \tau^*$ using the test statistic $T_{\text{t-stat-uneqvar}}$ for

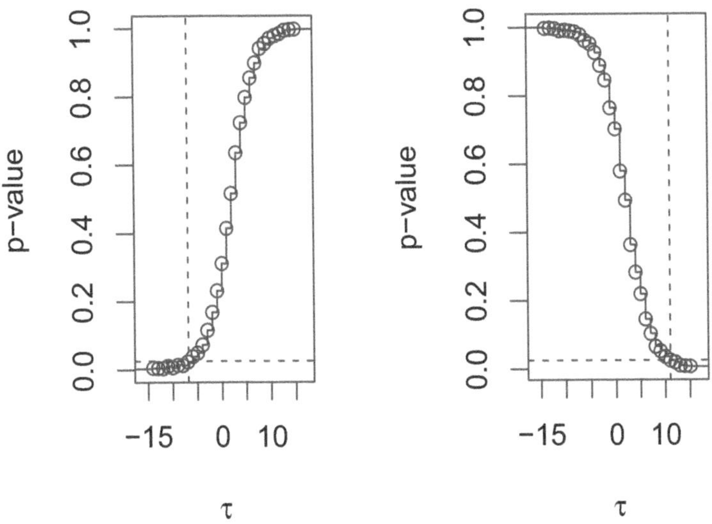

FIGURE 2.2
Computing Fisherian intervals for the tomato experiment using alternatives of the form $\tau_i > \tau^*$ (left panel) and $\tau_i < \tau^*$ (right panel) for $i = 1, \ldots, N$

$\tau^* \in \{-15, -14, \ldots, -2, -1, 0, 1, 2, \ldots, 14, 15\}$. For each τ^*, the p-value is approximated using 5000 draws from the distribution of $\mathbf{W}$. The plot of the non-decreasing function $p^+(\tau^*)$ obtained from these tests is shown in the left panel of Figure 2.2. From this plot, the value of τ that approximately corresponds to a p-value of 0.025 is $\tau_L = -6.88$ (shown by horizontal and vertical dotted lines). Similarly, the plot of the non-increasing function $p^-(\tau^*)$ is shown the right panel of Figure 2.2, from which the value of τ that approximately corresponds to a p-value of 0.025 is obtained as $\tau_U = 10.96$. Thus, an approximate 95% Fisherian interval for the treatment effect (assumed additive) from the tomato experiment is $[-7, 11]$. The results are obtained by using the R function INTERVALFISHER() shown below.

```
-------------------------------------------------------------------
# Inputs:
# w: N x 1 observed assignment vector
# y: N x 1 observed outcomes
# ITER: Number of draws from the distribution of W
# theta_min: Minimum value of tau^*
# theta_max: Maximum value of tau^*
# m: Number of values of tau^* at which the sharp null is assessed
# alpha: Confidence coefficient

INTERVALFISHER = function(y,w,ITER,theta_min, theta_max, m, alpha) {
```

```
int = (theta_max - theta_min)/(m-1)
taustarvec = seq(theta_min,theta_max,int)
pvalarray_g = numeric(m)
pvalarray_s = numeric(m)
for (i in 1:m) {
pvalarray_g[i] = FISHER_TEST(ytomato, wtomato, taustarvec[i],ITER)[1]
pvalarray_s[i] = FISHER_TEST(ytomato, wtomato, taustarvec[i],ITER)[2]
}

# Calculating lower limit

if (sum(pvalarray_g == alpha/2) == 1) {
tau_l = taustarvec[pvalarray_g==alpha/2]
}
else {
plbelow=max(pvalarray_g[pvalarray_g<alpha/2])
plabove=min(pvalarray_g[pvalarray_g>alpha/2])
tau_l_below = taustarvec[pvalarray_g==plbelow]
tau_l_above = taustarvec[pvalarray_g==plabove]
tau_l = tau_l_below +
(tau_l_above - tau_l_below)* (alpha/2 - plbelow)/(plabove-plbelow)
}

# Calculating upper limit

if (sum(pvalarray_s == alpha/2) == 1) {
tau_u = taustarvec[pvalarray_s==alpha/2]
}
else {
pubelow=max(pvalarray_s[pvalarray_s<alpha/2])
puabove=min(pvalarray_s[pvalarray_s>alpha/2])
tau_u_below = taustarvec[pvalarray_s==pubelow]
tau_u_above = taustarvec[pvalarray_s==puabove]
tau_u = tau_u_below +
(tau_u_above - tau_u_below)* (alpha/2 - pubelow)/(puabove-pubelow)
}

pminus=stepfun(taustarvec[-1],pvalarray_s)
pplus =stepfun(taustarvec[-1],pvalarray_g)
par(mfrow=c(1,2))
plot(pplus, main="Lower limit", xlab=expression(tau), ylab="p-value")
abline(v=tau_l, lty=2)
abline(h=alpha/2, lty=2)
plot(pminus, main="Upper limit", xlab=expression(tau), ylab="p-value")
abline(v=tau_u, lty=2)
abline(h=alpha/2, lty=2)

return(c(tau_l, tau_u))
```

```
}

# Implementing for the tomato experiment
> INTERVALFISHER(ytomato,wtomato,5000,-15, 15, 31, .05)
[1] -6.884615 10.962264
------------------------------------------------------------------
```

2.6 Neymanian inference: Testing the null hypothesis on average treatment effects

Assessment of the null hypothesis of equal treatment effect naturally leads to the randomization test procedure decribed in the previous section. However, sometimes hypotheses involving the population-level causal effect is considered important. The null hypothesis

$$H_{0N} : \overline{\tau} = \tau^*, \tag{2.21}$$

or equivalently

$$H_{0\mathrm{N}} : \overline{Y}(1) - \overline{Y}(0) = \tau^* \tag{2.22}$$

is often referred to as the "Neymanian" null hypothesis that the average treatment effect equals τ^*. Step 1 of Procedure 1.1 for the Neymanian form of inference essentially entails postulating such a hypothesis.

Step 2 of Procedure 1.1 entails finding a suitable test statistic for assessing $H_{0\mathrm{N}}$. As in Section 2.5.1, a test statistic that is widely used to test $H_{0\mathrm{N}}$ is $T_{\text{avg-unegvar}}$ defined in (2.15). This statistic is the conventional t-statistic used to test the null hypothesis that two population means differ by τ^*, allowing for possibly unequal variances in the two groups:

$$T_{\text{avg-unegvar}} = \frac{\widehat{\overline{\tau}} - \tau^*}{\sqrt{s_0^2/N_0 + s_1^2/N_1}}, \tag{2.23}$$

where s_0^2 and s_1^2 are defined by (2.16) and (2.17) respectively. The square of the denominator, $s_0^2/N_0 + s_1^2/N_1$ is an estimator of the sampling variance of $\widehat{\overline{\tau}}$ with a non-negative bias unless the unit-level treatment effects are all equal

Another possible choice is the conventional t-statistic $T_{\text{avg-eqvar}}$ defined in (2.18) and used to test the null hypothesis of equal means with assumed *equal variances* in the two groups.

It is important to realize that in spite of their similarities with the t-statistic used frequently in traditional statistical inference, the randomization distributions of $T_{\text{avg-unegvar}}$ and $T_{\text{avg-eqvar}}$ defined above solely arise from the probability distribution of $\mathbf{W}$. For a large finite population, this randomization distribution under H_{0N} can be approximated by a standard normal distribution, making closed-form computation of the p-value easy. Mathematical details of the approximation, derivations and related insights can be found in Ding (2017) [26] and Li and Ding (2017) [51].

Here we show that these approximations can be reasonable, even with small population sizes. Consider the potential outcomes in Table 2.7 in Problem 2 at the end of the chapter. The true value of the average treatment effect in this artificial example is $\tau^* = 12.74 - 9.02 = 3.72$. Assuming a completely randomized treatment assignment with $N_0 = N_1 = 5$, there can be $(10)!/(5!)^2 = 252$ possible realizations of the assignment vector $\mathbf{W}$, leading to 252 possible values of the test statistics $T_{\text{avg-unegvar}}$and $T_{\text{avg-eqvar}}$ with $\tau^* = 3.72$. Figure 2.3 shows the histogram for each statistic constructed from these possible values. It is remarkable to see that even with $N = 10$, the randomization distributions of both of these statistics appear to be quite similar and normal approximations seem to be quite reasonable.

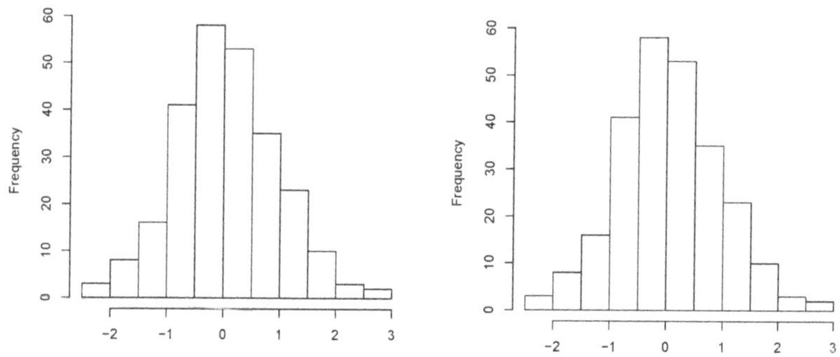

FIGURE 2.3
Randomization distributions of $T_{\text{avg-unegvar}}$ (left) and $T_{\text{avg-eqvar}}$ (right) with $\tau^* = 3.72$

Suppose in the tomato experiment we are interested in testing the null hypothesis that the average treatment effect is zero, i.e., $H_{0N} : \bar{\tau} = 0$ against the one-sided alternative $H_{1N} : \bar{\tau} > 0$ (i.e., fertilizer B produces higher average yield than fertilizer A). Then, substituting $\widehat{\tau} = 1.83$, $\tau^* = 0$, $s_0^2 = 49.43$ and $s_1^2 = 29.51$ into (2.15), we obtain

$$T^{\text{obs}}_{\text{avg-unegvar}} = \frac{1.83}{\sqrt{49.43/5 + 29.51/6}} \approx 0.48.$$

Using the normal approximation, the p-value is estimated as $\Pr(Z > 0.48) \approx 0.32$, where Z represents the standard normal variable. The large p-value suggests that the observed outcomes provide little evidence for rejecting the null hypothesis H_{0N}, and the change of fertilizer from A to B does not appear to have a significant causal effect on the *average yield* of tomatoes.

If, instead of the above one-sided alternative hypothesis, we are interested in testing the null against a two-sided alternative $\tau \neq 0$, then both large and small values of $T^{\text{obs}}_{\text{avg-unegvar}}$ will indicate departures from H_{0N}, and the corresponding p-value will be $Pr(|Z| > 0.48) \approx 0.63$.

Finding interval estimates of the population-level average causal effect using asymptotic inference in a Neymanian setting is straightforward. An approximate $100(1-\alpha)\%$ confidence interval for τ^*, the true value of $\overline{\tau}$, can be obtained as

$$\widehat{\overline{\tau}} \pm z_{\alpha/2}\sqrt{s_0^2/N_0 + s_1^2/N_1},$$

where $z_{\alpha/2}$ is the upper $\alpha/2$-th quantile of a standard normal distribution. For the tomato experiment, the 95% confidence interval for $\overline{\tau}$ using the normal approximation is $[-6, 9]$.

The inferential procedure explained above can be implemented in R using the function Neymanian_inference() shown below. The function takes $\mathbf{y}, \mathbf{w}, \alpha$ as the three arguments, and returns a list output of three elements – observed value of the test statistic, p-values (one-sided and two-sided) and a $100(1-\alpha)\%$ confidence interval.

```
-------------------------------------------------------------------
Neymanian_inference = function(y,w, alpha){
  N = length(y)
  N1 = sum(w)
  NO = N - N1
  tauhat = mean(y[w==1]) - mean(y[w==0])
  s0sq = var(y[w==0])
  s1sq = var(y[w==1])

Tobs = TUNEQUALCOMPUTE(y,w)
pval_minus = pnorm(Tobs)
pval_plus = 1 - pval_minus
pval_twosided = 2*(min(pval_minus, pval_plus))

taubarhat = mean(y[w==1]) - mean(y[w==0])

LL <- taubarhat - qnorm(1-alpha/2)*sqrt(s0sq/NO + s1sq/N1)
UL <- taubarhat + qnorm(1-alpha/2)*sqrt(s0sq/NO + s1sq/N1)

return(list(Tobs=Tobs,
            pvalues=c(pval_plus,pval_minus,
            pval_twosided),CI=c(LL,UL)))

}


> Neymanian_inference(ytomato, wtomato, 0.05)
$Tobs
[1] 0.4765076

$pvalues
```

```
[1] 0.3168564 0.6831436 0.6337128

$CI
[1] -5.707506  9.374173
-----------------------------------------------------------------
```

2.6.1 A bigger picture of Neymanian inference: choice of test statistic

So far we have discussed Neymanian inference along the lines of Procedure 1.1, in which step 1 involves postulating the Neymanian hypothesis $H_{0\mathrm{N}}$, steps 2 and 3 respectively involve identifying suitable test statistics like $T_{\text{avg-unequar}}$ or $T_{\text{avg-eqvar}}$ and computing their observed values. Finally, step 4 involves testing $H_{0\mathrm{N}}$ using the asymptotic randomization distribution of the test statistic. A crucial part in Neyman's form of randomization-based inference involves choice of an appropriate statistic. The statistic is typically a function of an estimator of the causal estimand of interest. For example, the statistics $\widehat{\overline{\tau}}$, $T_{\text{avg-unequar}}$ and $T_{\text{avg-eqvar}}$ are all functions of $\widehat{\overline{\tau}}$, an unbiased estimator of the causal estimand $\overline{\tau}$. Here we try to shed some light on the sampling distributions of these statistics under the Neymanian null hypothesis $H_{0\mathrm{N}} : \overline{\tau} = \tau^*$, because Neyman's inferential procedure uses knowledge of their large sample sampling distribution under $H_{0\mathrm{N}}$.

We have already mentioned in Section 2.5.1 that the estimator $\widehat{\overline{\tau}}$ defined by (2.12) is an unbiased estimator of the average causal effect $\overline{\tau}$. This provides motivation for using $\widehat{\overline{\tau}} - \tau^*$ or its scaled version as a suitable statistic, which would have zero expectation under $H_{0\mathrm{N}}$. An expression for sampling variance of $\widehat{\overline{\tau}}$ was derived by Neyman [65], who showed that under a completely randomized treatment assignment mechanism,

$$\mathrm{Var}(\widehat{\overline{\tau}}) = \frac{S_0^2}{N_0} + \frac{S_1^2}{N_1} - \frac{S_{\text{diff}}^2}{N}, \tag{2.24}$$

where

$$S_0^2 = \frac{1}{N-1}\sum_{i=1}^{N}\left(Y_i(0) - \bar{Y}(0)\right)^2, \tag{2.25}$$

$$S_1^2 = \frac{1}{N-1}\sum_{i=1}^{N}\left(Y_i(1) - \bar{Y}(1)\right)^2, \tag{2.26}$$

$$S_{\text{diff}}^2 = \frac{1}{N-1}\sum_{i=1}^{N}\left(\tau_i - \overline{\tau}\right)^2. \tag{2.27}$$

The quantities S_0^2 and S_1^2 defined by (2.25) and (2.26) above represent the variance (with divisor $N-1$) of all *potential outcomes* corresponding to levels 0 and 1 respectively. The quantity S_{diff}^2 represents the variance of the N unit-level treatment effects $\tau_1, \ldots, \tau_N$. The variances S_0^2, S_1^2 and S_{diff}^2 are shown in the second last row of Table 2.1.

The variance quantity S^2_{diff} measures the non-uniformity of unit-level causal effects $\tau_1, \ldots, \tau_N$, and reduces to zero if and only if

$$\tau_1 = \tau_2 = \cdots = \tau_N = \tau. \tag{2.28}$$

Condition (2.28) implies that for each unit, the potential outcomes under treatment levels 0 and 1 differ by an additive constant τ, that is,

$$Y_i(1) - Y_i(0) = \tau, \quad i = 1, \ldots, N. \tag{2.29}$$

Definition 2.1 (Additivity) *An $N \times 2$ matrix of potential outcomes is said to be* additive *if the equivalent conditions (2.28) and (2.29) hold.*

The assumption of additivity is often made to draw inference about population-level causal effects. Note that additivity also implies that the variances of potential outcomes for treatments 0 and 1 are equal, i.e., $S_0^2 = S_1^2$. However, the converse is not necessarily true, i.e., $S_0^2 = S_1^2$ does not necessarily imply additivity (see Problem 6).

Neyman's proof of (2.24) is complicated and somewhat difficult to follow. We provide a simpler proof in Appendix 11.5, which is along the lines of the proof is given in [44]. However, readers with less interest in technical derivations can skip the proof.

2.6.2 Estimation of sampling variance

To draw valid inference on $\overline{\tau}$, either in the form of interval estimators or in the form of a test of a statistical hypothesis, it is standard to obtain an estimator of $\text{Var}(\widehat{\overline{\tau}})$ given by (2.24). This estimation problem is arguably one of the most discussed and researched topic in Neymanian inference (see [71], for example). From (2.24), it is clear that to estimate $\text{Var}(\widehat{\overline{\tau}})$, one needs to estimate three terms – S_0^2, S_1^2 and S^2_{diff} . Let us first consider the estimation of S_0^2 and S_1^2.

Recall that s_0^2 and s_1^2 denote the sample variances of observed values for treatments 0 and 1 respectively, defined in (2.16) and (2.17) respectively. Then, by standard theory of finite population sampling (e.g., [18]), s_0^2 and s_1^2 are unbiased estimators of S_0^2 and S_1^2 respectively, i.e., $E\left(s_j^2\right) = S_j^2$ for $j \in \{0, 1\}$. However the variance term S^2_{diff} does not admit a meaningful unbiased estimator from the observed data, because both potential outcomes, and hence the unit-level treatment effects, are not observable for *any* of the N units.

Neyman (1923) [65] argued that the quantity $s_0^2/N_0 + s_1^2/N_1$ overestimates, in expectation, the true sampling variance of $\widehat{\overline{\tau}}$, because

$$E\left(s_0^2/N_0 + s_1^2/N_1\right) = \frac{S_0^2}{N_0} + \frac{S_1^2}{N_1} \geq \text{Var}(\widehat{\overline{\tau}}), \tag{2.30}$$

by equation (2.24). As explained in the previous paragraph, the term S^2_{diff} cannot be unbiasedly estimated without further assumptions. However, under the assumption of additivity of effects of treatments 0 and 1, i.e., under (2.28) or (2.29), it follows that $S^2_{\text{diff}} = 0$. In that case, $s_0^2/N_0 + s_1^2/N_1$ is an unbiased estimator of $\text{Var}(\hat{\tau})$. If s_0^2 and s_1^2 differ substantially, it is evidence against additivity.

These arguments, along with the asymptotic derivations mentioned earlier provide some justification for scaling the statistic $\hat{\bar{\tau}}$ by the natural estimate of the standard error, that is, the square root of the estimated sampling variance which equals $\sqrt{s_0^2/N_0 + s_1^2/N_1}$.

2.7 Optimal allocation of units to treatments

So far in this chapter, we have considered situations where N_1, the number of units to be exposed to treatment and N_0, the number of units to be exposed to control are pre-determined. However, in reality, whereas the total number of units available for the experiment may be determined by resource constraints, the experimenter does have the option of determining the allocation of these N units to the treated and control groups. A "balanced" design that allocates an equal number of units to the treatment and control groups, i.e., $N_0 = N_1 = N/2$ may be appealing but not necessarily the best choice. For example, in a situation where the variance of potential outcomes in the treated group is expected to be large compared to the variance of potential outcomes in the control group, perhaps more units should be allocated to the treated group than the control group. The question we address here is, "If one has such prior information on variances of potential outcomes in the form of estimates from previous or pilot studies, how can this information be used to improve the allocation of units into the two groups be made?"

Recall the expression of $\text{Var}(\hat{\bar{\tau}})$ from (2.24). Because the sampling variance of $\hat{\bar{\tau}}$ is a function of N_0 and N_1, we can aim at choosing N_0 and N_1 in a way that minimizes this sampling variance, leading to the most *precise* inference (e.g., shortest confidence intervals) for $\bar{\tau}$. Thus we can define the optimum allocation (N_0^*, N^*) as

$$\begin{aligned} N_0^* &= \arg\min_{N_0} \text{Var}(\hat{\bar{\tau}}) = \arg\min_{N_0} \left(\frac{S_0^2}{N_0} + \frac{S_1^2}{N - N_0} \right), \\ N_1^* &= N - N_0^*. \end{aligned} \tag{2.31}$$

Note that the third term S^2_{diff}/N in the right hand side of (2.24) can be ignored because N is fixed. After using a little calculus (exercise), it can be seen that the optimal allocation is

$$N_0^*/N_1^* = S_0/S_1,$$

i.e., proportional to the standard deviation of the potential outcomes in the control and treatment groups. The solution can also be written as

$$N_j^* = NS_j/(S_0 + S_1), \; j = 0, 1.$$

Suppose we plan to conduct a second round of tomato experiment on 20 new plots in the same location as those in the original experiment. Based on the previous experiment, we have estimates of S_0^2 and S_1^2 as $s_0^2 = 49.43$ and $s_1^2 = 29.51$ respectively, so that $s_0/(s_0 + s_1) = 0.56$. Thus the optimal allocation for the new experiment will be $N_0^* = 20 * .56 = 11$ and $N_1^* = 9$.

Recall from Definition 2.1 the case where all unit-level treatment effects are equal, leading to equality of the variances S_0^2 and S_1^2. Thus, under strict addivity, the optimal allocation of units into the two treatment groups is $N/2$. Consequently, under the assumption of strict addivity, or there is no reason to believe that the variances of potential outcomes in the two groups are different, a balanced allocation is optimal.

2.8 Experiments with binary outcomes

So far, much of the discussion in this chapter has been motivated by the case where the outcome variable is essentially continuous. In many situations, experimenters have to deal with binary outcome variables. For example, in the LEAP experiment introduced in Section 2.2, the primary outcome was whether the participants developed peanut allergy at 60 months of age, as determined by an oral food challenge. In this case each of the potential outcomes $Y(0)$ and $Y(1)$ can take two values : 1 for the outcome of passing the challenge (i.e., the participant would not develop a peanut allergy) and 0 for the outcome of failing the challenge (i.e., the participant would develop a peanut allergy).

In several applications of statistics, separate methods are used for analysis of continuous and binary (or categorical, in general) data. However, one beauty of randomization-based inference lies in the fact that all the methodology works irrespective of whether the outcome variable is continuous or binary. The same estimands, estimators and inferential methodology including the choice of test statistics as described earlier in this chapter are applicable to the case of binary response variables.

However, there are some interesting facts about experiments with binary responses that are worthwhile to note [27]. First, note that the averages of the potential outcomes in the treated and contol groups $\overline{Y}(1)$ and $\overline{Y}(0)$ are proportions of units that respectively achieve "success" (or pass) and "failure" in a certain event, where "success" is denoted by the outcome value 1 and "failure" by 0. For example, in the LEAP experiment, the proportion of infants who would not develop a peanut allergy if exposed to treatment can be denoted by $P_1 = \overline{Y}(1)$. Similarly the proportion who would not develop the allergy if exposed to control can be denoted by $P_0 = \overline{Y}(0)$. The average treatment effect can thus be interpreted as the difference between these two proportions, i.e., $\overline{\tau} = P_1 - P_0$.

The observed outcomes from such an experiment can be summarized in a 2×2 table as shown in Table 2.5. An unbiased estimator of the average treatment effect $\overline{\tau}$ is

$$\widehat{\overline{\tau}} = \widehat{P}_1 - \widehat{P}_0 = \frac{n_{11}}{N_1} - \frac{n_{01}}{N_0},$$

i.e., the difference between the estimated proportions of success in the treated and control groups.

TABLE 2.5
2×2 table of observed binary outcomes

	$y = 1$	$y = 0$	Total
$W = 1$	n_{11}	n_{10}	N_1
$W = 0$	n_{01}	n_{00}	N_0
Total	n_{+1}	n_{+0}	N

To assess the plausibility of the sharp null hypothesis of no treatment effect on any unit, i.e.,

$$H_{0F} : Y_i(0) - Y_i(1) = 0, \quad i = 1, \ldots, N,$$

as before, we can use $\widehat{\overline{\tau}}$ or its scaled versions $T_{\text{avg-uneqvar}}$ or $T_{\text{avg-eqvar}}$ as the test statistic. However, as shown by Ding and Dasgupta (2016) [27], under the sharp null hypothesis of no treatment effect, all of the above test statistics are monotonic functions of n_{11}, which means a randomization test based on any statistic will yield the same result as a randomization test using n_{11} as the test statistic. We demonstrate this aspect using the data from the LEAP experiment, that is summarized in Table 2.6.

TABLE 2.6
2×2 table of observed binary outcomes from the LEAP experiment

	$y = 1$	$y = 0$	Total
$W = 1$	262	5	267
$W = 0$	227	36	263
Total	489	41	530

Figure 2.4 shows the randomization distributions of n_{11} (left panel) and $|\widehat{\overline{\tau}}|$ (right panel) under H_{0F} constructed from 5000 draws of the assignment vector. Using the summarized information in Table 2.6, the observed values of n_{11} and $\widehat{\overline{\tau}}$ are respectively $n_{11}^{\text{obs}} = 262$ and $\widehat{\overline{\tau}} = \widehat{P}_1 - \widehat{P}_0 = 262/267 - 227/263 = 0.12$. Clearly, the two-sided p-value in either case is zero, providing the strongest possible evidence against the sharp null hypothesis of no treatment effect.

Neymanian inference can be conducted in the same way as outlined in Section 2.6 and is left as an exercise.

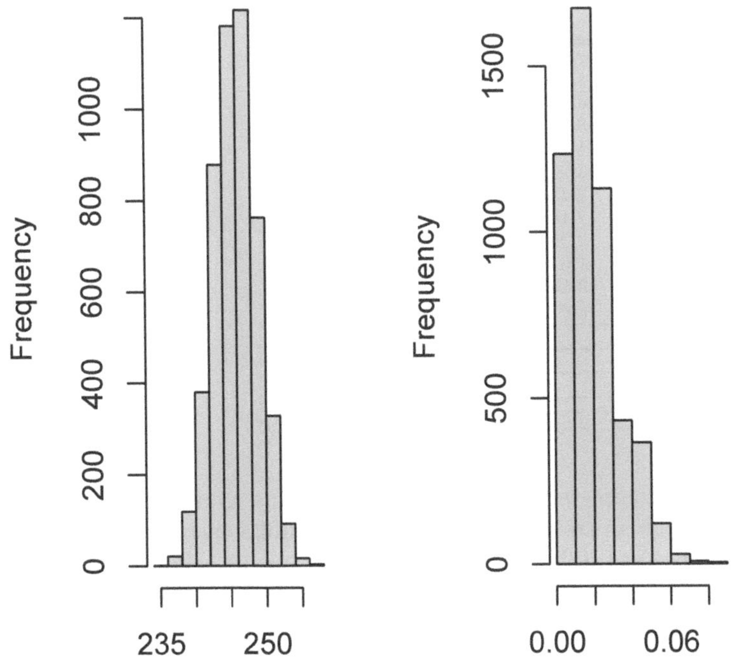

FIGURE 2.4
Randomization test for the LEAP experiment using statistic n_{11} (left panel) and $\widehat{\overline{\tau}}$ (right panel)

2.9 Problems

2.1 Discuss whether SUTVA is reasonable in the tomato experiment.

2.2 Three 10×2 tables of potential outcomes for ten units and two treatments are shown in Tables 2.7–2.9.

(a) From each table, calculate

(i) The missing summaries (average, median, variance and range) of each column.

(ii) The unit-level causal effects $\tau_1, \dots, \tau_{10}$.

(iii) The average treatment effect $\overline{\tau} = \overline{Y}(1) - \overline{Y}(0)$.

(iv) Two typical causal effects: median of unit-level causal effects: $\underset{i}{\text{Med}}(\tau_i)$ and the difference of the medians of the potential outcomes under treatments 0 and 1: $\underset{i}{\text{Med}}[Y_i(1)] - \underset{i}{\text{Med}}[Y_i(0)]$. Are they equal in any of the three cases? If so, why or why not?

TABLE 2.7

Unit	Treatment 0	Treatment 1
1	10.04	14.20
2	9.63	13.51
3	7.26	11.52
4	8.80	13.97
5	10.59	13.48
6	10.78	12.18
7	7.58	10.09
8	9.27	11.61
9	6.75	13.85
10	9.49	12.97
Avg		
Med		
Var		
Ran		

TABLE 2.8

Unit	Treatment 0	Treatment 1
1	2.77	4.77
2	2.26	4.26
3	0.69	2.69
4	1.49	3.49
5	3.65	5.65
6	4.01	6.01
7	0.81	2.81
8	1.89	3.89
9	0.53	2.53
10	2.10	4.10
Avg		
Med		
Var		
Ran		

TABLE 2.9

Unit	Treatment 0	Treatment 1
1	2.77	8.18
2	2.26	5.79
3	0.69	2.14
4	1.49	7.30
5	3.65	5.71
6	4.01	2.97
7	0.81	1.05
8	1.89	2.24
9	0.53	6.86
10	2.10	4.41
Avg		
Med		
Var		
Ran		

(b) If a completely randomized treatment assignment mechanism is used to assign treatments 0 and 1 to five units each, and if $\widehat{\overline{\tau}}$ denotes the observed difference of averages $\bar{y}(1) - \bar{y}(0)$, what is the value of the sampling variance of $\widehat{\tau}$ in each of the three cases?

(c) Does additivity hold in any of the three tables? If so, in which?

(d) For Table 2.7, using a Bernoulli treatment assignment mechanism (i.e., treatment assignment for each unit is determined by independent tosses of a fair coin), obtain a realization of the assignment vector $\mathbf{W} = (W_, \ldots, W_{10})^{\mathrm{T}}$ and thus show the observed and missing outcomes for each unit. Explain how you obtained the assignment (showing appropriate R codes).

(e) For Table 2.8, using a completely randomized treatment assignment mechanism to assign treatments 0 and 1 to five units each, obtain a realization of the assignment vector $\mathbf{W} = (W_, \ldots, W_{10})^{\mathrm{T}}$ and thus show the observed and missing outcomes for each unit. Explain how you obtained the assignment (showing appropriate R codes).

2.3 If a completely randomized treatment assignment mechanism is used to assign treatments 0 and 1 to four and six units respectively, and the observed assignment vector $\mathbf{w}$ is $(1,1,0,0,1,1,1,0,1,0)^{\mathrm{T}}$, create a table of observed and missing outcomes similar to Table 2.2 from the potential outcomes in Table 2.9 in Problem 2. Estimate the average causal effect $\overline{\tau}$ from the observed data. What is the TRUE sampling variance of this estimator? What will be the standard estimate of this sampling variance from the observed outcomes?

2.4 Prove equations (2.5)-(2.7).

2.5 Show that $\widehat{\overline{\tau}}$ defined by (2.12) is an unbiased estimator of $\overline{\tau}$ defined by (2.3).

2.6 Additivity implies that the variances of potential outcomes for treatments 0 and 1 are equal, i.e., $S_0^2 = S_1^2$. However the converse is not true. Construct a counter example where $S_0^2 = S_1^2$, but additivity does not hold.

2.7 For the tomato experiment, obtain 95% Fisherian intervals for the treatment effect (assumed additive) by inverting randomization tests conducted with the test statistics (i) $T_{\text{t-stat-eqvar}}$ given by (2.18) and (ii) T_{median} defined by (2.19).

2.8 Suppose, in the tomato experiment, the estimand of interest is the average treatment effect on $\sqrt{Y}$, i.e.,

$$N^{-1}\sum_{i=1}^{N}\sqrt{Y_i(1)} - N^{-1}\sum_{i=1}^{N}\sqrt{Y_i(0)}.$$

Find an unbiased estimator of this estimand, proving that it is unbiased.

2.9 Lyle et al. (1987) [56] reported the results of a randomized experiment on twenty one men conducted to investigate the effect of calcium on blood pressure in African American men. The treatment (a calcium supplement) was assigned to ten men selected at random from the twenty one, and the remaining eleven men instead received a placebo. All the men had their systolic blood pressures tested before and after the 12-week period. The data from the experiment are given in Table 2.10, and is also available at

`http://lib.stat.cmu.edu/DASL/Datafiles/Calcium.html.`

(a) Which of the following causal estimands do you care about more? Justify your answer.

(i) The difference between average post-treatment BP for the treatment and the control group.

(ii) The difference between average BP reduction for the treatment and the control group.

(b) For a particular person, the *relative* BP reduction is defined as $(Y - X)/X$. How do the estimands (i) and (ii) compare to (iii): the difference between the average relative BP reduction for the treatment and the control groups? Is there a situation where you would prefer estimand (iii) to estimands (i) and (ii)?

(c) Propose two unbiased estimators for estimand (i) defined in part (a) and explain why they are unbiased. Which of these two estimators is more efficient (i.e., should have smaller sampling variance) and why?

(d) Conduct a Fisher randomization test to compare the effects of the treatment on BP. Explain your choice of test statistic. What is the smallest p-value that can be obtained from such a test in this experiment?

TABLE 2.10
Effect of Calcium on BP

Treatment	BP before (X)	BP after (y)	Observed (negative) reduction ($y - X$)
Calcium	107	100	–7
Calcium	110	114	4
Calcium	123	105	–18
Calcium	129	112	–17
Calcium	112	115	3
Calcium	111	116	5
Calcium	107	106	–1
Calcium	112	102	–10
Calcium	136	125	–11
Calcium	102	104	2
Placebo	123	124	1
Placebo	109	97	–12
Placebo	112	113	1
Placebo	102	105	3
Placebo	98	95	–3
Placebo	114	119	5
Placebo	119	114	–5
Placebo	112	114	2
Placebo	110	121	11
Placebo	117	118	1
Placebo	130	133	3

(e) Obtain approximate 95% confidence intervals for estimands (i) and (ii).

2.10 Show, using calculus, that the solution to the optimization problem (2.31) is

$$N_0^* = N\frac{S_0}{S_0 + S_1}, \quad N_1^* = N\frac{S_1}{S_0 + S_1},$$

which means the optimal solution is to distribute N_0 and N_1 proportional to S_0 and S_1.

2.11 Show that for a binary outcome variable, the Neymanian variance estimator of the average treatment effect $\widehat{\overline{\tau}}$ is

$$\text{Var}(\widehat{\overline{\tau}}) = \frac{\widehat{P_1}(1 - \widehat{P_1})}{N_1 - 1} + \frac{\widehat{P_0}(1 - \widehat{P_0})}{N_0 - 1},$$

where $\widehat{P_1} = n_{11}/N_1$, $\widehat{P_0} = n_{01}/N_0$, and n_{11}, n_{01}, N_1, N_0 are as in Table 2.5.

2.12 Obtain a 95% confidence interval for the average treatment effect $\overline{\tau}$ for the LEAP experiment using the Neymanian inferential procedure described in Section 2.6.

3

Making better comparisons using blocking

3.1 Am I making a fair comparison? A hypothetical example

In Chapter 2, we introduced the design and analysis of experiments where the experimental units are assigned to two treatments using a completely randomized assignment mechanism. Intuitively, such an assignment mechanism splits (i.e., partitions) the population of experimental units into two groups in such a manner that they are balanced *in expectation* (or on average) with respect to their inherent characteristics (measured and unmeasured covariates). To explore what this means, consider the following simple hypothetical example:

Example 3.1 Suppose a new medication is being tested against a placebo control on a population of eight human subjects, four of whom are males and four are females. Suppose it is believed by some that the sex of a person can have an impact on the self-reported efficacy of the drug. A balanced completely randomized design will have $8!/(4!)^2 = 70$ possible allocations of the treatment assignment, with five different sex distributions over treatment and control group as shown in Table 3.1. The number of assignments out of the 70 that will lead to each of the five cases, i.e., specific sex distributions over the treatment and control groups, is also shown in Table 3.1.

Allocation 1 appears to be the best assignment, distributing an equal number of males and females in the treatment and control groups. However, such an allocation will actually occur only in roughly 50% of the cases (36 out of the 70 possible assignments) generated by the completely randomized mechanism. Allocations 4 and 5 are clearly undesirable, because in each of these allocations, the effect of the treatment on the outcome will be completely confounded (or inseparable) from the effect of sex. Each of these allocations, however, will occur with a low probability of 1/70 each. Allocations 2 and 3 are not ideal, but not as bad as allocations 4 and 5.

To quantify the "goodness" of a treatment assignment, we define a measure of covariate (sex) imbalance between the treatment and control groups as the difference in the proportions of treated males and untreated males. This measure, denoted by Δ, is shown in the last column of Table 3.1, and is a random variable taking values 0, 0.5, –0.5, 1, –1, with probabilities 36/70, 16/70, 16/70, 1/70 and 1/70 respectively. The expectation (avergae value with

DOI: 10.1201/9781003048794-3

TABLE 3.1
Covariate imbalance in a completely randomized design

Allocation	Treatment		Control		No. of	Imbalance
	Male	Female	Male	Female	possible assignments	measure (Δ)
1	2	2	2	2	36	0
2	3	1	1	3	16	0.5
3	1	3	3	1	16	–0.5
4	4	0	0	4	1	1
5	0	4	4	0	1	–1

respect to the probability distribution) of Δ is zero over all possible allocations, which means the treated and control groups have zero expected imbalance (i.e., are perfectly balanced in expectation. Note that this property would hold even if the sexes of the 8 subjects were not recorded or known before the experiment. This expected balance between treatment groups over observed *and* unobserved covariates is the key strength of randomized experiments.

However, an experimenter will conduct the experiment only once, observe only one of the 70 possible treatment assignments, and consequently will end up with only one of the five allocations shown in Table 3.1. An unlucky experimenter can therefore end up with the worst possible assignment. To summarize, the completely randomized experiment will be conducted with a non-ideal assignment about half the time.

3.2 Completely Randomized Resigns (CRD) to Randomized Block Designs (RBD): the motivation

Consider Example 3.1 described earlier. One straightforward way of ensuring that the design permits only one of the better randomizations with respect to sex is to use a randomized block design. This means stratifying the 8 experimental units into two strata (or blocks) of 4 each - one block consisting of the males and the other of females. Then a completely randomized assignment is used to assign the four units within each block equally to treatment and control. The number of assignments is $(4!/2!2!)^2 = 36$. Such a design is called a randomized block design (RBD) or a "stratified randomized design" as referred to in [44].

Let us revisit the part of the "partial Shadish experiment" reported by [85] and introduced in Chapter 2, in which 235 undergraduate students from introductory psychology classes at a large U.S. public university received either the treatment (vocabulary training) or control (math training) through a completely randomized treatment assignment. The number of students

TABLE 3.2
Distribution of males and females in the "partial Shadish" experiment

Sex	Treatment	Control	Total
Female	81	88	169
Male	35	31	66
Total	116	119	235

assigned to treatment and control were 116 and 119, respectively. The outcome we focus on is the vocabulary test score after the experiment. One important covariate in the experiment was sex of the student, and Table 3.2 shows the distribution of the students into the treated and control groups by sex. It appears that the completely randomized assignment resulted in a relatively even distribution of 169 females and 66 males into the treated and control groups, respectively. However, this happened by chance. If sex was apriori was considered important in this experiment because of its possible effect on the outcome, a better strategy would have been to consider sex as a blocking factor, and randomly assign treatments within each block (sex) so that approximately the same proportion of females receive treatment and control, and the same holds for males. In this chapter we will assume that a hypothetical RBD was used in which it was pre-decided that 81 females would receive treatment and 88 control, and 35 males would receive treatement and 31 control. The experiment will be analyzed accordingly.

Example 3.2 Consider a hypothetical example in which we are interested in comparing two treatments with respect to their ability to reduce low-density lipoprotein (LDL) or "bad" cholesterol on twelve individuals. One of the two treatments is the standard treatment (labeled 0) and the other is new (labeled 1). The goal is to decide whether the new treatment is better than the old one. Twelve participants with very similar pre-treatment LDL cholesterol levels (around 180 mg/dL, considered high) are selected for the study. These twelve participants are numbered $1, \ldots, 12$ in increasing order of their ages in years (denoted by x), which are $20, 25, 30, 35, 40, 45, 50, 55, 60, 65, 70, 75$ respectively. Hypothetical potential outcomes for the twelve participants under the two treatments and the unit-level treatment effects $\tau_1, \ldots, \tau_{12}$ are shown in Table 3.3. The lower the unit-level treatment effects $\tau_i = Y_i(1) - Y_i(0)$ for $i = 1, \ldots, 12$, the better is the new treatment.

One of the goals of the experiment is to estimate the average treatment effect $\overline{\tau}$, the true (but unknown in reality) value of which is -0.5. Looking at the unit-level treatment effects in Table 3.3, it immediately becomes obvious that there is an association between the treatment effect and age. Specifically, the new treatment appears to be superior compared to the old treatment for younger individuals, but the superiority tends to be reduced for older individuals. This association can be observed clearly from Figure 3.1.

TABLE 3.3
Covariate values and all potential outcomes from hypothetical LDL experiment

Unit (i)	Age (x_i)	$Y_i(0)$	$Y_i(1)$	τ_i
1	20	100	97	–3
2	25	99	97	–2
3	30	99	97	–2
4	35	97	96	–1
5	40	102	101	–1
6	45	99	99	0
Block mean	32.5	99.3	97.8	–1.5
7	50	97	97	0
8	55	98	98	0
9	60	100	101	1
10	65	99	99	0
11	70	97	98	1
12	75	100	101	1
Block mean	62.5	98.5	99	0.5
Overall mean	47.5	98.9	98.4	–0.5

If this association was known to the experimenter prior to the experiment, it would be pragmatic to partition the units into two strata or blocks, where the first block consists of units $1, \dots, 6$ for which $\tau_i \leq 0$ and the second block consists of units $7, \dots, 12$ for which $\tau_i \geq 0$. As seen from Table 3.3, the true average treatment effect within block 1 is –1.5 whereas that within block 2 is 0.5. Although these block-level average treatment effects will not be

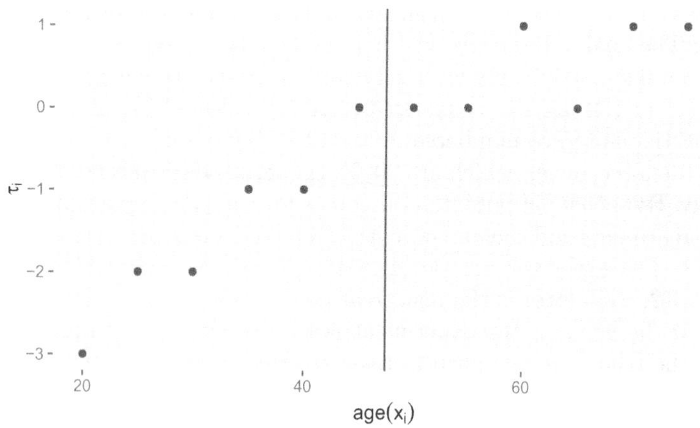

FIGURE 3.1
Treatment effect versus age from Table 3.3

known to the experimenter apriori, the association between age and treatment can possibly be intuited by a domain expert (a medical practitioner in this example) from prior experience, thereby leading to stratification of the twelve units into those two blocks before assigning treatments.

3.3 Blocks, potential outcomes and causal estimands–general notation

As in Chapter 2, consider an experiment with N units and two treatments (i.e., one factor with two levels) labeled 0 and 1. Assuming SUTVA, let $Y_i(j)$ denote the potential outcome corresponding to unit i, when assigned to treatment j, $j = 0, 1$. Suppose the N units are partitioned into B blocks indexed by $1, \ldots, B$, each of size $M_1, \ldots, M_B$ respectively on the basis of pre-treatment covariates, where $M_b \geq 2$ for $b = 1, \ldots, B$. In Example 3.1, $N = 8$, $B = 2$. Assuming that the first four subjects were female and the last four male, the set $\{5, 6, 7, 8\}$ represents block 1, i.e., the block of male units, and $\{1, 2, 3, 4\}$ represents block 2, i.e., the block of female units, and $M_1 = M_2 = 4$. In Example 3.2, $N = 12$, $B = 2$, $M_1 = M_2 = 6$.

The condition $M_b \geq 2$ indicates that the number of units within each block is at least as large as the number of treatments. This condition guarantees that within each block, at least one unit is assigned to each treatment. An RBD with this condition is referred to, in the traditional experimental design literature, as a *complete block design*.

Assuming, for notational simplicity, that the first M_1 units belong to block 1, the next M_2 units belong to block 2, and so on, Table 3.4 shows the potential outcomes for the N experimental units. As before, note that only one of the two potential outcomes from each experimental unit (i.e., in each row of Table 3.4) can be *actually observed*.

We now introduce notation to define various causal estimands (i.e.., treatment effects at various levels - unit, block and population). We will also introduce notation for non-causal estimands (e.g., averages of potential outcomes within blocks, block effects).

For $i = 1, \ldots, N$, we can define unit-level treatment effects, τ_i, in the same way as in Chapter 2, i.e., $\tau_i = Y_i(1) - Y_i(0)$ as per definition (2.1). Next we define block-level treatment effects $\overline{\tau}_b$ by averaging the unit-level treatment effects over the units in block b, $b = 1, \ldots, B$, i.e.,

$$\overline{\tau}_b = \frac{1}{M_b} \sum_{i \in \text{ block } b} \tau_i = \overline{Y}_b(1) - \overline{Y}_b(0), \tag{3.1}$$

where for $j = 0, 1$,

$$\overline{Y}_b(j) = \frac{1}{M_b} \sum_{i \in \text{ block } b} Y_i(j) \tag{3.2}$$

TABLE 3.4
Potential outcomes for the N experimental units in B blocks

Block	Unit	Treatment 0	Treatment 1	Block average	Treatment effect Unit-level	Treatment effect Block-level
1	1 ... M_1	$Y_1(0)$ ... $Y_{M_1}(0)$	$Y_1(1)$ ... $Y_{M_1}(1)$	$\overline{Y}_1(\cdot)$	τ_1 ... τ_{M_1}	$\overline{\tau}_1$
... ...	... ...	... ...	... ...	... ...	... ...	... ...
B	$N-M_B+1$ ... N	$Y_{N-M_B+1}(0)$ ... $Y_N(0)$	$Y_{N-M_B+1}(1)$ ... $Y_N(1)$	$\overline{Y}_B(\cdot)$	τ_{N-M_B+1} τ_N	$\overline{\tau}_B$
Treatment average		$\overline{Y}.(0)$	$\overline{Y}.(1)$	Grand avg $\overline{Y}.(\cdot)$	Simple avg $\overline{\tau}$	Wghtd avg $\overline{\tau}$

represents the average of all potential outcomes for all units corresponding to treatment j in the bth block. Note that the bar in the notation $\overline{\tau}_b$ represents an average treatment effect, and the subscript b indicates that the average is taken over all units in block b.

Finally, the population-level treatment effect is defined exactly as in (2.3), and can also be expressed in terms of the block-level treatment effects as follows:

$$\overline{\tau} = \frac{1}{N}\sum_{i-1}^{N}\tau_i = \frac{1}{N}\sum_{b=1}^{B} M_b\overline{\tau}_b. \tag{3.3}$$

Having defined treatment effects at the block level and at the population level, it is sometimes important to understand if these effects are *similar* or *consistent* across blocks. To measure such similarity, recall that the block-level effect $\overline{\tau}_b$, defined in (3.1), compares the two treatments within block b, and the population-level contrast $\overline{\tau}$ defined in (3.3) compares the treatments in the entire population. Thus, the quantity

$$\overline{\gamma}_b = \overline{\tau}_b - \overline{\tau} \tag{3.4}$$

compares the treatment effect within block b to the average of all unit-level treatment effects in the entire population. The quantity $\overline{\gamma}_b$ is called the *Interaction effect* or *mean-centered interaction effect* between block b and the treatment. A small value of $|\overline{\gamma}_b|$ or its square suggests the treatment effect within block b differs little from the average treatment effect across all blocks. It can be seen that $\sum_{b=1}^{B}\overline{\gamma}_b = 0$, i.e., the mean-centered interaction effects sum up to zero (Exercise 3.2).

In Example 3.1, the 12 units are divided into two blocks based on age as seen in Table 3.3. The average treatment effects in block 1 and block 2 are $\overline{\tau}_1 = -1.5$ and $\overline{\tau}_2 = 0.5$ respectively. The population-level treatment effect $\overline{\tau} = -0.5$. The true interaction effects are

$$\overline{\gamma}_1 = \overline{\tau}_1 - \overline{\tau} = -1, \ \overline{\gamma}_2 = \overline{\tau}_2 - \overline{\tau} = 1,$$

which sum up to zero.

3.4 Assignment mechanism and observed outcomes in RBD

As in chapter 2, for $i = 1, \ldots, N$, define assignment variables W_i such that

$$W_i = \begin{cases} 1 & \text{if the } i\text{th unit is assigned to treatment 1} \\ 0 & \text{if the } i\text{th unit is assigned to treatment 0.} \end{cases}$$

Further, let $\mathbf{W} = (W_1, \ldots, W_N)^{\mathrm{T}}$ denote the $N \times 1$ vector of treatment assignments.

Returning to Example 3.2, consider first a Bernoulli assignment mechanism in which the assignments of twelve units to one of the two treatments are determined by independent tosses of a fair coin. The total number of allocations possible under this mechanism is 2^{12}, two of which will allocate all units to a single treatment, generating assignment vectors $(0, 0, \ldots, 0)$ or $(1, 1, \ldots, 1)$. To avoid such scenarios, the experimenter can use a completely randomized treatment assignment mechanism where it is decided apriori that six units will be assigned to each treatment. Among the $\binom{12}{6}$ possible randomizations that can be generated by this design, the worst possible realizations of $\mathbf{W}$ are $(0, 0, 0, 0, 0, 0, 1, 1, 1, 1, 1, 1)$ and $(1, 1, 1, 1, 1, 1, 0, 0, 0, 0, 0, 0)$. The former assigns the six youngest participants to treatment 0 and the six oldest participants to treatment 1, and the latter does just the opposite. In either case, the effect of age on the outcome (which was seen to be strong) will be confounded with the effect of the treatment.

To avoid such undesirable allocations, we can use a stratified assignment mechanism, also referred to as a *randomized block design* in the literature. In this design, the 12 participants are split into two strata or blocks based on their ages. Thus, block 1 and block 2 respectively consist of the first six and the last six participants. Then a completely randomized assignment mechanism with 3 units assigned to each treatment group is used to assign treatments within each block. The total number of randomizations possible with such a design is $\binom{6}{3} \times \binom{6}{3} = 400$.

To get a deeper insight into why this change of assignment mechanism helps, we recall the imbalance measure between the treatment and control groups (Δ) introduced earlier in this chapter. Denoting the covariate value (age) for unit i by x_i, the average covariate values in the treated and control groups are respectively:

$$\begin{aligned} \overline{x}(1) &= \frac{1}{6} \sum_{i:W_i=1} x_i = \frac{1}{6} \sum_{i=1}^{12} W_i x_i, \\ \overline{x}(0) &= \frac{1}{6} \sum_{i:W_i=0} x_i = \frac{1}{6} \sum_{i=1}^{12} (1 - W_i) x_i. \end{aligned}$$

Then the imbalance measure, $\Delta = \overline{x}(1) - \overline{x}(0)$ can be used to assess the "goodness" of a specific realization of an assignment vector $\mathbf{W}$ and can be computed for any such vector (The readers are asked to compute Δ for the worst possible assignment vectors generated by the Bernoulli assignment mechanism and the balanced completely randomized assignment mechanism in Exercise 3.3). An *ideal* assignment is one for which the criterion is equal to zero.

Under a balanced completely randomized assignment, the expectation of the criterion $\overline{x}(1) - \overline{x}(0)$ is zero (Exercise 3.4), which means a completely randomized assignment mechanism will perfectly balance the groups of units assigned to the two treatments *on an average*. However, one specific realization (e.g., $(0, 0, 0, 0, 0, 0, 1, 1, 1, 1, 1, 1)$ in this example) can be very bad. Using a randomized block design can prevent such undesirable allocations.

Figure 3.2 shows a comparison of the distributions of all possible values of the criterion Δ in Example 3.2 generated by a completely randomized design and a randomized block design with two blocks. Both distributions are centered around zero, but the dispersion is higher for CRD, indicating that the RBD "shrinks" the distribution of the balance measure towards the ideal value of zero.

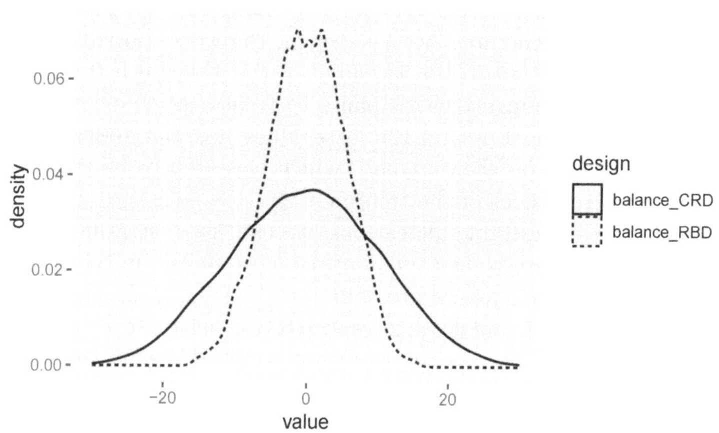

FIGURE 3.2
Distributions of Δ for CRD (solid curve) versus RBD (dotted curve) over all possible randomizations

We now introduce some general notation to describe the assignment mechanism for an RBD. Consider a randomized block design with N units partitioned into B blocks. For $j = 0, 1$ and $b = 1, \ldots, B$, let N_{bj} denote the predetermined number of units in block b to be assigned to treatment j. Consistent with the notation used in Chapter 2, let $\sum_{b=1}^{B} N_{bj} = N_j$ denote the total number of units assigned to treatment j. Because the N_{bj}'s are predetermined for all b and $j = 0, 1$, so are the N_j's. Also, note that $N_{b0} + N_{b1} = M_b$, the size (number of units in) of the bth block. In Example 3.1, $N = 8$, $B = 2$, $M_1 = M_2 = 4$, $N_0 = N_1 = 4$ and $N_{10} = N_{11} = N_{20} = N_{21} = 2$. In the hypothetical partial Shadish experiment, from Table 3.2, $N = 235$, $B = 2$, $M_1 = 169$, $M_2 = 66$, $N_0 = 119$, $N_1 = 116$, $N_{10} = 88$, $N_{11} = 81$, $N_{20} = 31$ and $N_{21} = 35$. In Example 3.2, $N = 12$, $B = 2$, $M_1 = M_2 = 6$, $N_0 = N_1 = 6$ and $N_{10} = N_{11} = N_{20} = N_{21} = 3$.

Note that there are $M_b!/\left(N_{b0}!N_{b1}!\right)$ possible ways in which M_b experimental units in the bth block can be randomly allocated to the two treatments so that N_{bj} units are assigned to treatment $j = 0, 1$. Assuming that the assignment of treatments to units within each block is independent of the assignment of treatments to units in all other blocks, the total number of treatment assignments is therefore the product across the blocks:

$$\frac{M_1!}{N_{10}!N_{11}!} \times \ldots \times \frac{M_B!}{N_{B0}!N_{B1}!} = \prod_{b=1}^{B} \frac{M_b!}{N_{b0}!N_{b1}!}.$$

Thus, in the partial Shadish experiment, a completely randomized treatment assignment mechanism would choose one from 235!/(116!119!) equally likely assignments, whereas a randomized block design with sex as the blocking factor would choose one allocation from 169!/(88!81!) × 66!/(31!35!) equally likely allocations.

Assuming that the randomized assignments are independent across blocks, the joint distribution of $\mathbf{W}$ can be written as

$$Pr(\mathbf{W} = \mathbf{w}) = \begin{cases} \prod_{b=1}^{B} \frac{N_{b1}!N_{b0}!}{M_b!} & \text{if} \quad \sum\limits_{i \in \text{ block } b} W_i = N_{b1} \text{ for all } b, \\ 0 & \quad \text{otherwise.} \end{cases},$$

The constraint $\sum\limits_{i \in \text{ block } b} W_i = N_{b1}$ reflects that within block b, exactly N_{b1} units receive the treatment.

An R code that can be used to generate an assignment vector from an RBD in Example 3.2 is shown below. The algorithm simply generates two separate random permutations of a binary vector with three zeros and three ones, and then concatenates them.

```
---------------------------------------------------------------
set.seed(9999)
> (W_B1 <- sample(c(rep(0,3),rep(1,3))))
[1] 1 0 1 0 0 1
> (W_B2 <- sample(c(rep(0,3),rep(1,3))))
[1] 1 1 0 0 0 1
> (W = c(W_B1, W_B2))
 [1] 1 0 1 0 0 1 1 1 0 0 0 1
---------------------------------------------------------------
```

As earlier, for $i = 1, \ldots, N$, let

$$y_i = W_i Y_i(1) + (1 - W_i) Y_i(0)$$

denote the observed outcome for unit i, and let $\mathbf{y}$ denote the N-vector of observed outcomes. The observed and missing outcomes generated from Table 3.3 using the assignment vector $\mathbf{W} = (1, 0, 1, 0, 0, 1, 1, 1, 0, 0, 1)$ are shown in Table 3.5.

TABLE 3.5
Observed and missing potential outcomes from hypothetical LDL experiment

Block (Age)	Unit (i)	$Y_i(0)$	$Y_i(1)$	W_i	y_i
$1(\leq 45)$	1	?	97	1	97
	2	99	?	0	99
	3	?	97	1	97
	4	97	?	0	97
	5	102	?	0	102
	6	?	99	1	99
Block mean		$\overline{y}_1(0) = 99.3$	$\overline{y}_1(1) = 97.7$		
Block variance		$s^2_{10} = 6.3$	$s^2_{11} = 1.3$		
$2\ (> 45)$	7	?	97	1	97
	8	?	98	1	98
	9	100	?	0	100
	10	99	?	0	99
	11	97	?	0	97
	12	?	101	1	101
Block mean		$\overline{y}_2(0) = 98.7$	$\overline{y}_2(1) = 98.7$		
Block variance		$s^2_{20} = 2.3$	$s^2_{21} = 4.3$		

A few observed outcomes for the hypothetical blocked partial Shadish experiment are shown in Table 3.6.

3.4.1 Estimators of treatment effects

From the definitions of block-level and population-level treatment effects in (3.2) and (3.3), it is clear that unbiased estimators of these can be obtained by substituting unbiased estimators of $\overline{Y}_b(j)$ for $j = 0, 1$ in (3.2) and those of $\overline{Y}(j)$ for $j = 0, 1$ in (3.3). To find these unbiased estimators, we define the following observed treatment averages within and across the blocks. Let

$$\overline{y}.(1) = \frac{1}{N_{b1}} \sum_{i \in \text{ block } b} W_i y_i, \quad \overline{y}.(0) = \frac{1}{N_{b0}} \sum_{i \in \text{ block } b} (1 - W_i) y_i \tag{3.5}$$

denote the average of observed outcomes for treatments 1 and 0 respectively within block b, and for $j = 0, 1$, let

$$\overline{y}.(j) = \frac{1}{N} \sum_{b=1}^{B} M_b \overline{y}_b(j), \tag{3.6}$$

TABLE 3.6
Data from partial Shadish experiment

Block (sex)	Unit (i)	$Y_i(0)$	$Y_i(1)$	W_i	y_i
1 (Female)	1	?	17	1	17
	2	?	16	1	16
	⋮	⋮	⋮	⋮	⋮
	169	7	?	0	7
Block mean		$\overline{y}_1(0) = 8.023$	$\overline{y}_1(1) = 15.926$		
Block variance		$s_{10}^2 = 8.643$	$s_{11}^2 = 14.794$		
2 (Male)	1	?	20	1	20
	2	11	?	0	11
	⋮	⋮	⋮	⋮	⋮
	66	?	12	1	12
Block mean		$\overline{y}_2(0) = 8.226$	$\overline{y}_2(1) = 16.800$		
Block variance		$s_{20}^2 = 9.247$	$s_{21}^2 = 13.047$		

denote the weighted average of $\overline{y}_b(j)$'s over blocks. Then, using the results of Chapter 2, it is straightforward to see that an unbiased estimator of $\overline{\tau}_b$ is

$$\widehat{\overline{\tau}}_b = \overline{y}_b(1) - \overline{y}_b(0). \tag{3.7}$$

Substitution of (3.7) into (3.3) yields the following unbiased estimator of the average unit-level treatment effect $\overline{\tau}$:

$$\widehat{\overline{\tau}} = \frac{1}{N} \sum_{b=1}^{B} M_b \widehat{\overline{\tau}}_b. \tag{3.8}$$

From Table 3.5, point estimates of the average treatment effects within block 1 and block 2 can be computed as follows:

$$\begin{aligned} \widehat{\overline{\tau}}_1 &= \overline{y}_1(1) - \overline{y}_1(0) = 97.7 - 99.3 = -1.6, \\ \widehat{\overline{\tau}}_2 &= \overline{y}_2(1) - \overline{y}_2(0) = 98.7 - 98.7 = 0. \end{aligned}$$

A point estimate of the overall average treatment effect is:

$$\widehat{\overline{\tau}} = \frac{\widehat{\overline{\tau}}_1 + \widehat{\overline{\tau}}_2}{2} = \frac{-1.6}{2} = -0.8.$$

The mean-centered interaction effects can be estimated as:

$$\begin{aligned} \widehat{\overline{\gamma}}_1 &= \widehat{\overline{\tau}}_1 - \widehat{\overline{\tau}} = -1.6 + 0.8 = -0.8, \\ \widehat{\overline{\gamma}}_2 &= \widehat{\overline{\tau}}_2 - \widehat{\overline{\tau}} = 0 + 0.8 = 0.8, \end{aligned}$$

which sum up to zero. Note that when there are only two blocks, as in this example,

$$\widehat{\overline{\gamma}}_1 = \frac{\widehat{\overline{\tau}}_1 - \widehat{\overline{\tau}}_2}{2}, \quad \widehat{\overline{\gamma}}_2 = -\frac{\widehat{\overline{\tau}}_1 - \widehat{\overline{\tau}}_2}{2}.$$

Thus, we can use a single quantity $\widehat{\overline{\tau}}_1 - \widehat{\overline{\tau}}_2$ to unbiasedly estimate the interaction between the block and the treatment.

3.5 Testing Fisher's sharp null hypothesis in an RBD

A randomized block design permits us to test the sharp null hypothesis of a *constant* treatment effect for *all* experimental units within a specific block $b = 1, \ldots, B$. Such a hypothesis for block b can be formally expressed as

$$H_{0F}^b : Y_i(1) - Y_i(0) = \overline{\tau}_b^*. \quad i \in \text{ block } b, \tag{3.9}$$

As in the case of a completely randomized treatment assignment, the plausibility of the above sharp null hypothesis can be assessed using a randomization test with a suitable test statistic. The difference between the unbiased estimator $\widehat{\overline{\tau}}_b$ defined in (3.7) and its true hypothesized value $\overline{\tau}_b^*$, or the following scaled version is a resonable statistic to test this hypothesis:

$$T_{b,\text{avg-unegvar}} = \frac{\widehat{\overline{\tau}}_b - \overline{\tau}_b^*}{\sqrt{s_{b0}^2/N_{b0} + s_{b1}^2/N_{b1}}}, \tag{3.10}$$

where for $j = 0, 1$,

$$s_{bj}^2 = \frac{1}{N_{bj} - 1} \sum_{i \in \text{ block } b} W_i \left(y_i - \overline{y}_b(j)\right)^2, \tag{3.11}$$

is the sample variance of the N_{bj} observed outcomes in the bth block for treatment j. Note that this is essentially the same test statistic used to assess the sharp-null hypothesis of a constant treatment effect under a completely randomized assignment, and was defined earlier in (2.15) of Chapter 2. The only difference here is the subscript b representing the block for which the assessment is being made.

Because under the sharp null hypothesis (3.9), the variances of the potential outcomes for treatments 0 and 1 are equal, another reasonable test statistic is

$$T_{b,\text{avg-eqvar}} = \frac{\widehat{\overline{\tau}}_b - \overline{\tau}_b^*}{s_{b,\text{pooled}}\sqrt{1/N_{b0} + 1/N_{b1}}}, \tag{3.12}$$

where

$$s_{b,\text{pooled}}^2 = \frac{(N_{b0} - 1)s_{b0}^2 + (N_{b1} - 1)s_{b1}^2}{N_{b0} + N_{b1} - 2}, \tag{3.13}$$

is the pooled variance within block b. The statistic (3.12) is essentially the same test statistic defined earlier in (2.18) of Chapter 2, with an additional subscript b representing block b.

As in Section 2.5, we can also test the sharp null hypothesis of a constant treatment effect for *all* of the N experimental units. Note that we can still use the test statistic $T_{\text{avg-unegvar}}$ defined in (2.15) to test this hypothesis, keeping in mind that in this case the randomization distribution of this test statistic under the sharp null hypothesis H_{0F} will be obtained by drawing assignment vectors $\mathbf{w}_{\text{new}}$ from the distribution of $\mathbf{W}$ that follows from the blocked assignment mechanism defined in Section 3.4, rather than the completely randomized assignment mechanism of Chapter 2.

Another reasonable test statistic to test this hypothesis uses a scaling factor based on a weighted average of the sampling variances:

$$T_{\text{wt-avg-uneqvar}} = \frac{\widehat{\overline{\tau}} - \overline{\tau}^*}{N^{-1}\sqrt{\sum_{b=1}^{B} M_b^2\left(s_{b0}^2/N_{b0} + s_{b1}^2/N_{b1}\right)}}. \tag{3.14}$$

A scaling factor that uses pooled variances within each block can also be used:

$$T_{\text{wt-avg-eqvar}} = \frac{\widehat{\overline{\tau}} - \overline{\tau}^*}{N^{-1}\sqrt{\sum_{b=1}^{B} M_b^2 s_{b,\text{pooled}}^2\left(1/N_{b0} + 1/N_{b1}\right)}}, \tag{3.15}$$

where $s_{b,\text{pooled}}^2$ is defined in (3.13). Heuristic justifications for using the above two test statistics are provided in Section 3.5.1.

We now demonstrate the randomization test procedure in an RBD using the observed data for the hypothetical LDL experiment (Example 3.2) shown in Table 3.5. First, let us test whether the data provides evidence against the sharp null hypothesis of no treatment effect within block 1, i.e., $H_{0\text{F}}^1 : Y_i(1) - Y_i(0) = 0$ for $i = 1, \ldots, 6$ using Procedure 2.1. We choose the test statistic $T_{1,\text{avg-uneqvar}}$ defined in (3.10). From Table 3.5, $\widehat{\overline{\tau}}_1 = -1.6$, and consequently the observed value of the test statistic is

$$T^{\text{obs}} = \frac{-1.6}{\sqrt{6.3/3 + 1.3/3}} = -1.$$

Assuming that the experimenters are looking for evidence that the new treatment (1) is better than the old treatment (0), which means LDL cholesterol is expected to be lower under treatment 1 than under treatment 0, a reasonable alternative would be $H_{1F}^1 : Y_i(1) < Y_i(0)$. For this alternative, lower values of the test statistic would be considered extreme. Using the function FISHER_TEST() introduced in Section 2.5.5 of Chapter 2 that is based on the test statistic $T_{1,\text{avg-uneqvar}}$ with 10000 draws, we obtain the desired p-value as 0.35 as shown below.

```
--------------------------------------------------------------
y1 <- c(97,99,97,97,102,99) # Observed outcomes in block 1
W_B1 <- c(1,0,1,0,0,1) # Assignment vector in block 1
> FISHER_TEST(y1, W_B1, 0, ITER=10000)
[1] 0.9471 0.3546 0.6981
--------------------------------------------------------------
```

Note that, instead of obtaining 10000 draws from the distribution of **W**, we can enumerate all the 20 possible assignment vectors using the unique(permn()) function. We show the five of these 20 assignments below, extracted from the list AllrandB1. Calculating the exact one-sided p-value based on these 20 assignments is left as an exercise.

```
------------------------------------------------------------
library(combinat)
AllrandB1 <- unique(permn(W_B1))
> AllrandB1[1:5]
[[1]]
[1] 1 0 1 0 0 1

[[2]]
[1] 1 0 1 0 1 0

[[3]]
[1] 1 0 1 1 0 0

[[4]]
[1] 1 1 0 1 0 0

[[5]]
[1] 1 0 0 1 0 1
------------------------------------------------------------
```

Next, let us test the sharp null hypothesis of no treatment effect on *any* of the N experimental units using the test statistic $T_{\text{wt-avg-eqvar}}$ defined in (3.15). Using the summary statistics in Table 3.7, we calculate the denominator of the observed value of the test statistic as

$$\frac{1}{12}\sqrt{6^2 \times 3.8\,(1/3+1/3) + 6^2 \times 3.3\,(1/3+1/3)} = \frac{6}{12}\sqrt{(7.6/3+6.6/3)}$$
$$= 1.09.$$

TABLE 3.7
Block-wise summary statistics from hypothetical LDL experiment

b	M_b	$\widehat{\overline{\tau}}_b$	s^2_{b1}	s^2_{b0}	$s^2_{b,\text{pooled}}$
1	6	–1.6	6.3	1.3	3.8
2	6	0	4.3	2.3	3.3
Overall	12	–0.8	3.6	2.6	

Thus the observed value of the test statistic is

$$T^{\text{obs}} = \frac{\widehat{\overline{\tau}}}{1.09} = \frac{-0.8}{1.09} = -0.73.$$

The following R function takes $\mathbf{y}$, $\mathbf{W}$ and the block index vector (in this case (1,1,1,1,1,1,2,2,2,2,2,2) and computes the observed value of $T_{\text{wt-avg-eqvar}}$:

```
------------------------------------------------------------
TwtavgEVCOMPUTE = function(y,W,block,taustar=0) {
  # Computes t statistic with unequal variances
```

```
  N = length(y)
  B = max(unique(block)) # number of blocks
  N1 = sum(W)
  N0 = N - N1
  Nb1 = Nb0 = Mb = taubarhatb = s0sqb = s1sqb = sbsqpooled = numeric(B)
  for (b in 1:B){
  yb = y[block==b]
  Mb[b] = length(yb)
  Wb = W[block==b]
  Nb1[b] = sum(Wb)
  Nb0[b] = Mb[b] - Nb1[b]
  taubarhatb[b] = mean(yb[Wb==1]) - mean(yb[Wb==0])
  s0sqb[b] = var(yb[Wb==0])
  s1sqb[b] = var(yb[Wb==1])
  sbsqpooled[b] = ((Nb1[b]-1)*s1sqb[b] + (Nb0[b]-1)*s0sqb[b])/(Nb0[b]
  +Nb1[b]-2)
  } # end b
taubarhat = sum(Mb*taubarhatb)/N
num = taubarhat - taustar
den = 0
for (b in 1:B){
  den = den + (Mb[b])^2 * sbsqpooled[b] * (1/Nb0[b] + 1/Nb1[b])
   }
den = sqrt(den)
den = den/N
return(num/den)
}
-----------------------------------------------
```

Computation of the observed value of $T_{\text{wt-avg-eqvar}}$ from the data in Table 3.5 using the above function is shown below:

```
------------------------------------------------------------
y = c(97,99,97,97,102,99,97,98,100,99,97,101)
W = c(1,0,1,0,0,1,1,1,0,0,0,1)
block = c(1,1,1,1,1,1,2,2,2,2,2,2)
> TwtavgEVCOMPUTE(y,W,block,taustar=0)
[1] -0.7624929
------------------------------------------------------------
```

Next, to obtain the randomization distribution of the test statistic under $H_{0F} : Y_i(0) = Y_i(1)$ for $i = 1, \ldots, N$, we impute the missing outcomes in each row of Table 3.5 with the observed outcome. This entails use of two functions introduced in Section 2.5.5 of Chapter 2: Nbytwo() that converts the vectors $(\mathbf{y}, \mathbf{w})$ into an $N \times 2$ table of observed and missing potential outcomes, and then imputeundersharpnull() to impute the missing outcomes under the sharp null.

```
-------------------------------------------------------------
Y = Nbytwo(y,W)  # Convert (y,W) to an Nx2 matrix

Yimp <- imputeundersharpnull(Y,taustar=0) # Impute missing outcomes
> Yimp
      [,1] [,2]
 [1,]   97   97
 [2,]   99   99
 [3,]   97   97
 [4,]   97   97
 [5,]  102  102
 [6,]   99   99
 [7,]   97   97
 [8,]   98   98
 [9,]  100  100
[10,]   99   99
[11,]   97   97
[12,]  101  101
-------------------------------------------------------------
```

The next step is the key difference between the randomization test procedure described in Chapter 2 for a completely randomized assignment and that adopted here for randomized block design. To generate the randomization distribution of the test statistic, we repeatedly generate new draws of the assignment vector $\mathbf{W} = (W_1, \ldots, W_{12})$ in a manner consistent with the design. This means we generate two *independent* permutations of the vector $(0, 0, 0, 1, 1, 1)$ for the two blocks and concatenate them into a single assignment vector $Wnew$. Each new assignment vector generates new observed outcomes and calculates a new value of the test statistic. The function FISHER_TEST_RBD() shown below takes $\mathbf{y}$, $\mathbf{w}$, vector *block* of block indices, $\overline{\tau}^*$ and the number of iterations (ITER) as arguments and returns three p-values similar to the FISHER_TEST() function in Chapter 2.

```
-------------------------------------------------------------
FISHER_TEST_RBD = function(y,w,block, taustar=0, ITER) {
# Tests sharp null H0:tau_i=taustar against tau_i>taustar
N = length(y) # Number of units
B = max(unique(block)) # Number of blocks
Ypot = Nbytwo(y,w) # Convert data into Nx2 potential outcomes table
Tobs = TwtavgEVCOMPUTE(y,w,block,taustar) # Calculate observed value
of statistic
Yimp = imputeundersharpnull(Ypot, taustar)
# Impute missing potential outcomes

Tarray = NULL
for (i in 1:ITER) {
wnew = NULL
wstart = w[block==1] # assignment vector of block 1
for (b in 1:block) {
```

```
wnew = c(wnew, sample(wstart))
}
ynew = numeric(N)
for (i in 1:N){
ynew[i]=Yimp[i,wnew[i]+1]
}
Tnew =TwtavgEVCOMPUTE(ynew,wnew,block,taustar)
Tarray = c(Tarray, Tnew)
}

hist(Tarray, main=NULL, xlab=NULL)
abline(v=Tobs, lty=2, lwd=3.0)

# Calculate p-value for greater than alternative
pval_plus = sum(Tarray >= Tobs)/ITER

# Calculate p-value for smaller than alternative
pval_minus = sum(Tarray <= Tobs)/ITER

# Calculate p-value for two-sided alternative
pval_twosided = sum(abs(Tarray) >= abs(Tobs))/ITER

return(c(pval_plus, pval_minus, pval_twosided))
}
-------------------------------------------------------------
```

Applying the function on the current example, i.e., on the data in Table 3.5, we obtain a p-value of 0.25 against the one-sided alternative hypothesis $H_{1F}^1 : Y_i(1) < Y_i(0)$. The randomization distribution of the test statistic under the sharp null hypothesis of no treatment effect is shown in Figure 3.3, with the observed value of the test statistic superimposed on the histogram.

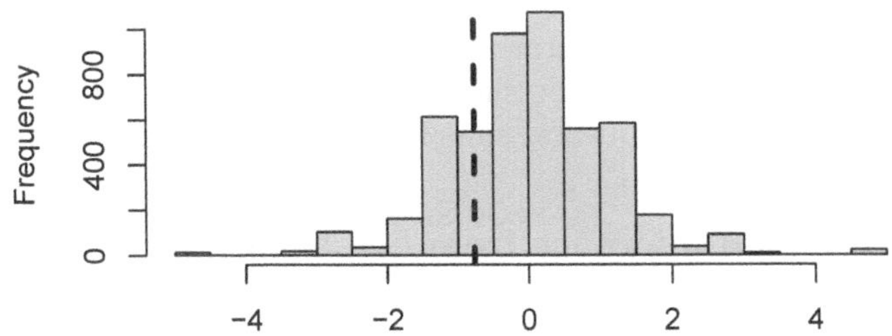

FIGURE 3.3
Randomization tests for the hypothetical LDL RBD experiment with test statistic ($T_{\text{wt-avg-eqvar}}$)

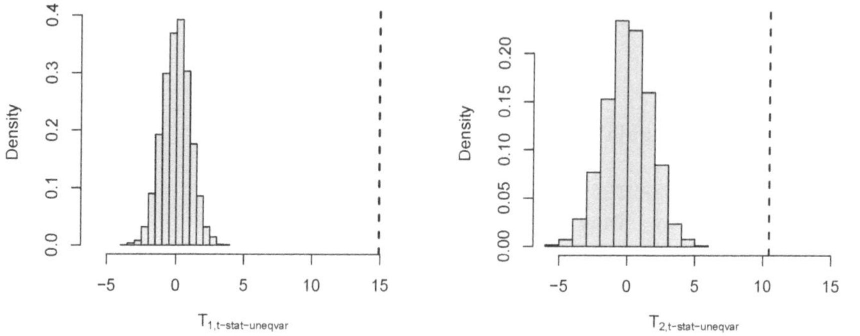

FIGURE 3.4
Randomization tests from partial Shadish experiment (for females on left and males on right)

```
-------------------------------------------------------------
y = c(97,99,97,97,102,99,97,98,100,99,97,101)
W = c(1,0,1,0,0,1,1,1,0,0,0,1)
block = c(1,1,1,1,1,1,2,2,2,2,2,2)
> FISHER_TEST_RBD(y,W,block, taustar=0, ITER=5000)
[1] 0.8072 0.2544 0.5102
-------------------------------------------------------------
```

For the partial Shadish experiment with hypothetical blocking by sex, the observed values of $T_{1,\text{avg-unegvar}}$, $T_{2,\text{avg-unegvar}}$ are 14.913 and 10.467 respectively. The randomization distributions of these two test statistic under the sharp null hypothesis of zero treatment effect within each block are shown in Figure 3.4, with the observed values of the test statistics superimposed on the histograms. Each of the two histrograms shown in the figure are based on 5000 random draws from the distribution of $\mathbf{W}$. Clearly, the estimated one-sided p-value is zero in both cases, thereby suggesting strong evidences aganist the sharp null hypotheses of zero treatment effects within each block.

The randomization test for the sharp null hypothesis of zero treatment effects for *all* of the N experimental units is left as an exercise (problem 3.7).

3.5.1 Neymanian inference from two-armed RBD experiments

We now discuss (asymptotic) Neymanian inference of block-level and population-level treatment effects $\overline{\tau}_b$ defined in (3.7) and (3.8) respectively. Because the treatment assignment within each block is completely randomized, the results of Section 2.6 are directly applicable for drawing inference on $\overline{\tau}_b$. Along the lines of the argument in Section 2.6.1, we can estimate the sampling variance of the estimator $\widehat{\overline{\tau}}_b$ by

$$\widehat{\text{Var}}\left(\widehat{\overline{\tau}}_b\right) = s_{b0}^2/N_{b0} + s_{b1}^2/N_{b1}, \tag{3.16}$$

where for $j = 0, 1$, s_{bj}^2 defined by (3.11) is the sample variance of the N_{bj} observed outcomes in the bth block for treatment j.

As explained in Section 2.6.1, $\widehat{\text{Var}}\left(\widehat{\overline{\tau}}_b\right)$ is an unbiased estimator of the true sampling variance of $\widehat{\overline{\tau}}_b$ if strict additivity holds within block b, i.e., for all $i \in$ block b, $Y_i(1) - Y_i(0)$ is a constant. Thus, to test the Neymanian hypothesis that the average treatment effect in block b is equal to $\overline{\tau}_b^*$, i.e.,

$$H_{0N}^b : \overline{\tau}_b = \overline{\tau}_b^*, \tag{3.17}$$

one can use the statistic

$$T_{b,\text{avg-unegvar}} = \frac{\widehat{\overline{\tau}}_b - \overline{\tau}_b^*}{\sqrt{\widehat{\text{Var}}\left(\widehat{\overline{\tau}}_b\right)}} = \frac{\widehat{\overline{\tau}}_b - \overline{\tau}_b^*}{\sqrt{s_{b0}^2/N_{b0} + s_{b1}^2/N_{b1}}},$$

which can considered a *standardized version* of the statistic $\widehat{\overline{\tau}}_b$. Using approximate normality of this statistic under the null hypothesis H_{0N}^b, the one-side p values under the alternatives $\overline{\tau}_b > \overline{\tau}_b^*$ and $\overline{\tau}_b < \overline{\tau}_b^*$ are, respectively $\text{pr}(Z \geq T_{b,\text{avg-unegvar}}^{\text{obs}})$ and $\text{pr}(Z \leq T_{b,\text{avg-unegvar}}^{\text{obs}})$, where $T_{b,\text{avg-unegvar}}^{\text{obs}}$ is the observed value of the statistic and Z is the standard normal variable. The two-sided p-value for testing this hypothesis is $\text{pr}(|Z| \geq |T_{b,\text{avg-unegvar}}^{\text{obs}}|)$.

An approximate $100(1-\alpha)\%$ confidence interval for the true value of $\overline{\tau}_b$, can be obtained as

$$\widehat{\overline{\tau}}_b \pm z_{\alpha/2}\sqrt{s_{b0}^2/N_{b0} + s_{b1}^2/N_{b1}},$$

where $z_{\alpha/2}$ is the upper $\alpha/2$th quantile of a standard normal distribution.

For the partial Shadish experiment with hypothetical blocking by sex, from Table 3.6, we obtain the estimates of the block-level treatment effects as $\widehat{\overline{\tau}}_1 = 7.903$ and $\widehat{\overline{\tau}}_2 = 8.574$ for females and males respectively, and their estimated variances respectively as $s_{10}^2/N_{10} + s_{11}^2/N_{11} = 0.281$ and $s_{20}^2/N_{20} + s_{21}^2/N_{21} = 0.671$. Consequently, 95% large sample confidence intervals for the block-level treatment effects are obtained as [6.864, 8.942] and [6.969, 10.180] for females and males respectively.

Now returning to the population-level treatment effect, from (3.8), it follows that

$$\text{Var}\left(\widehat{\overline{\tau}}\right) = \frac{1}{N^2}\sum_{b=1}^{B} M_b^2 \text{Var}\left(\widehat{\overline{\tau}}_b\right). \tag{3.18}$$

The above result follows from the independence of treatment assignment across blocks, making the estimators $\widehat{\overline{\tau}}_1, \ldots, \widehat{\overline{\tau}}_B$ mutually independent. Substituting the estimator of $\text{Var}\left(\widehat{\overline{\tau}}_b\right)$ from (3.16) in the right hand side of (3.18), we obtain the Neymanian estimator of the variance of $\widehat{\overline{\tau}}$ as

$$\widehat{\mathrm{Var}}\left(\widehat{\overline{\tau}}\right) = \frac{1}{N^2}\sum_{b=1}^{B} M_b^2\left(s_{b0}^2/N_{b0} + s_{b1}^2/N_{b1}\right), \tag{3.19}$$

Thus, to test the Neymanian hypothesis that the overall average treatment effect is equal to $\overline{\tau}^*$, one can use the statistic

$$T_{\text{wt-avg-uneqvar}} = \frac{\widehat{\overline{\tau}} - \overline{\tau}^*}{\sqrt{\widehat{\mathrm{Var}}\left(\widehat{\overline{\tau}}\right)}},$$

where $\widehat{\mathrm{Var}}\left(\widehat{\overline{\tau}}\right)$ is the Neymanian variance estimator in (3.19) . Using approximate normality of the statistic, under the null hypothesis H_{0N}, the one-side p values under the alternatives $\overline{\tau} > \overline{\tau}^*$ and $\overline{\tau} < \overline{\tau}$ are respectively $\mathrm{pr}(Z \geq T^{\text{obs}}_{\text{avg-unegvar}})$ and $\mathrm{pr}(Z \leq T^{\text{obs}}_{\text{avg-unegvar}})$, where $T^{\text{obs}}_{\text{avg-unegvar}}$ is the observed value of the statistic and Z is the standard normal variable. The two-sided p-value for testing this hypothesis is $\mathrm{pr}(|Z| \geq |T^{\text{obs}}_{\text{avg-unegvar}}|)$.

An approximate $100(1-\alpha)\%$ confidence interval for $\overline{\tau}^*$, can be obtained as

$$\widehat{\overline{\tau}} \pm z_{\alpha/2}\sqrt{\widehat{\mathrm{Var}}\left(\widehat{\overline{\tau}}\right)},$$

where $z_{\alpha/2}$ is the upper $\alpha/2$th quantile of a standard normal distribution and $\widehat{\mathrm{Var}}\left(\widehat{\overline{\tau}}\right)$ is as defined in (3.19).

For the partial Shadish experiment with hypothetical blocking by sex, we obtain $\widehat{\overline{\tau}} = 8.092$ and

$$\widehat{\mathrm{Var}}\left(\widehat{\overline{\tau}}\right) = \frac{1}{235^2}(169^2 * 0.281 + 66^2 * 0.671) = 0.198.$$

Thus, an asymptotic 95% confidence interval for the population-level treatment effect is [7.22, 8.96].

Another Neymanian hypothesis of interest is the equality of treatment effects in two blocks b and b'. i.e.,

$$H_{0N}^{b,b'} : \overline{\tau}_b = \overline{\tau}_b'. \tag{3.20}$$

An unbiased estimator of the difference $\overline{\tau}_b - \overline{\tau}_b'$ is $\widehat{\overline{\tau}_b} - \widehat{\overline{\tau}_b'}$. The sampling variance of this estimator can be estimated by

$$\widehat{\mathrm{Var}}\left(\widehat{\overline{\tau}_b} - \widehat{\overline{\tau}_b'}\right) = \frac{s_{b0}^2}{N_{b0}} + \frac{s_{b1}^2}{N_{b1}} + \frac{s_{b'0}^2}{N_{b'0}} + \frac{s_{b'0}^2}{N_{b'0}}. \tag{3.21}$$

Under what conditions will this estimator be an unbiased estimator of the true sampling variance $\mathrm{Var}\left(\widehat{\overline{\tau}_b} - \widehat{\overline{\tau}_b'}\right)$? The answer is left as an exercise, but obviously assumptions about additive effects will have a role to play.

To test (3.20), one can use the statistic

$$T_{b,b',\text{avg-uneqvar}} = \frac{\widehat{\tau}_b - \widehat{\tau}'_b}{\sqrt{\frac{s^2_{b0}}{N_{b0}} + \frac{s^2_{b1}}{N_{b1}} + \frac{s^2_{b'0}}{N_{b'0}} + \frac{s^2_{b'0}}{N_{b'0}}}}. \tag{3.22}$$

Use of this test statistic to assess whether the treatment effect in the partial Shadish experiment differs between female and male subjects is left as an Excercise.

3.6 Matched-pair designs

Recall the hypothetical experiment in Example 3.1 in which eight individuals - four males and four females - are to be exposed either to a new drug or a placebo. Suppose that we know two out of four males are old (e.g., above seventy) and the other two are young (say, below forty). Similarly two out of four females are old and two are young. It is believed that age can have an inpact on the outcome of interest. Then an RBD with two blocks based on sex may not be the best design anymore, because within either of the two blocks, both of the old individuals can be assigned to the same treatment group, which is undesirable.

One way to avoid such assignments is to create four pairs (blocks) of individuals - old males, young males, old females and young females - and assign the treatment and control randomly between the two individuals within each pair. This is a special case of an RBD discussed in Section 3.2 in which the number of blocks B is half the number of units, and each block is a pair, i.e., has two individuals one of whom receives treatment and the other control. Such a design is called a matched-pair design. These designs find a special place in experimental design and causal inference literature due to the special appeal that arises from *pairing* and their simple applicability.

Similarly, in the hypothetical cholesterol treatment experiment described in Example 3.2, a worst treatment allocation in the RBD with two blocks comprising units $(1, 2, 3, 4, 5, 6)$ and $(7, 8, 9, 10, 11, 12)$ is $(0, 0, 0, 1, 1, 1, 0, 0, 0, 1, 1, 1)$, which means within each block, younger individuals receive the standard treatment and the older individuals receive the new treatment, confounding the estimated treatment effect within each block with the effect of age. To avoid such confounding, we can split the 12 individuals numbered $\{1, \ldots, 12\}$ into six pairs: (1,2), (3,4), (5,6), (7,8), (9,10) and (11,12). Within each pair, the existing treatment is assigned to one unit and the new treatment to the other unit at random.

3.6.1 Some examples of matched-pair experiments

1. In a classic experiment conducted in 1934, Youden and Beale [95] "wished to find out if two preparations of a virus would produce different effects on tobacco plants". Half a leaf of a tobacco plant was rubbed with cheesecloth soaked in one preparation of the virus extract, and the second half was rubbed similarly with the second extract (Page 86, [89]). The outcome was the number of lesions on each half of each leaf. Although it is not exactly clear from this description how the two extracts were assigned to the two halves, assume that this assignment was done randomly, i.e., which half of a leaf would receive which extract was determined by the toss of a fair coin. This experiment was also reported in Snedecor and Cochran (1980) [89] and by Pruzek and Helmreich (2009) [70], and the data can be found in the R package "PairedData" [16].

2. In a medical experiment reported in [49], 18 asthma patients were grouped into nine pairs on the basis of covariates. One of the two patients within each pair was randomly chosen to be treated with a presumably active drug; the other was given a placebo. The outcome, asthmatic relief index, was measured and reported for each patient. Data from this experiment can be found in the R package "BSDA" [2].

3. In a dermatological investigation of the impact of shaving on human hair growth reported in [57], five healthy young men shaved one leg weekly for several months and used the other leg as unshaved control. Although the mechanism for choosing the leg to be shaved is not reported in the study, assume that the investigators assigned the treatment (shaving) to the right or the left leg of each subject randomly. The response was the weight of hairs shaved from specified locations (10 by 10 cm square area on both legs) after the experiment.

3.6.2 Potential outcomes and causal effects

The three matched-pair experiments described in Section 3.6.1 have the same objectives as the experiments (CRD or RBD) described earlier: compare two treatments based on the same fixed number, N, of experimental units. Therefore, we can use the same notation for potential outcomes used so far, (i.e., $Y_i(j)$, $i = 1, \dots, N$, $j \in \{0, 1\}$) to denote the potential outcomes of the ith location when exposed to treatment j. However, because the pairing of the $N = 2B$ units is known in advance, a more explicit way to write the table of potential outcomes is to incorporate the block (pair) information as shown in Table 3.8. As done throughout this chapter, we denote the B pairs (blocks) by $b = 1, \dots, B$.

The unit-level causal effect (treatment difference) for unit i is

$$\tau_i = Y_i(1) - Y_i(0). \tag{3.23}$$

TABLE 3.8
Potential outcomes in a matched-pairs design

Pair/Block (b)	Unit (i)	Treatment 0	Treatment 1	Causal Effect
1	1	$Y_1(0)$	$Y_1(1)$	τ_1
	2	$Y_2(0)$	$Y_2(1)$	τ_2
	Pair Mean	$\overline{Y}_1(0)$	$\overline{Y}_1(1)$	$\overline{\tau}_1(\cdot)$ (Pair-level)
...	...	...	...	...
...	...	...	...	...
B	$N-1$	$Y_{N-1}(0)$	$Y_{N-1}(1)$	τ_{N-1}
	N	$Y_N(0)$	$Y_N(1)$	τ_N
	Pair Mean	$\overline{Y}_B(0)$	$\overline{Y}_B(1)$	$\overline{\tau}_B(\cdot)$ (Pair-level)
Grand Mean		$\overline{Y}(0)$	$\overline{Y}(1)$	$\overline{\tau}$ (Population-level)

The average causal effect for the bth pair is

$$\overline{\tau}_b = \frac{1}{2}\left(\tau_{2b-1} + \tau_{2b}\right) = \overline{Y}_b(1) - \overline{Y}_b(0), \quad b = 1, \ldots, B, \tag{3.24}$$

where, for $j \in \{0, 1\}$

$$\overline{Y}_b(j) = \frac{1}{2}\left(Y_{2b-1}(j) + Y_{2b}(j)\right). \tag{3.25}$$

Finally, the population-level average causal effect is:

$$\overline{\tau} = \frac{1}{B}\sum_{b=1}^{B}\overline{\tau}_b = \frac{1}{N}\sum_{i=1}^{N}\tau_i = \overline{Y}(1) - \overline{Y}_(0), \tag{3.26}$$

where

$$\overline{Y}(j) = \frac{1}{N}\sum_{i=1}^{N} Y_i(j) \tag{3.27}$$

is the average of all N potential outcomes for treatment j, $j = 0, 1$. As discussed earlier, there are other causal estimands, such as the median of the unit-level causal effects, or differences of the medians of potential outcomes corresponding to treatments 1 and 0, but the traditional focus is on the means because the algebraic development is substantially simpler.

3.6.3 Assignment mechanism in matched-pair experiments

Typically in a matched-pair design, each unit within each pair is assigned to either of the two treatments with equal probability (1/2). This means the total number of equally likely possible assignments is 2^B. The observed outcomes will have an expression exactly analogous to that used earlier, i.e.,

$$y_i = W_i Y_i(1) + (1 - W_i) Y_i(0),$$

where W_i is the binary treatment assignment variable taking value 1 if unit i receives treatment and 0 otherwise. However, the key difference here from Chapter 2 lies in the distribution of the W_i's. Letting $\mathbf{W} = (W_1, \ldots, W_N)$ denote the vector of assignment variables, we can write

$$Pr(\mathbf{W} = \mathbf{w}) = \begin{cases} \frac{1}{2^B} & \text{if} \sum_{i \in \text{ block } b} w_i = 1, \text{ for } b = 1, \ldots, B, \\ 0 & \text{otherwise.} \end{cases}, \tag{3.28}$$

The constraint $\sum_{i \in \text{ block } b} w_i = 1, \quad b = 1, \ldots, B$ enforces the assignment of one unit within each pair to treatment and one to control. An R function that can be used to generate matched-pair design with B blocks or pairs is shown below. The function takes B as an input and returns a binary assignment vector $\mathbf{W}$. An example with $B = 6$ pairs is shown below the function where the matched pair assignment vector generated by the function is $\mathbf{w}_M = (1, 0, 0, 1, 0, 1, 1, 0, 1, 0, 1, 0)$.

```
------------------------------------------------------------
MPassignment = function(B) {
W_matched <- NULL
for (i in 1:B) {
W_B = sample(c(0,1))
W_matched <- c(W_matched, W_B)
}
return(W_matched)
}

# Example with 12 units

> set.seed(9999)
> w_M <- MPassignment(6)
> w_M
 [1] 1 0 0 1 0 1 1 0 1 0 1 0
------------------------------------------------------------
```

Table 3.9 shows the observed and missing outcomes from a hypothetical matched-pair cholesterol treatment experiment generated from Table 3.3 using assignment vector $\mathbf{w}_M = (1, 0, 0, 1, 0, 1, 1, 0, 1, 0, 1, 0)$.

The observed data from the asthma treatment experiment, i.e., Example 2 of Section 3.6.1, are shown in Table 3.10. Representation of outcomes of matched-pair experiment in this format is very common in statistics textbooks. However, such a representation of the data does not contain information about the treatment assignments, i.e., does not explicitly show which unit (one of the two paired asthma patients) received the treatment, whereas a representation like the one in Table 3.11 shows the (hypothetical) treatment assignment. Although the information available from Table 3.10 will allow us to perform a valid randomization analysis, assuming that treatments were assigned to the two units within each pair with equal probability, representations

TABLE 3.9
Observed and missing outcomes from the hypothetical LDL treatment matched-pair experiment

Pair (Age)	Unit (i)	$Y_i(0)$	$Y_i(1)$	W_i	y_i
1	1	?	97	1	97
	2	99	?	0	99
2	3	99	?	0	99
	4	?	96	1	96
3	5	102	?	0	102
	6	?	99	1	99
4	7	?	97	1	97
	8	98	?	0	98
5	9	?	101	1	101
	10	99	?	0	99
6	11	?	98	1	98
	12	100	?	0	100

like the ones in 3.9 and 3.11 provides the analyst with a clear understanding of the assignment mechanism, estimands and estimates.

3.6.4 Fisherian inference from matched-pair experiments

As in Section 2.5, the sharp null hypothesis of no treatment difference for *any* experimental unit can be formally expressed as $H_{0F} : Y_i(0) = Y_i(1), \quad i = 1, \ldots, N$, which in the case of matched-pair experiments, has the equivalent

TABLE 3.10
Observed data from the asthma treatment experiment: asthmatic relief index for 18 patients matched in nine pairs

Pair	Observed asthmatic relief index of patients receiving		Difference
	Placebo (0)	Drug (1)	
1	32	28	–4
2	33	31	–2
3	19	17	–2
4	26	22	–4
5	17	12	–5
6	30	32	2
7	26	24	–2
8	19	18	–1
9	25	25	0
Avg	25.22	23.22	–2
Var	34.44	44.69	4.75

TABLE 3.11
Representation of the asthma treatment data showing the potential outcomes and the treatment assignment

Pair (b)	Unit (i)	w_i	Potential outcomes $Y_i(0)$	$Y_i(1)$	Treatment effect for pair b ($\overline{\tau}_b$)	Observed pair difference $\widehat{\overline{\tau}}_b$
1	1	$w_1 = 0$	32	?	$\overline{\tau}_1$	-4
	2	$w_2 = 1$	?	28		
2	3	$w_3 = 1$	?	31	$\overline{\tau}_2$	-2
	4	$w_4 = 0$	33	?		
	...	...	...	...	...	...
	...	...	...	...	...	...
9	17	$w_{17} = 1$	?	25	$\overline{\tau}_9$	0
	18	$w_{18} = 0$	25	?		
Average			$\overline{Y}(0)$	$\overline{Y}(1)$	$\overline{\tau}$	$\widehat{\overline{\tau}}$

representation

$$H_{0F} : Y_i(0) = Y_i(1), \quad i \in \text{ block } b, \quad b = 1, \ldots, B.$$

As before, the plausibility of the above sharp null hypothesis can be assessed by using a randomization test. As test statistic, we first consider a natural estimator of $\overline{\tau}_b$, which is the difference between the two observed outcomes in the bth pair, i.e., units $i \in \{2b-1, 2b\}$:

$$\widehat{\overline{\tau}}_b = y_{2b} - y_{2b-1}. \tag{3.29}$$

It is easy to see (exercise) that $\widehat{\overline{\tau}}_b$ is an unbiased estimator of $\overline{\tau}_b$, that is, the average value of $\widehat{\overline{\tau}}_b$ over all possible equally likely assignment vectors is $\overline{\tau}_b$.

A natural estimator of the population-level causal effect $\overline{\tau}$, defined by (3.26), is the average of $\widehat{\overline{\tau}}_1, \ldots, \widehat{\overline{\tau}}_B$, and hence we can write

$$\widehat{\overline{\tau}} = \frac{1}{B} \sum_{b=1}^{B} \widehat{\overline{\tau}}_b. \tag{3.30}$$

Note that (3.30) is a special case of (3.8) with $N = 2B$ and the block size $M_b = 2$ for all $b = 1, \ldots, B$. The estimator $\widehat{\overline{\tau}}$ is an unbiased estimator of the population-level causal effect $\overline{\tau}$ and would thus be a natural statistic for testing H_{0F}. As seen earlier, another common practice is to scale the test statistic by a measure of variability, for which a popular choice is

$$T_{\text{avg-paired}} = \frac{\widehat{\overline{\tau}}}{s_{\widehat{\overline{\tau}}}/\sqrt{B}}, \tag{3.31}$$

where

$$s^2_{\widehat{\overline{\tau}}} = \frac{1}{B-1}\sum_{b=1}^{B}\left(\widehat{\overline{\tau}}_b - \widehat{\overline{\tau}}\right)^2 \tag{3.32}$$

denotes the sample variance of $\widehat{\overline{\tau}}_1, \ldots, \widehat{\overline{\tau}}_B$. If the outcome is ordinal (e.g., rating on a 1–5 scale), then the rank-based statistic T_{rank} defined in (2.20) may be more appropriate. We demonstrate the randomization test for the asthma treatment experiment in the following subsection. The randomization test for the hypothetical LDL cholesterol treatment matched-pair experiment is left as an exercise (Problem 3.14).

3.6.5 Analysis of the asthma treatment experiment

The asthma treatment experiment data in Table 3.10 can be loaded in R as follows:

```
-------------------------------------------------------------
library(BSDA)
Asthmati

# A tibble: 9 × 3
   drug placebo difference
  <int>   <int>      <int>
1    28      32          4
2    31      33          2
3    17      19          2
4    22      26          4
5    12      17          5
6    32      30         -2
7    24      26          2
8    18      19          1
9    25      25          0
-------------------------------------------------------------
```

Note that in the data frame "Asthmati" shown above, the signs of the differences are reversed from those in Table 3.10, thereby representing negative values of $\widehat{\tau}_b$ defined in (3.29). Thus, we have $\widehat{\overline{\tau}} = -2$, $s^2_{\widehat{\overline{\tau}}} = 4.75$ and consequently using (3.31), the observed value of the test statistic is obtained as $T^{\text{obs}}_{\text{avg-paired}} = -2/\sqrt{4.75/8} = -2.753$. These values can be calculated using the following lines of code in R:

```
-------------------------------------------------------------
> (taubarhat_b = -Asthmati$difference)
[1] -4 -2 -2 -4 -5  2 -2 -1  0
> (taubarhat = mean(taubarhat_b))
[1] -2
```

```
> (var_taubarhat = var(taubarhat_b))
[1] 4.75
> (tpaired_obs = taubarhat/sqrt(var_taubarhat/9))
[1] -2.752989
}
------------------------------------------------------------
```

To conduct Fisher's randomization test, we need to generate values of the test statistic under the sharp null hypothesis for the $2^9 = 512$ possible treatment assignments. As will be obvious from Table 3.11, under the sharp null hypothesis of no treatment effect, a new assignment within each pair simply flips the sign of the observed pair differences $\widehat{\overline{\tau}}_b$'s. There are 2^9 ways of choosing positive or negative signs for $\widehat{\overline{\tau}}_1, \ldots, \widehat{\overline{\tau}}_9$, and each of them will generate a new value of the test statistic, and thereby its randomization distribution. A histogram of the absolute value of the paired t-statistic obtained this way is shown in Figure 3.5. In the R code below, instead of completely enumerating the 2^9 vectors of signs, we obtain 1000 random draws from this set of vectors.

```
------------------------------------------------------------
set.seed(9999)
ITER=1000
tarray=NULL
for (count in 1:ITER){
sign=numeric(9)
for (i in 1:9){
sign[i]=sample(c(-1,1),1)
}
new_taubarhat_b = sign*taubarhat_b
new_tpaired = mean(new_taubarhat_b)/sqrt(var(new_taubarhat_b)/9)
tarray=c(tarray, new_tpaired)
}
hist(tarray,freq=FALSE, main=NULL, xlab="Paired t statistic")
abline(v=tpaired_obs, lwd=3, lty=3)
pval=sum(tarray > tpaired_obs)/ITER
> pval
[1] 0.975
------------------------------------------------------------
```

A one-sided alternative is more appropriate here, because the natural question to ask is whether the drug is more effective than the placebo. The one-sided p-value defined by $\text{pr}(T_{\text{avg-paired}} \geq T^{\text{obs}}_{\text{avg-paired}})$ is 0.975, suggesting no evidence for a significant effect of the asthma drug. In fact, the conclusion is obvious from the fact that the observed value of the test statistic is negative, indicating that the average observed performance under the drug was worse than that under the placebo.

3.6.6 Neymanian inference from matched-pair experiments

The Neymanian inference from matched-pair experiments is based on the sampling properties of the unbiased estimator $\widehat{\overline{\tau}}$, which has been extensively

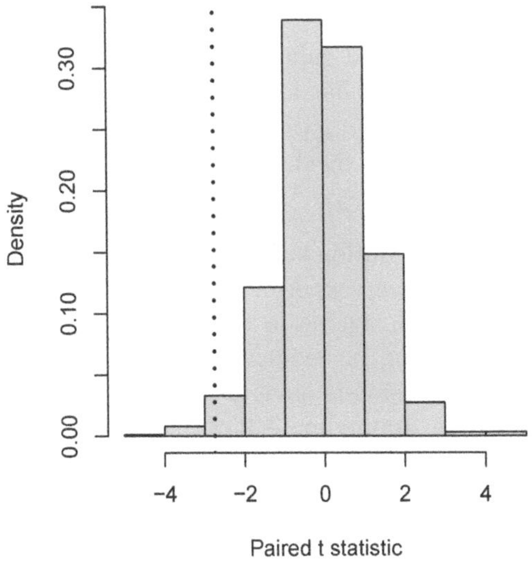

FIGURE 3.5
Randomization test from asthma experiment

studied by researchers (e.g., [36], [43]). After some surprisingly tedious algebraic manipulations, it can be established that the sampling variance of $\widehat{\overline{\tau}}$ with respect to its randomization distribution is:

$$\text{Var}(\widehat{\overline{\tau}}) = \frac{1}{4B^2} \sum_{b=1}^{B} \sum_{i,\tilde{i} \in \Omega_b} \left[(Y_i(1) + Y_i(0)) - (Y_{\tilde{i}}(1) + Y_{\tilde{i}}(0))\right]^2, \qquad (3.33)$$

As earlier, $s^2_{\widehat{\overline{\tau}}}$ is the sample variance of $\widehat{\overline{\tau}}_1, \ldots, \widehat{\overline{\tau}}_B$ defined in (3.32). The expectation of the estimator $s^2_{\widehat{\overline{\tau}}}/B$ was calculated by Imai(2008) [43] as:

$$E[s^2_{\widehat{\overline{\tau}}}/B] = \text{Var}(\widehat{\overline{\tau}}) + \frac{1}{B(B-1)} \sum_{b=1}^{B} (\overline{\tau}_b - \overline{\tau})^2. \qquad (3.34)$$

Equation (3.34) implies that $s^2_{\widehat{\overline{\tau}}}/B$ is an unbiased estimator of $\text{Var}(\widehat{\overline{\tau}})$ *if* $\overline{\tau}_b = \overline{\tau}$ for all b, which means that the treatment effect is the same in each pair, but otherwise $s^2_{\widehat{\overline{\tau}}}/B$ overestimates the actual sampling variance of $\widehat{\overline{\tau}}$.

Thus, from a Neymanian perspective, as discussed in Section 2.6, the estimator $s^2_{\widehat{\overline{\tau}}}/B$ can be called a "statistically conservative" estimator of $\text{Var}(\widehat{\overline{\tau}})$. Hence it can be used to construct asymptotic confidence intervals around $\overline{\tau}$ and also to test the Neymanian null hypothesis of zero average treatment effect, i.e., $H_{0\text{N}} : \overline{\tau} = 0$. against the one-sided alternative $\overline{\tau} > 0$. A natural test statistic for testing $H_{0\text{N}}$ is the statistic defined by (3.31), which can be used to calculate the one-sided p-value defined by $\Pr\left(T_{\text{avg-paired}} \geq T^{\text{obs}}_{\text{avg-paired}}\right)$, where $T^{\text{obs}}_{\text{avg-paired}}$ denotes the observed value of $T_{\text{avg-paired}}$. Using an asymptotic argument [26], it can be shown that for large B, the above probability can be approximated by $\Pr(Z > T^{\text{obs}}_{\text{avg-paired}})$, where Z denotes a standard normal random variable and $T^{\text{obs}}_{\text{avg-paired}}$ is fixed. An approximate $100(1-\alpha)\%$ one-sided confidence interval for $\overline{\tau}$ is $(0, \widehat{\overline{\tau}} + z_\alpha \left(s_{\widehat{\overline{\tau}}}/\sqrt{B}\right)$, where z_α denotes the point on a standard normal distribution such that the area above it is α.

For the asthma treatment experiment, the approximate p-value for the Neymanian hypothesis H_{0N} against a two-sided alternative is thus $\Pr(Z > -2.753) \approx 0.997$, indicating no evidence that the drug is better than the placebo. Again, this conclusion is obvious from the negative value of the observed test statistic. An approximate 95% confidence interval for the true mean difference $\overline{\tau}$ is $-2 \pm 1.96 \times \sqrt{4.75/8}$, i.e., $[0, -0.58]$, which also suggests that use of the drug does not produce any benefits.

3.7 Problems

3.1 Consider an RBD with B blocks, where M_b is the number of units in block B and $N = \sum_b M_b$ is the total number of units. For $i = 1, \dots, N$, let $Y_i(0)$ and $Y_i(1)$ denote the potential outcomes of the ith unit when exposed to treatments 0 and 1 respectively. Define assignment variables W_i satisfying

$$W_i = \begin{cases} 1 & \text{if the } i\text{th unit is assigned to treatment} \\ 0 & \text{otherwise.} \end{cases}$$

Calculate the following:

(i) $\mathbb{E}(W_i)$ and $\text{Var}(W_i)$ for $i \in$ block b and $b \in \{1, \dots, B\}$.

(ii) $\text{Cov}(W_i, W_{\tilde{i}})$ for $i \neq \tilde{i}$, $i, \tilde{i} \in$ block b, $b \in \{1, \dots, B\}$.

(iii) $\text{Cov}(W_i, W_{\tilde{i}})$ for $i \in$ block b, $\tilde{i} \in$ block $\tilde{b}$, $b \neq \tilde{b}$, $b \in \{1, \dots, B\}$.

3.2 Prove that the mean-centered interaction effects $\overline{\gamma}_b$, defined in (3.4), sum to zero, i.e., $\sum_{b=1}^{B} \gamma_b = 0$.

3.3 In Example 3.2, recall that $\overline{x}(1)$ and $\overline{x}(0)$ denote the average ages of participants who receive treatments 1 and 0 respectively and $\Delta = \overline{x}(1) - \overline{x}(0)$ is a measure of imbalance between the two treatment groups.

(i) Write down two assignment vectors $(W_1, \dots, W_{12})$, **other than the worst ones**, that are possible under the Bernoulli assignment but not under the balanced CRD and calculate Δ for each of these assignments.

(ii) Calculate Δ for the two worst assignment vectors $(W_1, \ldots, W_{12})$ under the balanced CRD.

(iii) Write down two assignment vectors $(W_1, \ldots, W_{12})$, other than the worst ones, that are possible under the balanced CRD but not under a balanced RBD with 6 units per block and 3 units assigned at random to each treatment within each block, and calculate the Δ for each of these assignments.

(iv) What is the worst possible assignment vector under the RBD? Calculate Δ for this assignment vector. If there are multiple such "worst" vectors that are equally bad, write all of them down and calculate Δ for each.

3.4 Let $x_1, \ldots, x_N$ denote the pre-measure values of a covariate x for N experimental units. For a given assignment vector $(W_1, \ldots, W_N)$ generated by a completely randomized design with N_1 and N_0 units assigned to treatments labeled 1 and 0 respectively, let

$$\overline{x}(1) = \frac{1}{N_1} \sum_{i=1}^{N} W_i x_i$$

and

$$\overline{x}(0) = \frac{1}{N_0} \sum_{i=1}^{N} (1 - W_i) x_i$$

respectively denote the average covariate value in the two groups assigned to treatments 1 and 0. We define $\overline{x}(1) - \overline{x}(0)$ as the measure of balance for the assignment vector $(W_1, \ldots, W_N)$. Prove that $\mathbb{E}\left[\overline{x}(1) - \overline{x}(0)\right] = 0$.

3.5 What is the total number of possible allocations of the 235 subjects into the two treatment groups in the partial Shadish experiment with a hypothetical randomized treatment assignment with sex as the blocking factor, where the allocation of units into the treated and control groups is predetermined according to Table 3.2. What percentage of all possible allocations in a completely randomized design would be discarded in this hypothetical experiment?

3.6 For the partial Shadish experiment with hypothetical blocking by sex, Figure 3.4 suggests that the treatment effects for females are distributed much more tightly compared to that for females due. Provide an explanation for this phenomenon.

3.7 Assess the plausibility of the sharp null hypothesis of zero treatment effect for *all* of the N experimental units in the hypothetical RBD version of partial Shadish experiment using (a) the $T_{\text{avg-uneqvar}}$ statistic defined in (2.15) and (b) The $T_{\text{wt-avg-uneqvar}}$ defined in (3.14).

3.8 In the hypothetical RBD version of partial Shadish experiment, consider testing whether the average treatment effects in the two blocks (female and male) are the same.

(a) State the conditions under which the estimator defined in (3.21) is an unbiased estimator of the true sampling variance $\text{Var}\left(\widehat{\overline{\tau}}_1 - \widehat{\overline{\tau}}_2\right)$, where $\widehat{\overline{\tau}}_b$ denotes the estimator of the treatment effect in block b. Justify your answer with proper mathematical arguments.

(b) Use the test statistic defined in (3.22) to determine whether the treatment effect in the partial Shadish experiment differs between female and male subjects.

3.9 In a matched-pair design with B pairs indexed by $1, \ldots, B$ and $N = 2B$ units denoted by $i = 1, \ldots, N$, let $Y_i(0)$ and $Y_i(1)$ denote the potential outcomes of the ith unit when exposed to treatments 0 and 1 respectively. Let W_i denote the binary treatment assignment variable defined in Problem 3.1.

Calculate the following:

(i) $\mathbb{E}(W_i)$ and $\text{Var}(W_i)$ for $i \in$ block b and $b \in \{1, \ldots, B\}$.

(ii) $\text{Cov}(W_i, W_{\tilde{i}})$ for $i \neq \tilde{i}$, $i, \tilde{i} \in$ block b, $b \in \{1, \ldots, B\}$.

(iii) $\text{Cov}(W_i, W_{\tilde{i}})$ for $i \in$ block b, $\tilde{i} \in$ block $\tilde{b}$, $b \neq \tilde{b}$, $b \in \{1, \ldots, B\}$.

3.10 Prove that the pair-level difference, $\widehat{\overline{\tau}}_b$ defined by (3.29), is an unbiased estimator of the pair-level causal effect $\overline{\tau}_b$, and the estimator $\widehat{\overline{\tau}}$, defined by (3.30), is an unbiased estimator of the population-level average causal effect $\overline{\tau}$.

3.11 A scientist wants to test the effect of a diet A (treatment) versus a more traditional diet B (control). He has 20 animals available for the experiment. The animals differ with respect to age, sex and other characteristics.

(i) What are the drawbacks of using a completely randomized treatment assignment?

(ii) Suppose the experimenter forms ten pairs of animals, so that within each pair the animals appear "similar to each other" to the experimenter, although the experimenter does not share the precise information that lead to the pairing. Within each pair, an animal gets either diet A or diet B and the other animal gets the other diet; allocation is decided by a flip of a fair coin. The response is the percentage decrease in the animals weight after two months of treatment. The results are shown in Table 3.12. What is the total number of randomizations possible in this experiment?

(iii) Give an example of a randomization that is possible in a completely randomized design but not with this design.

TABLE 3.12
Observed results of the diet experiment

Group	1	2	3	4	5
A	13.2 (1)	8.2 (2)	10.9 (1)	14.3 (1)	10.7 (2)
B	14.0 (2)	8.8 (1)	11.2 (2)	14.2 (2)	11.8 (1)
Group	6	7	8	9	10
A	6.6 (2)	9.5 (1)	10.8 (2)	8.8 (2)	13.3 (1)
B	6.4 (1)	9.8 (2)	11.3 (1)	9.3 (1)	13.6 (2)

(iv) Express the data in the form of a table that shows the observed and the missing potential outcomes and also reveals the observed assignment vector $\boldsymbol{w}$.

(iv) Perform a randomization test to assess Fisher's sharp null hypothesis of no treatment effect. Clearly state the test statistic you use, report the p-value and state your conclusions. Include a histogram that shows your results.

(v) Obtain an approximate 95% confidence interval for the average causal effect using Neyman's method.

3.12 Download data "Tobacco" obtained from the tobacco experiment conducted by Youden and Beale [95], described in Section 3.6.1, from the R package "PairedData" [16]. Perform a randomization test to assess Fisher's sharp null hypothesis of no treatment effect. Clearly state the test statistic you use, report the p-value and state your conclusions.

3.13 Partial data from the shaving experiment described in Section 3.6.1 is shown in Table 3.13. Although the assignment of the treatment (shaving) to the legs (left or right) for the five individuals participating in the study was not reported, assume that the investigators assigned the treatment (shaving) to the right or the left leg of each subject with equal probability

TABLE 3.13
Partial data from shaving experiment with hypothetical assignments

Individual	Leg	Y_W		Y_L	
		Treatment	Control	Treatment	Control
Z	L	−3	?	−0.02	?
	R	?	−1	?	−0.02
F	L	?	−8	?	−0.01
	R	0	?	−0.03	?
M	L	+10	?	−0.02	?
	R	?	+9	?	+0.02
S	L	+10	?	+0.01	?
	R	?	−3	?	−0.03
F	L	?	−35	?	+0.06
	R	−11	?	+0.04	?

(e.g., determined by toss of a fair coin). Table 3.13 shows a set of hypothetical assignments and two responses: difference between median width (Y_W) and median length(Y_L) of hairs sampled from specified locations on the legs before and after the experiment. Assess the sharp null hypotheses of no effect of shaving on the length or width of hair by performing randomization tests separately for the two responses. Comment on the interpretations of these p-values.

3.14 Using the observed outcomes from the hypothetical matched-pair LDL cholesterol experiment shown in Table 3.9,

(i) Conduct a Fisher randomization test to assess whether there is any differential effect between the two treatments on *any* patient.

(ii) Obtain an asymptotic confidence interval for the average treatment effect.

4

Beyond blocking: acceptable versus unacceptable randomizations

4.1 Acceptable randomizations: a stem-cell experiment

As noted in Chapter 1, and expounded in Chapter 3, as we try to protect our chosen allocations of units to treatments more and more from potential confounding effects of covariates, we move from a completely randomized design to a randomized block design, and to the special case of a matched-pair design. In an experiment with, say, 12 experimental units, a balanced completely randomized design has $12!(6!)^2 = 924$ possible allocations of the units into the two treatment groups. In a balanced randomized block design with two blocks of size 6 each, only $\{(6!(3!)^2\}^2 = 400$ of these allocations are allowed (see Exercise 3.2, for an example). A matched pair further reduces the number of allowable randomizations to $2^6 = 64$. Thus, blocked or matched pair designs can be viewed as designs that do not allow certain unacceptable randomized allocations to be used.

Stem cell research has received a vast attention in the early 21st century. Harnessing the ability of human embryonic stem cells to form nearly all cell types in the body is one of the important applications of stem cell therapy. One such application is treatment of Type I diabetes, where the goal is to convert embryonic stem cells to insulin generating pancreatic β cells that diabetic individuals lack. Thus, directing stem cells to tranfom into β cells through a process called "differentiation" yields potential treatment for type I diabetes, which can be triggered through application of small molecule chemical modulators.

Scientists have been conducted experiments to assess whether one or more chemical modulators (or combinations of modulators) can trigger differentiation of a stem cell into a β cell. In one such experiment, a population of healthy stem cells are distributed in physical wells arranged in a rectangular array, with each well comprising approximately 200,000 stem cells. For design purposes, each well can be considered as an experimental unit. For ease of exposition, consider a simplified version of the experiment where we have 12 wells arranged in a 3×4 rectangular array and there are two chemical modulators (treatment factors) whose effects on cell conversion are to be assessed. The modulators will be dispensed into the 12 wells through a robotic

DOI: 10.1201/9781003048794-4

workstation based on a stipulated assignment of treatments to wells. After a specified number of days, the percentage of cells in each well that expressed specific proteins, which are important markers of pancreatic β cells, are measured. Each such percentage is a response in this experiment.

Thus, the design problem is to assign six wells to the active treatment (i.e. dispensing a new modulator) and six to the control (dispensing an existing modulator). The researchers are mindful of spatial effects that may interfere with the outcomes of experiments. For example, wells located on the perimeter of the rectangular array may react differently from wells located at the center. There may also be unmeasured factors capable of affecting the response, that change across rows and columns. Therefore, a completely randomized assignment is not a good solution to this design problem because it may yield an allocation like the one shown in the left panel of Figure 4.1, which is undesirable due to possible confounding with column effects. However, a scientist may consider the allocation shown in the right panel of Figure 4.1 as an acceptable allocation because it spreads treated and control units somewhat uniformly across the rectangular array. However, it is difficult to avoid an unacceptable allocation by simple blocking.

1	1	0	0
1	1	0	0
1	1	0	0

0	1	0	0
0	1	1	0
1	0	1	1

FIGURE 4.1
Unacceptable allocation (left) and acceptable allocation (right)

In this case, each unit has two pre-measured covariates - the row number denoted by x_1 and the column number denoted by x_2. Table 4.1 shows the covariate values of the 12 units arranged by row. The last two columns show the two assignment vectors corresponding to the two allocations in Figure 4.1, the first one denoted by $\mathbf{w}_1$ and the second allocation by $\mathbf{w}_2$.

Each assignment vector partitions the 12 units into two groups. Let $\overline{x}_1(0)$ and $\overline{x}_1(1)$ denote the averages of x_1 in the control and treated groups respectively. Similarly let $\overline{x}_2(0)$ and $\overline{x}_2(1)$ denote the averages of x_2 in the control and treated groups respectively. Define vectors of covariate means in the treatment and control groups as:

$$\overline{\mathbf{x}}(1) = \begin{pmatrix} \overline{x}_1(1) \\ \overline{x}_2(1) \end{pmatrix} \text{ and } \overline{\mathbf{x}}(0) = \begin{pmatrix} \overline{x}_1(0) \\ \overline{x}_2(0) \end{pmatrix}$$

Then the 2-dimensional vector

$$\overline{\mathbf{x}}(1) - \overline{\mathbf{x}}(0) = \begin{pmatrix} \overline{x}_1(1) - \overline{x}_1(0) \\ \overline{x}_2(1) - \overline{x}_2(0) \end{pmatrix}$$

TABLE 4.1
Units, covariates and assignments

Unit	x_1	x_2	$\mathbf{w}_1$	$\mathbf{w}_2$
1	1	1	1	0
2	1	2	1	1
3	1	3	0	0
4	1	4	0	0
5	2	1	1	0
6	2	2	1	1
7	2	3	0	1
8	2	4	0	0
9	3	1	1	1
10	3	2	1	0
11	3	3	0	1
12	3	4	0	1

represents the difference of the covariate means between the treatment and control groups. For the assignment vectors $\mathbf{w}_1$ and $\mathbf{w}_2$, $\overline{\mathbf{x}}(1)-\overline{\mathbf{x}}(0) = (0,-2)^{\mathrm{T}}$ and $(2/3,0)^{\mathrm{T}}$ respectively (see R code below).

```
---------------------------------------------------------
x1<- c(1,1,1,1,2,2,2,2,3,3,3,3)
x2 <- rep(seq(1,4,1),3)
w1 <- rep(c(1,1,0,0),3)
w2 <- c(0,1,0,0,0,1,1,0,1,0,1,1)

# Design 1
(x1D1 = mean(x1[w1==1]) - mean(x1[w1==0]))
(x2D1 = mean(x2[w1==1]) - mean(x2[w1==0]))
> (x1D1 = mean(x1[w1==1]) - mean(x1[w1==0]))
[1] 0
> (x2D1 = mean(x2[w1==1]) - mean(x2[w1==0]))
[1] -2

# Design 2
(x1D2 = mean(x1[w2==1]) - mean(x1[w2==0]))
(x2D2 = mean(x2[w2==1]) - mean(x2[w2==0]))
> (x1D2 = mean(x1[w2==1]) - mean(x1[w2==0]))
[1] 0.6666667
> (x2D2 = mean(x2[w2==1]) - mean(x2[w2==0]))
[1] 0
---------------------------------------------------------
```

Figure 4.2 shows the 2-dimensional vector $\overline{\mathbf{x}}(1) - \overline{\mathbf{x}}(0)$ as points on a 2-dimensional plot for all treatment assignment vectors under a completely randomized allocation of 12 units into two groups. Each point in this plot corresponds to one specific allocation, (i.e., treatment assignment vector), with the ones corresponding to $\mathbf{w}_1$ and $\mathbf{w}_2$ shown on the plot. The size of each point is proportional to the percentage of points that have the same values of $\overline{\mathbf{x}}(1) - \overline{\mathbf{x}}(0)$. For example, 5% of the assignments yield the vector $(0, 0)$, which represents the "best" assignments in the sense that they perfectly balance the treated and control groups with respect to covariate averages.

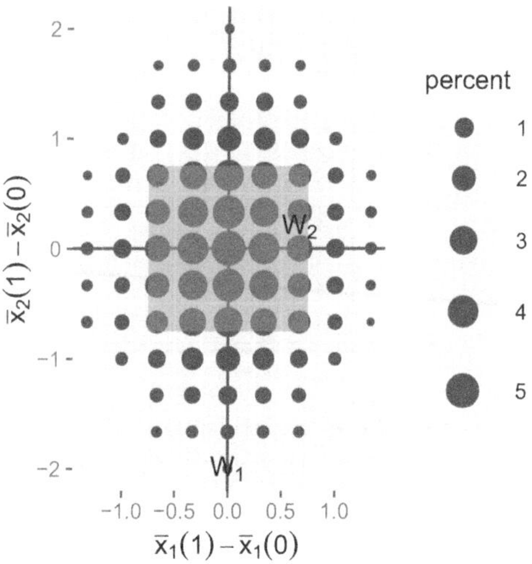

FIGURE 4.2
Difference of covariate means for all assignments

Points close to the origin $(0, 0)$ of the graph represent assignment vectors that are *good* in the sense that they balance the treated and control groups well with respect to the covariates, with points at the origin representing perfect covariate balance. The farther points are from the origin, the less balanced are the groups. As shown in Figure 4.2, an acceptable set of assignment vectors can be represented by all the points within a square that permits a largest deviation of a in any direction from the origin. Thus, any assignment $\mathbf{w}$ satisfying both

$$|\overline{x}_1(1) - \overline{x}_1(0)| \leq a_1, \quad \text{and} \quad |\overline{x}_2(1) - \overline{x}_2(0)| \leq a_2,$$

where a_1 and a_2 are two predefined positive real numbers, is acceptable; others are not. In the figure, the thresholds a_1 and a_2 are both taken to be equal to $3/4$, making assignment vector $\mathbf{w}_1$ unacceptable but $\mathbf{w}_2$ acceptable, because the point corresponding to this assignment vector lies within the square. Choosing $a_1 = a_2 = 1/3$ would make both $\mathbf{w}_1$ and $\mathbf{w}_2$ unacceptable.

With $a_1 = a_2 = 3/4$, the design, or the treatment assignment mechanism, is defined by an equally likely draw from the set of "acceptable" assignment vectors that produce a point within the square. Such a design can be constructed by an acceptance-rejection algorithm, i,e, randomly drawing a sample **w** from the set of 924 possible vectors under a complete randomized assignment mechanism and retaining the drawn allocation only if the covariate balance satisfies the acceptability criteria. If not, another sample is drawn, and the process continues until an acceptable **w** is obtained.

An alternative acceptance criterion

Whereas a measure of imbalance based on the difference of covariate means between the treatment and control groups is often natural and appealing, an alternative way of measuring imbalance can be motivated by the desire to "spread out" the treatment and control locations across the 3×4 array of wells. Thus, we can consider the vector $\left(\overline{d}(0), \overline{d}(1)\right)^{\mathrm{T}}$ of average pairwise Euclidean distance between the units in the control and treatment groups respectively as a measure of imbalance between the two groups. According to this definition, the larger the elements of this vector, the better is the allocation in terms of covariate balance.

To illustrate this idea, denote the covariates of the ith unit by (x_{1i}, x_{2i}). Recall that the Euclidean distance between unit i and $\widetilde{i}$ is

$$d(i,\widetilde{i}) = \sqrt{\left(x_{1i} - x_{1\widetilde{i}}\right)^2 + \left(x_{2i} - x_{2\widetilde{i}}\right)^2}.$$

Then the average pairwise Euclidean distances between units in the control group and the treatment group are, respectively,

$$\overline{d}(0) = \frac{1}{15} \sum_{\substack{i,\widetilde{i}:w_i=w_{\widetilde{i}}=0 \\ i\neq\widetilde{i}}} d(i,\widetilde{i}), \quad \overline{d}(1) = \frac{1}{15} \sum_{\substack{i,\widetilde{i}:w_i=w_{\widetilde{i}}=1 \\ i\neq\widetilde{i}}} d(i,\widetilde{i}),$$

where as before, w_i denotes the binary assignment variable for unit i.

Consider the assignment vector $\mathbf{w}_1$ shown in Table 4.1. The Euclidean distance matrix for the 15-pairs of units in the control group is shown in Table 4.2, and the average Euclidean distance in the control group is $\overline{d}(0) = 1.409$. The R function used to obtain these outputs is shown below:

```
-------------------------------------------------------
# Given any (X,w), the following function computes the distance matrix
# and average distance between points in treated and control groups
avg_dist = function(X,w) {
X_c<- X[w==0,]
X_t <- X[w==1,]
return(list(dist(X_c), dist(X_t), mean(dist(X_c)),mean(dist(X_t))))
}

X = cbind(x1,x2)

> round(avg_dist(X,w1)[[1]],3)
      1     2     3     4     5
```

```
2 1.000
3 1.000 1.414
4 1.414 1.000 1.000
5 2.000 2.236 1.000 1.414
6 2.236 2.000 1.414 1.000 1.000

> (dw1=avg_dist(X,w1))[[3]] # average distance for control group
 under w1
[1] 1.408599
-------------------------------------------------------
```

The matrix representation of assignment $\mathbf{w}_1$ in the left panel of Figure 4.1 implies that the average distance among the treated and control units are the same for this allocation, i.e., $\overline{d}(0) = \overline{d}(1) = 1.409$.

TABLE 4.2
Euclidean distances between pairs of units in the control group

Coordinates (x_1, x_2)	(1,3)	(1,4)	(2,3)	(2,4)	(3,3)
(1,4)	1				
(2,3)	1	1.414			
(2,4)	1.414	1	1		
(3,3)	2	2.236	1	1.414	
(3,4)	2.236	2	1.414	1	1

Figure 4.3 shows the 2-dimensional vector $(\overline{d}(0), \overline{d}(1))$ as points on a 2-dimensional plot for all treatment assignment vectors under a completely randomized allocation of 12 units into two groups. Each point in this plot corresponds to one specific treatment assignment vector, with the one corresponding to $\mathbf{w}_1$ shown on the plot.

Contrary to Figure 4.2, in Figure 4.3, points close to a straight line $\overline{d}(0) = \overline{d}(1)$ that are distant from the origin $(0, 0)$ of the graph represent assignment vectors that are *good* in the sense that they spread out the points both in the treated and control groups. Points close to the origin indicate represent allocations that do not spread the units out in either group. As shown in Figure 4.2, an acceptable set of assignment vectors can be represented by all the points within a square that permits a largest deviation of a in any direction from the origin. Thus, assignments $\mathbf{w}$ satisfying

$$\min\left(\overline{d}(0), \overline{d}(1)\right) \geq a$$

can be considered acceptable. In the figure, the threshold a is taken to be 1.8, generating the rectangular region in the upper right corner of the graph bounded by the dotted lines. Under this criterion, assignment vector $\mathbf{w}_1$ is unacceptable, despite lying on the 45-degree line. The readers are asked to verify the acceptability of assignment vector $\mathbf{w}_2$.

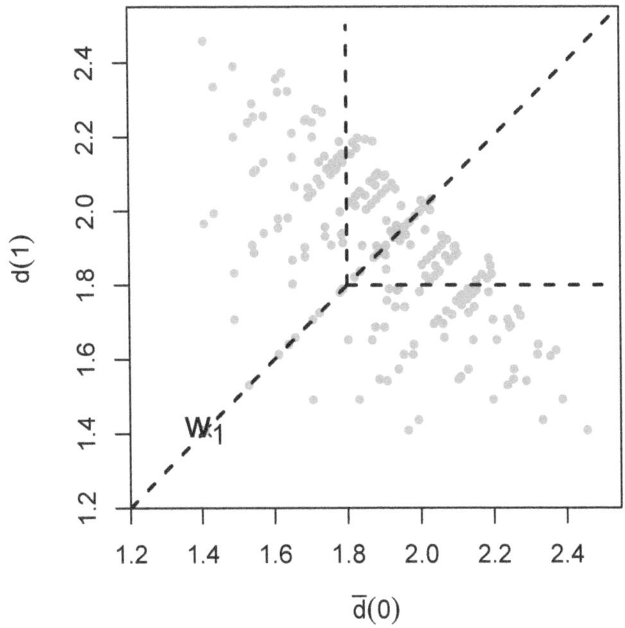

FIGURE 4.3
Difference of covariate means for all assignments

4.2 Framework for design and analysis

The idea presented in Section 4.1 formalizes the unified framework introduced in Chapter 1 for experiments designed using a pre-defined criterion for acceptable allocations in terms of covariate balance. The Fisherian mode of inference is a natural choice in this framework. We start with a finite population of N units with pre-measured covariates $x_1, \ldots, x_k$ for each unit. Table 4.3 shows the covariate values and the fixed potential outcomes in this setting.

TABLE 4.3
Units, covariates and potential outcomes

Unit	Covariates			Potential outcomes	
	x_1	$\cdots$	x_k	$Y(0)$	$Y(1)$
1	x_{11}	$\cdots$	x_{k1}	$Y_1(0)$	$Y_1(1)$
2	$\vdots$	$\vdots$	$\vdots$	$\vdots$	$\vdots$
N	x_{1N}	$\cdots$	x_{kN}	$Y_N(0)$	$Y_N(1)$

The steps involved in designing the experiment are:

Step 1: Specify balance criterion determining when a randomized allocation (or an assignment vector) is acceptable.

Step 2: Obtain a realization of the randomly drawn assignment vector $\mathbf{w}$ that satisfies the criterion for acceptance defined in step 1. A naive but natural way to do this is to generate a vector $\mathbf{w}$ without restriction under a completely randomized mechanism and *reject and re-randomize* if the draw does not satisfy the acceptance criterion. Such an acceptance-rejection rule was termed "rerandomization" by Morgan and Rubin (2012) [61] and has received considerable attention in literature since then. Figure 4.4 depicts the steps involved in rerandomization.

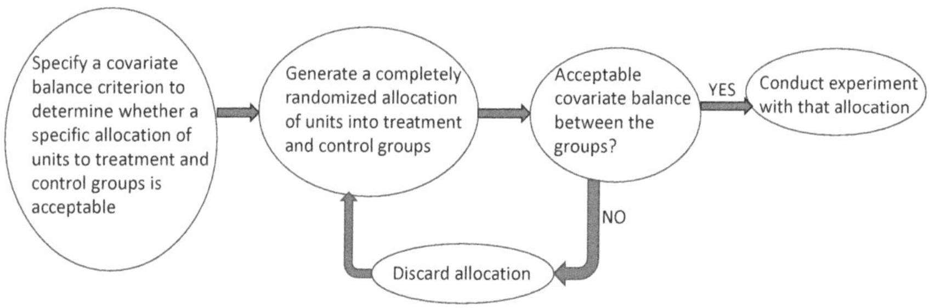

FIGURE 4.4
Designing a rerandomized experiment

Analysis

The randomized design with pre-defined acceptability criterion permits us to a randomization test to compute a p-value to assess the sharp null hypothesis of a constant treatment effect on each unit. By this time, the readers should be quite familiar with the mechanics of the randomization test. As in Chapters 2 and 3, any test statistic can be used. After imputation of the missing potential outcomes under the sharp null, the randomization distribution of the test statistic T under the sharp null can be obtained by generating $\mathbf{w}$'s *that satisfy the same acceptance criterion used to design the experiment*, and then the p-value can be used to determine the extremeness of the observed value of the test statistic with respect to its randomization distribution under that sharp null.

We illustrate the randomization test through a hypothetical version of the simplified stem cell experiment described in Section 4.1. Assume that two modulators - one that is already known to be useful (indicated by 0) and a new untested modulator (indicated by 1) - are being compared with respect to their ability to convert embryonic stem cells to insulin generating pancreatic β-cells. Consider the acceptance criterion

$$|\overline{x}_1(1) - \overline{x}_1(0)| \leq 3/4, \quad |\overline{x}_2(1) - \overline{x}_2(0)| \leq 3/4.$$

Then the assignment vector $\mathbf{w}_2$ shown in Table 4.1 is acceptable, and can be used to conduct the experiment. The R function sample_acceptable() shown below can be used to check if an assignment vector meets the acceptance criteria that the difference between the vectors of covariate averages in the treated and control groups lies within a hypercube (generalization of a square) centered at the origin with vertices located at a units from the center along each axis. As seen in the examples below the function, the function rejects assignment vector $\mathbf{w}_1$ (the output is FALSE) but accepts $\mathbf{w}_2$ (the output is TRUE).

```
-------------------------------------------------------
####### Function to draw from acceptable randomization from a hypercube
## hypercube centered at the origin with vertices "a" units from center
## inputs are (a) a binary assignment vector W of N0 zeros and N1 ones
## (b) an Nxp matrix X of covariates

sample_acceptable = function(X,W,a) {
  flag = FALSE
  Xt = X[W==1,]
  Xc = X[W==0,]
  Xtbar = as.vector(apply(Xt,2,mean))
  Xcbar = as.vector(apply(Xc,2,mean))
  if (max(abs(Xtbar - Xcbar)) <= a) {
    flag = TRUE
  }
  return(flag)
}

> sample_acceptable(X,w1,3/4)
[1] FALSE
> sample_acceptable(X,w2,3/4)
[1] TRUE
-------------------------------------------------------
```

The outcome variable in this experiment is the proportion of embryonic stem cells in each well that are converted to β-cells. The observed assignment and outcome are shown in Figure 4.5

We can use the test statistic $T_{\text{avg-eqvar}}$ defined in (2.18) to assess the sharp null hypothesis of no difference between the treatments for any unit. The observed value of the statistic is 2.43. To obtain its randomization distribution under the sharp null of no treatment different, we only need to make one subtle change to the procedure (and R functions) outlined in Section 2.5.5. The function Fisher_test() needs two more inputs - the covariate matrix $\mathbf{X}$ and the

0	1	0	0
0	1	1	0
1	0	1	1

0.10	0.13	0.00	0.04
0.23	0.34	0.18	0.16
0.29	0.27	0.56	0.53

FIGURE 4.5
Observed assignment (left) and outcomes (right) in the hypothetical simplified stem cell experiment

acceptance threshold a. A realization of a completely randomized assignment mechanism would be a vector that is a random permutation of six zeros and six ones. Every time such a "trial vector" wtry is obtained, it is checked using the function sample_acceptable(), and if found acceptable, it is considered as a new acceptable assignment wnew under the stated criterion. Thus, among the total number of iterations ITER specified by the user, only a fraction would be used to generate the randomization distribution of the test statistic under the sharp null hypothesis. This modified R function FISHER_TEST_hypercube is shown below and is used to conduct the re-randomization test with the data in Figure 4.5.

```
------------------------------------------------------
# Fisher re-randomization test of sharp null tau_i = tau^*
# using test statistic with equal variances
#########################################
FISHER_TEST_hypercube = function(y,w,X,a,taustar,ITER) {
  # Tests sharp null H0:tau_i=taustar against one or two-sided
  alternatives
  # Inputs: observed outcomes y, observed assignment w.
  # covariate matrix X, hypercube size from origin a
  N = length(y) # Number of units
  Ypot = Nbytwo(y,w)
  Tobs = TEQUALCOMPUTE(y,w, taustar) # Calculate observed value of
  statistic
  Yimp = imputeundersharpnull(Ypot, taustar)
# Create imputed matrix of potential outcomes
  Tarray = NULL

  for (i in 1:ITER) {
    wtry = sample(w) # Try any permutation
    if (sample_acceptable(X,wtry,a) == TRUE){  # if acceptable
    wnew = wtry
    ynew = numeric(N)
    for (i in 1:N){
      ynew[i]=Yimp[i,wnew[i]+1]
    }
    Tnew =TEQUALCOMPUTE(ynew,wnew, taustar)
    Tarray = c(Tarray, Tnew)
    } # if acceptable
```

```
  } # end ITER

  acc = length(Tarray) # number of randomizations accepted
  hist(Tarray, main=NULL, xlab=NULL)
  abline(v=Tobs, lty=2, lwd=3.0)

  # Calculate p-value for greater than alternative
  pval_plus = sum(Tarray >= Tobs)/ITER

  # Calculate p-value for smaller than alternative
  pval_minus = sum(Tarray <= Tobs)/ITER

  # Calculate p-value for two-sided alternative
  pval_twosided = sum(abs(Tarray) >= abs(Tobs))/ITER

  return(list(acc/ITER,c(pval_plus, pval_minus, pval_twosided)))
}


# Test for simplified stem cell experiment
w2 = c(0,1,0,0,0,1,1,0,1,0,1,1)
x1<- c(1,1,1,1,2,2,2,2,3,3,3,3)
x2 <- rep(seq(1,4,1),3)
Xstem = cbind(x1,x2)
ystem = c(0.10,0.13,0.00,0.04,0.23,0.34,0.18,0.16,0.29,0.27,0.56,0.53)

> set.seed(100)
> FISHER_TEST_hypercube(ystem,w2,Xstem, 2/3, taustar=0,ITER=10000)
[[1]]
[1] 0.7088

[[2]]
[1] 0.0037 0.7070 0.0085
-------------------------------------------------------
```

Using 10000 draws from the distribution of assignment vector $\mathbf{w}$ under a completely randomized assignment mechanism, we find from the output of the above function that roughly 70% satisfy the acceptance criterion

$$|\overline{x}_1(1) - \overline{x}_1(0)| \leq 3/4, \quad |\overline{x}_2(1) - \overline{x}_2(0)| \leq 3/4.$$

The histogram of the values of the test statistic obtained from these acceptable draws is shown in Figure 4.6. The estimated p-value for the one-sided alternative $\tau_i > 0$ for all $i = 1, \ldots, N$ is $\text{pr}(|T_{\text{avg-eqvar}}| \geq 2.43)$ is 0.0037, suggesting that the hypothesis of no treatment effect is not likely.

We end this section with the observation that the rerandomization procedure can be computationally heavy because if the proportion of acceptable randomizations is π, then following the rules of geometric distribution, the expected number of draws required to obtain one acceptable randomization

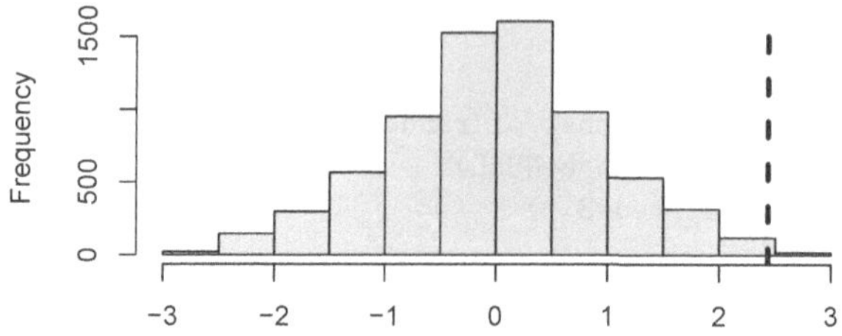

FIGURE 4.6
Rerandomization test for hypothetical stem cell experiment

is $1/\pi$. Thus in order to obtain one acceptable draw from a rule that allows 20% of all draws, one is expected to generate five draws. More generally, in order to generate r acceptable draws from a distribution with proportion π of acceptable randomizations, one needs to draw r/π random samples on an average. This follows from the negative binomial law. See Exercise problem 4.4 for details.

4.3 Mahalanobis distance as the acceptance criterion

Although there are many options for measuring covariate distance between treatment and control groups as seen in Section 4.1 (also see Problem 4.3), balancing means is typically a preferred strategy because it has the benefit of simple interpretability. As mentioned in [61], it also has several statistical advantages, including use of the finite population central limit theorem, making theoretical results straightforward.

Recall that in the simplified stem call experiment introduced in Section 4.1 with covariate mean vectors in the treatment and control groups, $\overline{\mathbf{x}}(1)$ and $\overline{\mathbf{x}}(0)$, and with the 2-dimensional vector $\overline{\mathbf{x}}(1) - \overline{\mathbf{x}}(0)$ of difference of covariate means between the treatment and control groups as a measure of covariate balance, acceptable allocations were defined by placing bounds $|\overline{x_1}(1) - \overline{x_1}(0)| \leq a_1$ and $|\overline{x_2}(1) - \overline{x_2}(0)| \leq a_2$ on the two individual elements of $\overline{\mathbf{x}}(1) - \overline{\mathbf{x}}(0)$, to ensure that each covariate difference was within its specified caliper.

This strategy resulted in the region of acceptable allocations taking the form of a box or rectangle (or square when $a_1 = a_2$) as seen in Figure 4.2. As suggested by Morgan and Rubin (2012), a better strategy may be to use a univariate measure of multivariate (bivariate in the current example) balance. The most natural measure is the norm (or length) of the vector $\overline{\mathbf{x}}(1) - \overline{\mathbf{x}}(0)$ that measures the Euclidean distance (D_{E}) between the two vectors. Thus, we can define an acceptance criterion of the form

$$D_{\mathrm{E}} \equiv ||\overline{\mathbf{x}}(1) - \overline{\mathbf{x}}(0)|| \leq a, \tag{4.1}$$

where a is a predetermined positive real number. Because (4.1) is equivalent to

$$(\overline{x}_1(1) - \overline{x}_1(0))^2 + (\overline{x}_2(1) - \overline{x}_2(0))^2 \leq a^2,$$

the acceptance region represents a circle on the two-dimensional plot of $(\overline{x}_1(1) - \overline{x}_1(0), \overline{x}_2(1) - \overline{x}_2(0))$ with center at the origin and radius a, as shown in Figure 4.7, where $a = 3/4$. Note that the assignment vector $\mathbf{w}_1$ is still deemed unacceptable by this criterion, whereas the assignment vector $\mathbf{w}_2$, lying within the circle, is considered acceptable.

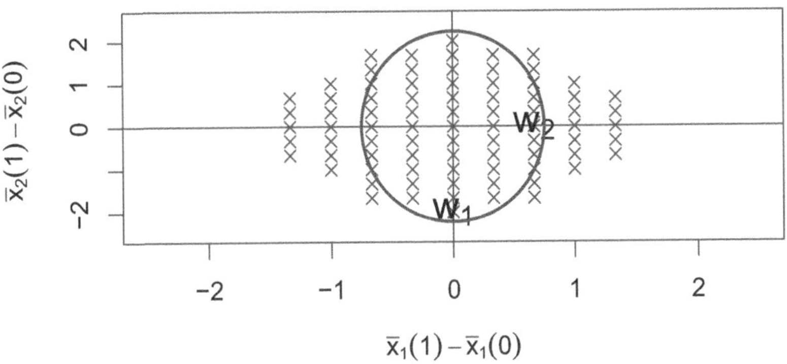

FIGURE 4.7
Acceptance region based on Euclidean distance for the simplified stem cell experiment

However, this measure based on the Euclidean distance does not reflect the correlation between the vectors $\overline{\mathbf{x}}(1)$ and $\overline{\mathbf{x}}(0)$. Morgan and Rubin [61] suggest using a measure that takes into consideration the joint distribution of the covariates. One such measure is the Mahalanobis distance [58] that standardizes the vector of differences in covariate means by the inverse of its covariance matrix. This measure has the advantage of automatically placing each covariate on the same scale and accounting for correlations between covariates.

The Mahalanobis distance between the covariate mean vectors $\overline{\mathbf{x}}(1)$ and $\overline{\mathbf{x}}(0)$ is defined as

$$D_{\mathrm{M}} = (\overline{\mathbf{x}}(1) - \overline{\mathbf{x}}(0))^{\mathrm{T}} \Sigma^{-1} (\overline{\mathbf{x}}(1) - \overline{\mathbf{x}}(0)), \tag{4.2}$$

where Σ is the covariance matrix of $(\overline{\mathbf{x}}(1), \overline{\mathbf{x}}(0))$ under complete randomization. Skipping some technical details from [61], it can be shown that if p_1

and p_0 represent the proportion of units in the treatment and covariate group respectively, then the covariate matrix Σ in (4.2) is

$$\Sigma = \frac{1}{Np_1p_0}\mathrm{Cov}(\mathbf{X}), \tag{4.3}$$

where Cov($\mathbf{X}$) is the sample covariance matrix of the covariate matrix

$$\mathbf{X} = \begin{pmatrix} x_{11} & x_{21} \\ \vdots & \vdots \\ x_{1N} & x_{2N} \end{pmatrix},$$

and can be expressed as the 2×2 matrix

$$\begin{pmatrix} \frac{1}{N-1}\sum_{i=1}^{N}(x_{1i}-\overline{x}_1)^2 & \frac{1}{N-1}\sum_{i=1}^{N}(x_{1i}-\overline{x}_1)(x_{2i}-\overline{x}_2) \\ \frac{1}{N-1}\sum_{i=1}^{N}(x_{1i}-\overline{x}_1)(x_{2i}-\overline{x}_2) & \frac{1}{N-1}\sum_{i=1}^{N}(x_{2i}-\overline{x}_2)^2 \end{pmatrix},$$

that can be computed by using the cov() function is R for a specified matrix of covariates.

Substitution of (4.3) into (4.2) results in the following form for the Mahalanobis distance between vectors $\overline{\mathbf{x}}(1)$ and $\overline{\mathbf{x}}(0)$:

$$D_{\mathrm{M}} = Np_1p_0\left(\overline{\mathbf{x}}(1)-\overline{\mathbf{x}}(0)\right)^{\mathrm{T}}\left[\mathrm{Cov}(\mathbf{X})\right]^{-1}\left(\overline{\mathbf{x}}(1)-\overline{\mathbf{x}}(0)\right), \tag{4.4}$$

which can easily be computed for any specific assignment vector $\mathbf{w}$. Note that in (4.4) N, p_0, p_1 and Cov($\mathbf{X}$) is the same for all assignment vectors; only the vector $(\overline{\mathbf{x}}(1)-\overline{\mathbf{x}}(0))$ depends on the specific treatment assignment which specifies which units receive treatment and which receive control. Finally, one can define an acceptance criterion as

$$D_{\mathrm{M}} \le a,$$

where a is a predefined positive real number.

Note that the second and third columns of Table 4.1 constitute the 12×2 covariate matrix $\mathbf{X}$ in the simplified stem cell experiment. It is easy to check (Problem 4.5) that in this case the Mahalanobis distance between $\overline{\mathbf{x}}(1)$ and $\overline{\mathbf{x}}(0)$ reduces to a scaled version of the Euclidean distance between them.

Example: Partial LEAP trial

To illustrate the idea of using Mahalanobis distance as a measure of balance between the treatment and control groups, we consider a subset of 20 subjects (infants) from the LEAP trial described in Chapter 2. The age in months (x_1) and sex (1 if female and 2 if male) of these 20 infants are the pre-treatment covariates shown in Table 4.4. Suppose the design problem is to assign 10 infants to the treatment consumption of dietary peanut) and 10 to control (avoidance of dietary peanut) so that the treatment and control groups are balanced with respect to the covariate averages.

TABLE 4.4
20 subjects in the LEAP trial

Unit (infant)	x_1 (age in months)	x_2 (sex)	Unit (infant)	x_1 (age in months)	x_2 (sex)
1	6.08	1	11	7.26	1
2	7.59	1	12	8.28	2
3	5.98	2	13	5.13	2
4	7.03	1	14	6.70	2
5	6.41	2	15	10.45	1
6	6.05	1	16	7.03	1
7	4.93	1	17	8.87	2
8	8.38	1	18	7.69	1
9	6.57	2	19	10.84	1
10	10.12	1	20	7.66	2

For a specific assignment vector, one can obtain the vectors $\overline{\mathbf{x}}(1)$ and $\overline{\mathbf{x}}(0)$ (covariate means in the treatment and control groups) in the same way as before. The covariance matrix of $\mathbf{X}$ and its inverse are respectively:

$$\text{Cov}(\mathbf{X}) = \begin{pmatrix} 2.73 & -0.21 \\ -0.21 & 0.25 \end{pmatrix}, \quad [\text{Cov}(\mathbf{X})]^{-1} = \begin{pmatrix} 0.39 & 0.33 \\ 0.33 & 4.23 \end{pmatrix}$$

Substituting $N = 20$, $p_0 = p_1 = 0.5$ and $[\text{Cov}(\mathbf{X})]^{-1}$, obtained above, into (4.4), we obtain the Mahalanobis distance $D_{\mathbf{M}}$ between the vectors $\overline{\mathbf{x}}(1)$ and $\overline{\mathbf{x}}(0)$ as

$$5\Big(\overline{x}_1(1) - \overline{x}_1(0), \overline{x}_2(1) - \overline{x}_2(0)\Big) \begin{pmatrix} 0.39 & 0.33 \\ 0.33 & 4.23 \end{pmatrix} \begin{pmatrix} \overline{x}_1(1) - \overline{x}_1(0) \\ \overline{x}_2(1) - \overline{x}_2(0) \end{pmatrix}$$
$$= 1.95\Big(\overline{x}_1(1) - \overline{x}_1(0)\Big)^2 + 3.3\Big(\overline{x}_1(1) - \overline{x}_1(0)\Big)\Big(\overline{x}_2(1) - \overline{x}_2(0)\Big) + 21.15\Big(\overline{x}_2(1) - \overline{x}_2(0)\Big)^2.$$

An acceptance region of the form $D_{\mathbf{M}} \leq a$ will consist of all points $\Big(\overline{x}_1(1) - \overline{x}_1(0), \overline{x}_2(1) - \overline{x}_2(0)\Big)$ that lie within the ellipse

$$1.95\Big(\overline{x}_1(1) - \overline{x}_1(0)\Big)^2 + 3.3\Big(\overline{x}_1(1) - \overline{x}_1(0)\Big)\Big(\overline{x}_2(1) - \overline{x}_2(0)\Big) + 21.15\Big(\overline{x}_2(1) - \overline{x}_2(0)\Big)^2 = a.$$

When $a = 2$, this ellipse is shown in Figure 4.8. All points within or on the boundary of the ellipse represent acceptable allocations, whereas points outside represent unacceptable randomizations.

Computation of Mahalanobis distance:

The function MD(X,W) shown below takes two input arguments: an $N \times k$ matrix of covariates $\mathbf{X}$ and an $N \times 1$ binary assignment vector $\mathbf{W}$, and computes the Mahalanobis distance between the vectors $\overline{\mathbf{x}}(1)$ and $\overline{\mathbf{x}}(0)$ using (4.4). Application of the function to calculate the Mahalanobis distance between the vectors $\overline{\mathbf{x}}(1)$ and $\overline{\mathbf{x}}(0)$ on the basis of the covariate matrix $\mathbf{X}$ in Table 4.4

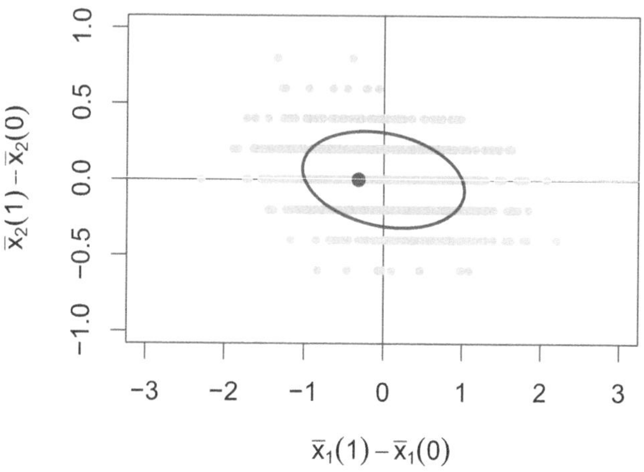

FIGURE 4.8
Acceptance region based on $D_{\mathrm{M}} \leq 2$ for 20 LEAP study subjects in Table 4.4. The black dot within the ellipse corresponds to assignment vector *wtest* (see computation example below)

for assignment vector $wtest = (1,1,1,1,0,0,0,1,0,1,0,0,1,1,0,1,1,0,0,0)$ is also shown below. The Mahalanobis distance of 0.20 suggests that *wtest* is an acceptable allocation if the threshold a is chosen as 2, which is visually confirmed by the black dot representing $(\overline{\mathbf{x}}(1), \overline{\mathbf{x}}(0))$ for the assignment vector *wtest* lying entirely within the ellipse for $a = 2$.

```
-------------------------------------------------------
MD = function(X,W) {
N1 = sum(W)
N0 = sum(1-W)
N = N1 + N0
sigmainv = solve(cov(X))
Xt = X[W==1,]
Xc = X[W==0,]
Xtbar = as.vector(apply(Xt,2,mean))
Xcbar = as.vector(apply(Xc,2,mean))

MD = (N0*N1)/N * t(Xtbar - Xcbar) %*% sigmainv %*% (Xtbar - Xcbar)
return(MD)
}

> MD(X,wtest)
          [,1]
[1,] 0.2029673
-------------------------------------------------------
```

Choice of a: Let p_a denote the proportion of acceptable randomized allocations that are deemed acceptable by the criterion $D_{\mathrm{M}} \leq a$. In order to choose the threshold a, one can specify an acceptable proportion p_a of allocations first, and then determine a from $\Pr(D_{\mathrm{M}} \leq a) = p_a$, where $\Pr(\cdot)$ represents the probability distribution of the Mahalanobis distance under all possible randomized allocations. Morgan and Rubin [61] showed that if the sample sizes are large enough to assume normality of the covariate normal means, then the Mahalanobis distance D_{M} is approximately distributed as a χ^2 random variable with k degrees of freedom, where k denotes the number of covariates. Thus a can be determined from $\Pr(\chi^2_k \leq a) = p_a$, using, for example, the qchisq(p=α, df=k) function in R. The choice of $a = 2$ in the partial LEAP experiment with 20 units leads to $p_a = Pr(\chi^2_2 \leq 2) = .63$, meaning 63% of the allocations will be acceptable under this criterion. This approximation, however, is not good, because with only 20 units the normality of covariate means is questionable, especially with one covariate (sex) being binary. This can also be seen in Figure 4.8.

The choice of p_a or a also has a direct connection to the reduction of variance of the covariate mean differences. Morgan and Rubin [61] proved that for a balanced design with $p_1 = p_0 = 1/2$, the covariance matrix of the covariate difference in means under the acceptance criterion $D_{\mathrm{M}} \leq a$ is simply the original covariance matrix (under pure randomization) of the covariate difference in means, scaled by a constant v_a, where

$$\mathrm{Cov}\left(\overline{\mathbf{x}}(1) - \overline{\mathbf{x}}(0) \middle| D_{\mathrm{M}} \leq a\right) = v_a \left(\mathrm{Cov}\left(\overline{\mathbf{x}}(1) - \overline{\mathbf{x}}(0)\right)\right), \tag{4.5}$$

where

$$v_a = \frac{\Pr(\chi^2_{k+2} \leq a)}{\Pr(\chi^2_k \leq a)}$$

is the ratio of the variance of the difference in covariate means under the acceptance criterion $D_{\mathrm{M}} \leq a$ to the variance of the difference in covariate means under pure randomization. This result tells us that the variance reduction for the covariate difference in means, or equivalently, the balance improvement, is the same and calculable in advance for each covariate. See problem 4.6 for application of this idea to understand the consequences of selecting a and p_a on the potential reduction in variance between the covariate means in the treatment and control groups.

We conclude this section by noting that although all the ideas, concepts and methodologies described in this chapter so far have been illustrated with a 2-dimensional covariate matrix $\mathbf{X}$, they can easily be generalized to the case of k covariates, where the covariate matrix $\mathbf{X}$ will be of dimension $N \times k$, $\overline{\mathbf{X}}(1)$ and $\overline{\mathbf{X}}(0)$ will be k-dimensional vectors and $\mathrm{Cov}(\mathbf{X})$ will be a $k \times k$ matrix. However, the Mahalanobis distance will still be a scalar quantity, D_{M}, and the acceptance criterion for an allocation will be of the form $D_{\mathrm{M}} \leq a$. We now illustrate such an application with the Shadish experiment.

4.4 Example: Shadish experiment

In Chapters 2 and 3, we introduced and discussed some partial and hypothetical versions of the experiment reported by Shadish et al. (2008) in [85]. This experiment was used by Morgan and Rubin (2012) [61] and Morgan (2012) [62] to illustrate the notion of acceptable criteria for randomized allocations in terms of Mahalanobis distance, because the example has a rich set of covariates. The full study was designed to examine whether observational studies can be analyzed to yield valid estimates of causal effects. The causal effects being estimated were the effect that taking a vocabulary training course has on vocabulary test scores, and the effect that taking a math training course has on math test scores.

Participants were first randomized to be in either a randomized experiment or an observational study. Those in the randomized experiment were then randomized to take either a vocabulary training course or a mathematics training course, whereas those in the observational study got to choose which course to take. The goal was to see whether the estimates of the causal effects computed from the observational study were close to the estimates computed from the randomized experiment, whose estimate was treated as the "gold standard".

As pointed out by Rubin (2008) [82] in a discussion to [85], such a design is implicitly relying on randomized experiments to give correct answers, which will only be the case if the randomization adequately balance the treatment groups across relevant covariates. In such instances in which we are treating randomized experiments as gold standards, certainly we would like to *ensure* a balanced randomization, and rerandomization provides a way to do this.

Morgan (2012) [62] examined the covariate balance resulting from the actual randomized allocations in the experiment. Note that there are two randomizations here, the first randomization to experiment or observational study, and then a second randomization to vocabulary or mathematics training. Focusing on the first of these randomizations in which among 445 participants, 235 were allocated to treatment (randomized experiment) and 210 to control (self-selection) using a completely randomized design, 10 covariates were selected on the basis of their perceived potential to be highly associated with scores on either a vocabulary or mathematics test, or with choosing to take either a vocab or a math course. The balance of these covariates in terms of standardized difference of means between the treatment and control groups is displayed in Figure 4.9, similar to the one in [62]. For a specific covariate x, the standardized difference in means between the treatment and control groups is defined as

$$\frac{\overline{x}(1) - \overline{x}(0)}{s_x\sqrt{1/N_1 + 1/N_0}},$$

where s_x denotes the sample standard deviation of x. In this example, $N_1 = 235$ and $N_0 = 210$.

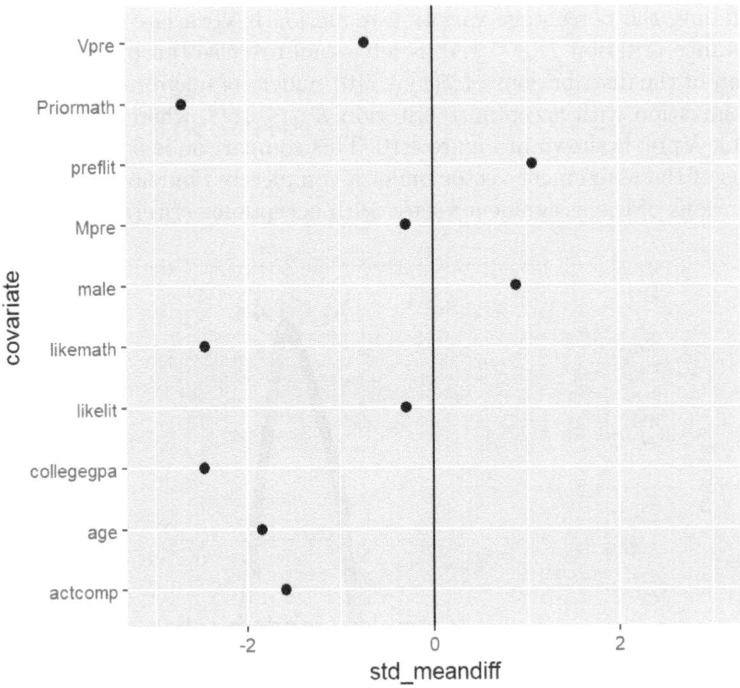

FIGURE 4.9
Standardized difference in covariate means in the Shadish (2008) experiment

The means for the number of math classes taken (priormath), liking math (likemath), and college GPA (collegegpa) are all significantly (at 5% level) lower in the students randomized to participate in the randomized experiment. Any of these could be related to the effect taking a course in either vocabulary or math will have on vocabulary or math test scores. Because the two groups differ substantially on these important background covariates, it is impossible to conclude whether a significant effect detected by a Fisher randomization test using simple or scaled difference in sample means is actually due to the difference between using a randomized experiment versus an observational study, or arose due to covariate imbalance in the two groups. Only if the two groups are similar to begin with can we make a convincingly fair comparison between them.

Based on available computing resources, it was felt that $p_a = 0.001$ or $p_a = 0.0001$ were reasonable choices. Because with a sample size of 445, a normal approximation is reasonable, we can use $\Pr(\chi^2_k \leq a) = p_a$ to determine a. For $p_a = 0.001$, a is obtained as 1.48, which means the acceptance criterion

$D_M \leq 1.48$. The Mahalanobis distance for the actual randomization in the Shadish experiment was computed by equation (4.4) as 21, the 98th percentile of D_M, making it an unusually unlucky randomization.

Finding the percentage variation reduction in covariate means under the acceptance criterion $D_M \leq 1.48$ is left as an exercise (problem 4.7). A comparison of the distributions of $\overline{x}(1) - \overline{x}(0)$ under complete randomization and randomization with acceptance criterion $D_M \leq 1.48$, where x represents the variable Vpre, is shown in Figure 4.10. This comparison is based on 500 simulations of the assignment vector under a completely randomized rule and 500 simulations of the assignment vector with acceptance criterion $D_M \leq 1.48$.

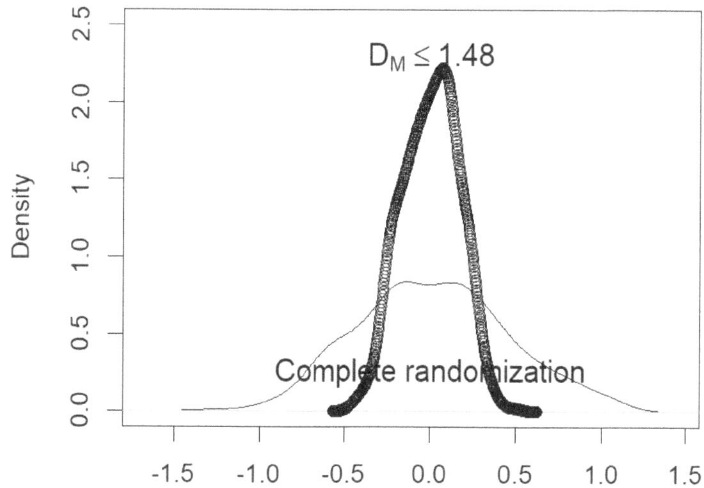

FIGURE 4.10
Difference between Vpre mean and reduction in variance under acceptance criterion $D_M \leq 1.48$

4.5 Summary

The act of randomization aims to balance pre-measured and unmeasured covariates between treatment groups, but only does that on average, and in any one experiment covariates may not be balanced. The chance of getting some observed covariate imbalance increases with the number of observed covariates. If the treatment groups are observed to be dissimilar with respect to covariates that can possibly have an impact on the outcome, inference about the causal effect of the treatment will be counfounded with the effects of the unbalanced covariates.

Defining a pre-defined acceptance criterion for assignment vectors (i.e., allocations) in a randomized experiment is a simple and intuitive way to improve covariate balance in randomized experiments. As we have already explained, the traditional technique of blocking is a special case of this approach and can be used to protect an experiment from an unlucky randomization. However, blocking is unintuitive when one is faced with a large number of covariates, especially if some are essentially continuous, and a criterion based on the Mahalanobis distance between treatment and control covariate means is simple, intuitive and flexible. Rerandomization, the idea of rejecting bad randomizations and selecting among the good ones is also intuitive and easy for practitioners to understand. In certain situations, a Mahalanobis distance based criterion, especially when combined with blocking, can be used to avoid undesirable allocations (see problem 4.9).

It has been pointed out in [62] that although it is nice to balance covariate means, it may make sense to balance higher moments as well. Higher moments can be balanced using the Mahalanobis metric by including transformations (e.g., higher powers) of the covariates in the matrix $\mathbf{X}$. The covariate matrix $\mathbf{X}$ may include, not only original covariates, but higher powers of covariates, interactions, other transformations, and anything else for which balance between the treatment and control groups is desired. For example, to balance both means and variances of the treatment and control groups for a specific covariate x, both x and x^2 should be included in the covariate matrix $\mathbf{X}$ when computing the Mahalanobis distance. Of course, deciding to balance on higher powers or certain interactions or transformations may be difficult before any outcome data is collected, and this decision will have to be based on another data source or subject matter knowledge.

4.6 Problems

4.1 In the simplified version of the stem-cell experiment described in this chapter that deals with allocation of treatment and control to wells in a 3×4 array of wells, consider the two allocations shown in Figure 4.11

0	1	0	0
0	1	1	1
1	0	1	0

1	0	0	1
1	0	0	1
0	1	0	1

FIGURE 4.11
Two allocations of units in a 3×4 array of experimental units to treatments 0 versus 1

For each of the two allocations, do (a) and (b) below.

(a) Compute the vector $\overline{\mathbf{x}}(1) - \overline{\mathbf{x}}(0)$ of differences of covariate means between the treatment and control groups and determine if the allocation is acceptable under the criterion

$$|\overline{x}_1(1) - \overline{x}_1(0)| \leq a, \quad |\overline{x}_2(1) - \overline{x}_2(0)| \leq a,$$

when $a = 1$ and $a = 1/3$.

(b) Compute the vector $\left(\overline{d}(0), \overline{d}(1)\right)^{\mathrm{T}}$ of average Euclidean distances among pairs of units in the control and treatment groups and determine whether the allocation is acceptable under the acceptance criterion $\max\left(\overline{d}(0), \overline{d}(1)\right) \geq 1.8$.

(c) Compute the vector $\left(\overline{d}(0), \overline{d}(1)\right)^{\mathrm{T}}$ of average Euclidean distances between pairs of units in the control and treatment groups for assignment vector $\mathbf{w}_2$ in Table 4.1 and determine whether the assignment is acceptable under the acceptance criterion defined in (b) above.

4.2 The Hamming distance between two vectors (x_{1i}, x_{2i}) and $(x_{1\tilde{i}}, x_{2\tilde{i}})$ is the number of elements that differ between the two vectors. For example, if $x_{1i} = x_{1\tilde{i}}$ but $x_{2i} \neq x_{2\tilde{i}}$, then the Hamming distance is 1. If $x_{1i} \neq x_{1\tilde{i}}$ and $x_{2i} \neq x_{2\tilde{i}}$, then the Hamming distance is 2. Let $\left(\overline{d}_{\mathrm{H}}(0), \overline{d}_{\mathrm{H}}(1)\right)^{\mathrm{T}}$ denote the vector of average Hamming distances among pairs of units in the control and treatment groups for a given treatment allocation. Create a two-dimensional plot of this vector (analogous to Figures 4.2 and 4.3) for all completely randomized allocations in the simplified version of the stem-cell experiment. Suggest an acceptable criterion for an allocation and determine if the two allocations $\mathbf{w}_1$, $\mathbf{w}_2$ in Table 4.1 and the two allocations in Problem 1, are acceptable under the stated criterion.

4.3 Perform the randomization test for the hypothetical stem cell experiment summarized in Figure 4.5, assuming that the acceptance criterion for an allocation is $\max\left(\overline{d}(0), \overline{d}(1)\right) \geq 1.8$, where $\overline{d}(0)$ and $\overline{d}(1)$ represent the average Euclidean distances between units in the control group and the treatment group respectively. Sepcifically, use the test statistic $T_{\text{avg-eqvar}}$ and generate 1000 draws from its randomization distribution under the sharp null hypothesis. Report the percentage of draws that had to be rejected by the rerandomization procedure in order to generate 1000 acceptable draws.

4.4 Consider a sequence of iid Bernoulli trials with probability of success π. Let X be a random variable denoting the number of trials required to obtain r successes. Then X is said to follow a negative binomial distribution with parameters r and π.

(a) Write down the probability mass function $\Pr(X = x)$ for $x = 1, 2, \ldots,$ and show that its expectation is $\mathbb{E}[X] = r/\pi$.

(b) An alternative representation of the negative binomial distribution is in terms of random variable Y that denotes the number of failures preceeding the rth success. Write down its probability mass function $\Pr(Y = y)$ for $y = 0, 1, \ldots$ and obtain its expectation.

(c) If an acceptance criterion for a randomized allocation permits 10% of all unrestricted randomized allocations, on an average how many draws will be rejected in a rerandomization procedure before obtaining 5000 acceptable allocations?

4.5 For the simplified stem cell experiment, the second and third columns of Table 4.1 constitute the 12×2 covariate matrix $\mathbf{X}$. Compute the covariance matrix $\text{Cov}(\mathbf{X})$, and show that in this case the Mahalanobis distance between the vectors $\overline{\mathbf{x}}(1)$ and $\overline{\mathbf{x}}(0)$ reduces to a scaled version of the Euclidean distance between them. Provide an intuitive explanation of this result.

4.6 For the subset of 20 infants in the LEAP trial shown in Table 4.4, what percentage of variance of the covariate mean differences $\overline{x}_1(1) - \overline{x}_1(0)$ and $\overline{x}_2(1) - \overline{x}_2(0)$ would be reduced if only treatment allocations that satisfy the acceptance criterion $D_{\text{M}} \leq a$ for $a = 1$ and $a = 2$ are allowed when designing the experiment? Assuming that the normal approximation for covariate means holds, what percentage of assignment vectors drawn under complete randomization will be accepted in these cases? Compare these approximations with those estimated by simulation.

4.7 Assume that the truncated version of the LEAP trial shown in Table 4.4 resulted in the following assignment vector with binary observed outcome (1 if the infant passed the test and 0 if not)

(a) Could the assignment vector $\mathbf{w}$ have been generated using an acceptance criterion $D_{\text{M}} \leq 1$?

(b) If the answer to part (a) is yes, conduct a randomization test to assess the sharp null hypothesis of no treatment effect on any infant assuming that the design was created using such an acceptance criterion. If the answer is no, conduct the test assuming that a balanced completely randomized assignment mechanism was used.

(c) Compute the standardized difference between covariate means for the given assignment.

4.8 In the Shadish experiment, find the percentage reduction in variance of each of the 10 covariate mean differences (standardized) between the treatment and control groups when the acceptance criterion $D_{\text{M}} \leq 1.48$ is used. Display these reductions on a single plot.

4.9 Consider, again, the truncated version of the LEAP trial shown in Table 4.4. Suppose that it was designed by blocking on covariate x_2 (sex) and

TABLE 4.5
Hypothetical data from truncated LEAP trial

Unit	Assignment (**w**)	Observed outcome (y)
1	1	1
2	1	1
3	1	1
4	1	1
5	0	1
6	0	1
7	0	1
8	1	0
9	0	0
10	1	1
11	0	1
12	0	1
13	1	0
14	1	1
15	0	1
16	1	1
17	1	1
18	0	1
19	0	1
20	0	1

using a criterion based on the Mahalanobis distance for covariate x_1 (age). The treatment assignments and observed outcomes for the 20 subjects are shown in Table 4.5.

(a) Write down the form of the Mahalanobis distance if $k = 1$, i.e., when we have a single pre-measured covariate.

(b) If x_2 is used as a blocking variable, so that an equal number of units are assigned to the treatment and control using a completely randomized mechanism, and if $D_{\mathrm{M}} \leq 2$ is used an acceptance criterion for x_1, what proportion of randomizations generated by a completely randomized allocation will be declared unacceptable? (You can provide an estimate based on simulations or provide an analytical solution using an approximation based on normality for the means of x_2 in the two groups).

(c) What will be the variance reduction in $\overline{x}_1(1) - \overline{x}_1(0)$ and $\overline{x}_2(1) - \overline{x}_2(0)$ compared to a completely randomized allocation if this design is used?

5

Randomized experiments with $J(> 2)$ treatment arms

We now discuss the design and analysis of randomized experiments with a single treatment factor at multiple levels applied to a population of experimental units (e.g., individual people). We motivate the material in this chapter with the following three examples:

Example 5.1 We first consider another simplified version of the stem cell experiment described in Chapter 4, where *three* modulators are compared with respect to their ability to convert embryonic stem cells to insulin generating pancreatic β cells. As before, the experimental units are 12 distinct physical wells on a plate arranged in a 3×4 rectangular array, where the experiment will be conducted using two such plates.

Example 5.2 We next consider a study conducted by Foa, Rothbaum, Riggs, and Murdock [34] to assess the effects of certain treatments (i.e., therapies) for *Post Traumatic Stress Disorder* (PTSD). Here, forty-five rape victims were randomly assigned to one of following four treatment groups: 1) Stress Inoculation Therapy (SIT), where subjects were taught a variety of coping skills; 2) Prolonged Exposure (PE), where subjects went over the rape in their mind repeatedly for seven sessions; 3) Supportive Counseling (SC), which was a standard-therapy control group; and 4) a Waiting List (WL) control. We restrict our discussion primarily to the post-treatment data on PTSD Severity, which can be considered as an overall measure of effectiveness of the treatments.

Example 5.3 In an experiment reported by Hennessy et al. (2016) [40], a company manufacturing consumer products conducted a product marketing experiment before launching a new dish-washing detergent. Eleven "test" detergents were manufactured. A total of 2,427 households were chosen as the experimental units for this study, and the eleven test products were assigned randomly to these households. Each household was mailed one of eleven test products, instructed to use the product and then to return a survey regarding the product's performance. The primary outcome of interest was the household's response to the survey question "How would you rate the product overall?" on a three-point ordinal scale with 1 denoting poor and 3 denoting excellent. The goal was to identify whether there were important differences

DOI: 10.1201/9781003048794-5

between the test products, and if so, which product version the subjects preferred. The survey also collected covariate information, such as income and ethnicity.

5.1 Potential outcomes and estimands

In the three examples described above, we have a finite population of N experimental units, each of which can be exposed to any one of J treatments. In Example 5.1, $N = 12 \times 2 = 24$ and $j = 3$; in Example 5.2, $N = 45$ and $J = 4$, whereas in Example 5.3, $N = 2427$ and $J = 11$. Let Y denote the response of interest (e.g., PTSD severity in Example 5.2) and let $Y_i(j)$ denote the potential outcome corresponding to unit i when assigned to treatment j, $j = 1, \ldots, J$. Note that this notation is adequate to represent all values that can be observed under SUTVA Rubin (1980) [81], explained in Chapter 2.

Table 5.1 shows possible pre-measured covariates and potential outcomes for the N experimental units. Note that only one of the J potential outcomes from each experimental unit (i.e., in each row of Table 5.1) can be *actually observed*. Let $\mathbf{Y}(j)$ denote the N-component column vector of all potential outcomes corresponding to treatment j, i.e.,$\mathbf{Y}(j) = (Y_1(j), \ldots, Y_N(j))^{\mathrm{T}}$.

TABLE 5.1
Covariates and Potential outcomes for the N experimental units

Experimental Unit	Covariates			Treatment				
	x_1	$\cdots$	x_k	1	2	$\cdots$	J	Average
1	x_{11}	$\cdots$	x_{k1}	$Y_1(1)$	$Y_1(2)$	$\cdots$	$Y_1(J)$	$\overline{Y}_1$
2	x_{12}	$\cdots$	x_{k2}	$Y_2(1)$	$Y_2(2)$	$\cdots$	$Y_2(J)$	$\overline{Y}_2$
$\vdots$	$\vdots$	$\vdots$	$\vdots$	$\vdots$	$\vdots$	$\vdots$	$\vdots$	$\vdots$
N	x_{1N}	$\cdots$	x_{kN}	$Y_N(1)$	$Y_N(2)$	$\cdots$	$Y_N(J)$	$\overline{Y}_N$
Average	$\overline{x}_1$	$\cdots$	$\overline{x}_k$	$\overline{Y}(1)$	$\overline{Y}(2)$	$\cdots$	$\overline{Y}(J)$	$\overline{Y}$
Variance	$s^2_{x_1}$	$\cdots$	$s^2_{x_k}$	S^2_1	S^2_2	$\cdots$	S^2_J	$S^2_{\overline{Y}}$

Unit-level causal estimands

The $J-1$ unit level treatment j minus treatment $\widetilde{j}$ differences are

$$\tau_i(j, \widetilde{j}) = Y_i(\widetilde{j}) - Y_i(j), \quad j \neq \widetilde{j}, j, \widetilde{j} \in \{1, 2, \ldots, J\}. \tag{5.1}$$

Next, consider the unit-level "mean-centered" treatment effects,

$$\tau_i(j) = Y_i(j) - \overline{Y}_i, \tag{5.2}$$

where $\overline{Y}_i = \sum_{j=1}^{J} Y_i(j)/J$ denotes the average of all potential outcomes for unit i, as shown in the last column of Table 5.1. Note that they satisfy a *zero-sum constraint*, i.e.,

$$\sum_{j=1}^{J} \tau_i(j) = 0,$$

For each unit, we can define linear functions of potential outcomes, called *contrasts* as follows:

$$\tau_i(\lambda_1, \ldots, \lambda_J) = \lambda_1 Y_i(1) + \ldots \lambda_J Y_i(J), \tag{5.3}$$

where the coefficients $\lambda_1, \ldots, \lambda_J$ sum to zero, but $|\lambda_1| + \ldots + |\lambda_J| \neq 0$, which means at least two of them are non-zero. For example, in the stem cell experiment, if $j = 1$ represents an existing modulator (control) and $j = 2, 3$ represnt two new modulators, then, the contrast $Y_i(1) - (Y_i(2) + Y_i(3))/2$ measures the difference between the control level and the average treatment level. In the PTSD experiment, the contrast $(Y_i(4) + Y_i(3))/2 - (Y_i(2) + Y_i(1))/2$ is the difference between the average outcome over levels (3,4) and the average outcome over levels (1,2) for unit i. Note that the unit-level adjusted treatment effects $\tau_i(j)$, defined in (5.2), are unit-level treatment contrasts (Problem 5.2).

Population-level causal estimands: The following population-level causal estimands can be obtained by averaging the unit-level estimands over the N units:

1. The $J - 1$ paired average differences

$$\overline{\tau}(j, \tilde{j}) = \frac{1}{N} \sum_{i=1}^{N} \tau_i(j, \tilde{j}) = \overline{Y}(\tilde{j}) - \overline{Y}(j), \tag{5.4}$$

   where $\overline{Y}(j) = \sum_{i=1}^{N} Y_i(j)$ is the average of all potential outcomes for treatment j and shown in the row of averages in Table 5.1.

2. Mean-centered treatment averages

$$\overline{\tau}(j) = \sum_{i=1}^{N} \tau_i(j)/N = \overline{Y}(j) - \overline{Y}, \tag{5.5}$$

   where $\overline{Y} = \sum_{i=1}^{N} \sum_{j=1}^{J} Y_i(j)/(NJ)$ is the average of all potential outcomes. Like their unit-level counterparts, these mean-centered treatment averages also satisfy the zero-sum constraint

$$\sum_{j=1}^{J} \overline{\tau}(j) = 0.$$

3. Linear contrasts of treatment means of the form

$$\tau(\lambda_1, \ldots, \lambda_J) = \frac{1}{N} \sum_{i=1}^{N} \tau_i(\lambda_1, \ldots, \lambda_J) = \lambda_1 \overline{Y}(1) + \ldots \lambda_J \overline{Y}(J), \tag{5.6}$$

where $\tau_i(\lambda_1, \ldots, \lambda_J)$ are the unit-level contrasts defined in (5.3), with the coefficients $\lambda_1, \ldots, \lambda_J$ satisfy $\lambda_1 + \ldots + \lambda_J = 0$ where at least two of them are non-zero. The population-level contrasts have similar interpretations as their unit-level counterparts.

As mentioned in the previous chapters, there is no specific reason other than algebraic simplicity to restrict the causal estimands to linear functions of potential outcomes. For example, one can be interested in the treatment medians and their differences. As noted in Chapter 2, $\text{Med}_i[\tau_i(j, \widetilde{j})]$ is generally not equal to $\text{Med}_i[Y_i(j)] - \text{Med}_i[Y_i(\widetilde{j})]$, unlike the corresponding two average difference estimands.

Before concluding the discussion on potential outcomes, we extend Definition 2.1 of additivity of potential outcomes for experiments with two treatment arms to the case with J treatment arms.

Definition 5.1 (Strict additivity) *An $N \times J$ matrix of potential outcomes is said to be* strictly additive *if, for any pair of treatments $(j, \widetilde{j})$, $j \neq \widetilde{j}$, $j, \widetilde{j} \in \{1, \ldots, J\}$, additivity holds for potential outcomes in columns j and $\widetilde{j}$, i.e.,*

$$Y_i(\widetilde{j}) - Y_i(j) = \tau(j, \widetilde{j}), \quad i = 1, \ldots, N. \tag{5.7}$$

Thus, under strict additivity, the potential outcomes in any column can be expressed as the sum of the corresponding potential outcomes in any other column and a constant. Consider a toy example with $N = 6$ individuals, $J = 3$ treatments and the potential outcomes given in Table 5.2.

TABLE 5.2
Table of potential outcomes for a toy example

Unit	Treatment 1	Treatment 2	Treatment 3
1	8	13	4
2	13	18	9
3	7	12	3
4	10	15	6
5	13	18	9
6	9	14	5
Mean	$\overline{Y}(1) = 10$	$\overline{Y}(2) = 15$	$\overline{Y}(3) = 6$

Note that the unit-level treatment effects $\tau_i(1,2) = Y_i(2) - Y_i(1) = 5$ and $\tau_i(1,3) = Y_i(3) - Y_i(1) = -4$ for all $i = 1, \ldots, 6$, making the 6×3 table of potential outcomes strictly additive. For such a matrix, all unit-level treatment contrasts will be equal and thus equal to the population-level contrast (see Problem 5.3).

5.2 Assignment mechanism: determining the appropriateacceptability criterion

We now discuss different assignment mechanisms, i.e., rules for assigning the J treatments to the N units that will help us avoid undesirable randomizations and consequently enable fair comparisons of the J treatments. Let us revisit the simplified stem cell experiment in Example 5.1, for which the table of covariates and potential outcomes can be written as in Table 5.3:

TABLE 5.3
Units, covariates and potential outcomes for Example 5.1

Unit	Covariates			Potential outcomes		
	b (Plate number)	x_1 (row)	x_2 (column)	$Y(1)$	$Y(2)$	$Y(3)$
1	1	1	1	$Y_1(1)$	$Y_2(1)$	$Y_1(3)$
⋮	⋮	⋮	⋮	⋮	⋮	⋮
12	1	3	4	$Y_{12}(1)$	$Y_{12}(2)$	$Y_{12}(3)$
13	2	1	1	$Y_{13}(1)$	$Y_{13}(1)$	$Y_{13}(3)$
⋮	⋮	⋮	⋮	⋮	⋮	⋮
24	2	3	4	$Y_{24}(1)$	$Y_{24}(2)$	$Y_{24}(3)$

The first aspect of the design that the experimenter must decide is the distribution of the 24 units to the three treatments. Whereas a balanced allocation of 8 units to each of the three treatments sounds like a reasonable choice, it need not be an optimal choice as explained in Section 2.7 of Chapter 2. There can also be practical or cost constraints influencing this choice, e.g., higher cost of applying newer treatments. We will re-visit this aspect later. For the time being, let us assume that the numbers of units $N_1, \ldots, N_J$ to be assigned to the J treatments have been decided.

In spite of having covariate information as in Table 5.3, an experimenter can decide to ignore such information and choose a completely randomized treatment assignment mechanism, which, henceforth, will be refered to as a completely randomized design (CRD). A CRD will ensure that the covariates are balanced over the treatment groups, but only on an average. For example, in this case, assuming that $N_1 = N_2 = N_3 = 8$, it is possible that a CRD leads to applying the "control" modulator 1 only to wells on plate 1, whereas all the 12 wells on plate 2 receive either of the "treatment" modulators 2 or 3. In that case, any comparison of control vs treatment (e.g., inference for contrast $\overline{\tau}(1, -1/2, -1/2)$) could be confounded with hidden variables capable of creating systematic differences in the outcomes observed on plate 1 and plate 2.

To avoid undesirable allocations, like the one just described, one can use the plate factor to define blocks and use a CRD to assign treatments to units within each block. But then undesirable allocations as described in Chapter 4 (e.g., the assignment in the left panel of Figure 4.1) may occur within each plate. If only column effects are anticipated within each, then covariate x_2 (column) can be used as a second blocking variable, and the three units in each column can be randomly assigned to the three treatments. If both row and column effects are anticipated, then a criterion for defining acceptable randomizations can be defined following the ideas presented in Chapter 4.

In the classical experimental design literature, the problem of assigning $J(>2)$ treatments to units when there are two blocking variables with levels $B_1(>2)$ and $B_2(>2)$ has been studied under the name "row-column designs" (Shrikande 1951 [87], Youden and Beale 1934 [88], Youden 1940 [94]). Such a name originates from agricultural experiments where experimental units were plots of land, as in the tomato experiment described in Chapter 1, and undesirable allocations were likely to make effect of treatments (e.g., types of fertilizers) indistinguishable from the effect of fertility gradients that could exist along two perpendicular directions (rows and columns). It is interesting to note that the simplified stem cell experiment confronts an analogous problem. Further, if $B_1 = B_2 = J$, i.e., each of the two blocking factors have the same number of levels as that of the treatment factor, then the problem reduces to assigning J levels of a treatment on a $J \times J$ square array. We will describe such designs later in Section 5.6, emphasizing that such designs are often too restrictive, and neither intuitive nor applicable when the number of covariates is large.

As earlier, let W_i denote the assignment variable for unit i that can take integer values $1, \ldots, J$, and let $\mathbf{W} = (W_1, \ldots, W_N)^{\mathrm{T}}$ denote the $N \times 1$ column vector of treatment assignments. The joint distribution of $\mathbf{W}$ will follow from the design (assignment mechanism). For example, if the treatments are assigned to the N units using a CRD, then all the $N!/(N_1!N_2! \cdots N_J!)$ possible allocations are equally likely, and thus

$$\Pr(\mathbf{W} = \mathbf{w}) = \begin{cases} \frac{N_1!N_2!\cdots N_J!}{N!} & \text{if } \sum_{1:W_i=j} W_i = N_j \\ 0 & \text{otherwise.} \end{cases},$$

To assign J treatments to N units such that treatments $1, \ldots, J$ are assigned to $N_1, \ldots, N_J$ units respectively, we can use the R function sample() introduced in earlier chapters. In this case the argument of the function would be an N-dimensional vector consisting of N_1 ones, N_2 twos, and continuing this way N_J J's. For example, to assign three treatments to the six units in the toy example in Table 5.2 so that each treatment is assigned to two units, we use:

```
-----------------------------------------------------------------
> sample(c(1,1,2,2,3,3))
[1] 1 2 2 1 3 3
-----------------------------------------------------------------
```

Now assume that the N units are grouped into B blocks indexed by $b = 1, \ldots, B$ with block sizes $M_1, \ldots, M_b$ such that $\sum_b M_b = N$. Suppose, the experimenter wants to use a randomized block design with a predetermined number N_{bj} of units in block b to receive treatment j, such that $\sum_{b=1}^{N} N_{bj} = N_j$. Because N_{bj}'s are predetermined for all b and j, so are the N_j's. Also, let $\sum_{j=1}^{J} M_{bj} = M_b$, the size of the bth block. To understand these notations, assume that in the simplified stem cell experiment, plate is used as a blocking variable, and within each plate a CRD is used for assigning the three treatments to the 12 units, so that four units receive each treatment. This is a randomized block design with $N = 24$ units, $B = 2$ blocks of sizes $M_1 = M_2 = 12$, $J = 3$ treatments, $N_{11} = N_{12} = N_{13} = 4$ units to receive each treatment within block 1, and $N_{21} = N_{22} = N_{23} = 4$ units to receive each treatment within block 2.

To implement such an assignment mechanism, we should assign the J treatments to the units within each block using a CRD. After generating B such assignment vectors independently for the B blocks, they should be concatenated to obtain the assignment vector $W_1, \ldots, W_N$). In the example given in the previous paragraph, for each of the two blocks we should independently assign 3 treatments to 12 units using a CRD. The R code used to generate such an assignment vector $(W_1, \ldots, W_{24})$ is shown below.

```
-----------------------------------------------------------------
> set.seed(9999)
> W_B1 <- sample(c(1,1,1,1,2,2,2,2,3,3,3,3))
> W_B1
 [1] 2 3 3 1 1 2 3 1 1 2 3 2
> W_B2 <- sample(c(1,1,1,1,2,2,2,2,3,3,3,3))
> W_B2
 [1] 1 3 1 2 3 3 2 3 2 2 1 1
> c(W_B1, W_B2)
 [1] 2 3 3 1 1 2 3 1 1 2 3 2 1 3 1 2 3 3 2 3 2 2 1 1
-----------------------------------------------------------------
```

Obtaining the total number of equally likely allocations and writing down the probability distribution $\Pr[\mathbf{W} = \mathbf{w}]$ are left as exercises (Problems 5.4 and 5.5).

5.3 Observed outcomes and estimators in a CRD

Let y_i denote the observed outcome for unit i for $i = 1, \ldots, N$, which is *one* of the J potential outcomes $Y_i(1), \ldots, Y_i(J)$ shown in Table 5.1, and can be expressed as a function of the (random) treatment received and the (fixed) potential outcomes:

$$y_i = Y_i(W_i) = \sum_{j=1}^{J} \mathbb{I}(W_i = j)Y_i(j), \tag{5.8}$$

where $\mathbb{I}(W_i = j)$ denotes an indicator random variable that takes the value one if $W_i = j$ and zero otherwise. Because unit i can receive one and only one treatment, only one of the indicator variables $\mathbb{I}(W_i = 1), \ldots, \mathbb{I}(W_i = J)$ in (5.8) will be one and the remaining $J - 1$ will be zeros. The potential outcomes for these $J - 1$ treatments will be missing for unit i. Let $\mathbf{y}$ denote the N-component column vector of observed outcomes.

Consider again the toy example in Table 5.2. Under completely randomized treatment assignment with two units assigned to each treatment, we have $6!/(2!)^3 = 90$ possible treatment assignments. The treatment assignment vector $\boldsymbol{W}$ can be any one of the 90 possible permutations of the six integers $(1, 1, 2, 2, 3, 3)$. Suppose the observed treatment assignment vector $\boldsymbol{w}$ is $(1, 2, 2, 1, 3, 3)^{\mathrm{T}}$. Then, the observed outcomes are shown in Table 5.4, where the "?" symbol represents the unobserved (or missing) potential outcomes.

TABLE 5.4
Table of observed and missing outcomes for the example in Table 5.2

Unit	Treatment 1	Treatment 2	Treatment 3	w
1	8	?	?	1
2	?	18	?	2
3	?	12	?	2
4	10	?	?	1
5	?	?	9	3
6	?	?	5	3
Mean	$\overline{y}_.(1) = 9$	$\overline{y}_.(2) = 15$	$\overline{y}_.(3) = 7$	

Now, we identify estimators of common estimands of interest defined in Section 5.1. A natural estimator of $\overline{Y}(j)$ is the average of all observed outcomes corresponding to treatment j:

$$\overline{y}.(j) = \frac{1}{N_j} \sum_{i:W_i=j} y_i == \frac{1}{N_j} \sum_{i=1}^{N} \mathbb{I}(W_i = j)Y_i(j) \quad j = 1, \ldots, J. \tag{5.9}$$

The dot in the suffix of $\overline{y}_.(j)$ in (5.9) indicates that the average is taken over all N units. The average of all N observations can similarly be defined as

$$\overline{y}.(\cdot) = \frac{1}{N} \sum_{i=1}^{N} y_i = \frac{1}{N} \sum_{j=1}^{J} N_j \overline{y}.(j). \tag{5.10}$$

Estimators of the population-level estimands defined in Section 5.1 can be obtained by substituting $\overline{y}.(j)$ and $\overline{y}.(\cdot)$ in place of $\overline{Y}(j)$ and $\overline{Y}$ respectively. Thus, estimators of the paired average differences $\overline{\tau}(j,\tilde{j})$ defined in (5.4), adjusted treatment averages $\overline{\tau}(j)$ defined in (5.5), and linear contrasts of treatment means $\tau(\lambda_1,\ldots,\lambda_J)$ defined in (5.6) are respectively

$$\widehat{\overline{\tau}}(j,\tilde{j}) = \overline{y}.(\tilde{j}) - \overline{y}.(j), \tag{5.11}$$

$$\widehat{\overline{\tau}}(j) = \overline{y}.(j) - \overline{y}.(\cdot), \tag{5.12}$$

$$\widehat{\tau}(\lambda_1,\ldots,\lambda_J) = \sum_{j=1}^{J} \lambda_j \overline{y}.(j). \tag{5.13}$$

For example, an unbiased estimate of the average difference between treatments 1 and 2 obtained from Table 5.4 is $\widehat{\overline{\tau}}(1,2) = \overline{y}_.(2) - \overline{y}_.(1) = 15 - 9 = 6$. Computing estimates of other causal estimands is left as an exercise (Problem 5.6).

5.4 Inference from CRD experiments

In this section, we discuss the two different forms of randomization-based inference from CRD experiments. First, in Section 5.4.1 we describe Fisherian inference to assess the sharp null hypothesis of the same treatment effects for all units. Next, we discuss Neymanian inference with the aim of obtaining large sample confidence intervals for population-level treatment effects and contrasts.

5.4.1 Fisherian inference with J treatments from a CRD

We now describe how to investigate whether one or more of the J treatments differ from the others in terms of their effects on the potential outcomes. The sharp null hypothesis of no treatment difference for *any* unit can be formally expressed as:

$$H_{0F} : Y_i(1) = Y_i(2) = \ldots = Y_i(J), \quad i = 1,\ldots,N. \tag{5.14}$$

At this point of time, the readers should be well-versed in the mechanics of the randomization test. Until Chapter 4, we have compared only two treatments, making the difference of observed outcomes in the two treatment groups (or its scaled version) a natural test statistic. We also discussed other statistics, like the rank statistic, in Chapter 2, which was also based on a comparison of observed outcomes in the two treatment groups. However, in this chapter we aim to compare more than two treatments, and will consider test statistics that can capture discrepancy from the sharp null hypothesis

(5.14). The orthogonal decomposition and analysis of variance discussed next will lead us to such a statistic.

5.4.1.1 Orthogonal decomposition and one-way ANOVA

We have the following mathematical identity among the observed values y_i (obtained by adding and subtracting terms on the right hand side that cancel each other):

$$y_i = \overline{y}.(\cdot) + \{\overline{y}.(W_i) - \overline{y}.(\cdot)\} + \{y_i - \overline{y}.(W_i)\}. \tag{5.15}$$

From (5.15), it follows that

$$y_i - \overline{y}.(\cdot) = \{\overline{y}.(W_i) - \overline{y}.(\cdot)\} + \{y_i - \overline{y}.(W_i)\}. \tag{5.16}$$

For example, the first observed outcome $y_1 = 8$ in Table 5.4 can be expressed as:

$$\begin{aligned} y_1 &= \overline{y}.(\cdot) + \{\overline{y}.(1) - \overline{y}.(\cdot)\} + \{y_i - \overline{y}.(1)\} \\ &= 31/3 + (9 - 31/3) + (8 - 9). \end{aligned}$$

Here, $31/3$ is the average of all the six observed outcomes, the difference (9-31.3) is the estimate $\widehat{\overline{\tau}}(1)$ of the mean-centered treatment effect $\overline{\tau}(1)$, and the difference (8–9) is the residual effect that is not explained by the treatment received by unit 1.

Squaring both sides of (5.16) and summing over $i = 1, \ldots, N$, we can show that the cross-product term on the right hand side reduces to zero (show this explicitly as an exercise in Problem 5.8), and arrive at the following decomposition:

$$\text{SSTot} = \text{SSTre} + \text{SSRes}, \tag{5.17}$$

where

$$\text{SSTot} = \sum_{i=1}^{N} \{y_i - \overline{y}.(\cdot)\}^2, \tag{5.18}$$

$$\begin{aligned} \text{SSTre} &= \sum_{i=1}^{N} \{\overline{y}.(W_i) - \overline{y}.(\cdot)\}^2 = \sum_{j=1}^{J} \sum_{i:W_i=j} \{\overline{y}.(j) - \overline{y}.(\cdot)\}^2 \\ &= \sum_{j=1}^{J} N_j \{\overline{y}.(j) - \overline{y}.(\cdot)\}^2 \\ &= \sum_{j=1}^{J} N_j \{\widehat{\overline{\tau}}(j)\}^2, \text{ by (5.12)} \end{aligned} \tag{5.19}$$

$$\text{SSRes} = \sum_{i=1}^{N} \{y_i - \overline{y}.(W_i)\}^2 = \sum_{j=1}^{J} \sum_{i:W_i=j} \{y_i - \overline{y}.(j)\}^2. \tag{5.20}$$

Here SSTot, SSTre and SSRes are acronyms for the "Total sum of squares", "Treatment sum of squares" and "Residual sum of squares". The decomposition (5.17) is referred to as the *Analysis of Variance* (ANOVA) decomposition in the statistical literature.

Returning to the example shown in Table 5.4, we see that the overall mean $\overline{y}.(\cdot) = 31/3$. Thus, from (5.18),

$$\text{SSTot} = (8 - 31/3)^2 + (10 - 31/3)^2 + \ldots + (9 - 31/3)^2 + (5 - 31/3)^2 = 97.33.$$

Note that SSTot is the sum of squares of six individual components $(8 - 31/3), \ldots, (5 - 31/3)$ that sum to zero. Thus, if five of these components are known, the sixth can be determined. The number of components in a sum of squares that can be chosen without any restriction or constraint is called its *degrees of freedom*. Thus the degrees of freedom associated with SSTot is five. In general, if the number of units is N, the degrees of freedom associated with SSTot will be $N - 1$.

Next, we can calculate the treatment sums of squares using (5.19):

$$\text{SSTre} = 2\{(9 - 31/3)^2 + (15 - 31/3)^2 + (7 - 31/3)^2\} = 69.33.$$

The treatment sum of squares comprises three components that sum up to zero. Thus SSTre has two degrees of freedom, which reflects two independent comparisons that can be made between three treatments. From any two of these comparisons, the third comparison can be immediately inferred. In general, for J treatments, the degrees of freedom associated with SSTre is $J - 1$.

Finally, the residual sum of squares can either be directly computed from (5.20), or by subtracting SSTre from SSTot, yielding SSRes = SSTot−SSTre = 28. By the same logic as above, the degrees of freedom associated with SSRes is $N - J = 3$, which can also be obtained by subtracting the degrees of freedom for SSTre from that of SSTot. The ANOVA decomposition thus also applies to the degrees of freedom, denoted by DF below:

$$\text{DF(SSTot)} = \text{DF(SSTre)} + \text{DF(SSRes)},$$

or equivalently,

$$(N - 1) = (J - 1) + (N - J).$$

The ANOVA decomposition provides us with a useful tool for comparing the estimated effects of several treatments on the outcome variable that we will discuss in Section 5.4.1.3. The following subsection (Section 5.4.1.2) gives some geometrical insights of the ANOVA decomposition and degrees of freedom and can be skipped by readers with little or no interest in such intuition.

5.4.1.2 ANOVA: A geometrical intuition

In the example shown in Table 5.4, the observed data vector $\mathbf{y}$ and the observed assignment vector $\mathbf{w}$ are $(8, 18, 12, 10, 9, 5)^{\mathrm{T}}$ and $(1, 2, 2, 1, 3, 3)^{\mathrm{T}}$

respectively. Thus, the decomposition in (5.15) can be written as

$$\begin{pmatrix} 8 \\ 18 \\ 12 \\ 10 \\ 9 \\ 5 \end{pmatrix} = \begin{pmatrix} 31/3 \\ 31/3 \\ 31/3 \\ 31/3 \\ 31/3 \\ 31/3 \end{pmatrix} + \begin{pmatrix} -4/3 \\ 14/3 \\ 14/3 \\ -4/3 \\ -10/3 \\ -10/3 \end{pmatrix} + \begin{pmatrix} -1 \\ 3 \\ -3 \\ 1 \\ 2 \\ -2 \end{pmatrix} \tag{5.21}$$

Reordering the vectors in (5.21) according to the observed treatment, i.e., writing $\tilde{\mathbf{y}} = (\tilde{y}_1, \ldots, \tilde{y}_6)^{\mathrm{T}}$ as a permutation $(y_1, y_4, y_2, y_3, y_5, y_6)^{\mathrm{T}}$ of the observed data vector $\mathbf{y}$, we can rewrite (5.21) as

$$\begin{pmatrix} 8 \\ 10 \\ 18 \\ 12 \\ 9 \\ 5 \end{pmatrix} = \begin{pmatrix} 31/3 \\ 31/3 \\ 31/3 \\ 31/3 \\ 31/3 \\ 31/3 \end{pmatrix} + \begin{pmatrix} -4/3 \\ -4/3 \\ 14/3 \\ 14/3 \\ -10/3 \\ -10/3 \end{pmatrix} + \begin{pmatrix} -1 \\ 1 \\ 3 \\ -3 \\ 2 \\ -2 \end{pmatrix} \tag{5.22}$$

Note that the three vectors on the right hand sides of (5.21) and (5.22) are orthogonal to each other, which means the inner product between any two of them is zero. Further, the first vector has only one unique element. The second vector has three unique elements –4/3, 14/3, and –10/3, that sum to zero, implying that it has only two unique elements and the third one can be determined from the other two using the fact that they sum to zero. The third vector has three pairs of elements (–1,1), (–2,2) and (–3,3) such that the two elements within each pair sum to zero. Consequently, one needs to know three unique elements (one within each pair) independently to complete the vector. Mathematically, the first, second and third vectors on the right hand sides of (5.21) and and (5.22) reside in one-dimensional, two-dimensional and three-dimensional vector spaces respectively, that are all orthogonal to each other. The number of unique elements in a vector that can be chosen without restrictions or constraints is known as its *degrees of freedom*. Based on the argument above, the first, second and third vectors on the right hand sides of (5.21) and (5.22) have degrees of freedom 1, 2 and 3 respectively.

Generalizing the above observation to N experimental units and J treatments, we can write (5.15) in vector form similar to (5.22) as

$$\widetilde{\mathbf{y}} = \mathbf{y}_1 + \mathbf{y}_2 + \mathbf{y}_3, \tag{5.23}$$

where

$$\widetilde{\mathbf{y}} = \begin{pmatrix} \widetilde{y}_1 \\ \cdots \\ \widetilde{y}_{N_1} \\ \hline \vdots \\ \hline \widetilde{y}_{N_{J-1}+1} \\ \cdots \\ \widetilde{y}_{N_J} \end{pmatrix}$$

represents the ordered vector of observations stacked by treatments $1, \ldots, J$, and

$$\mathbf{y}_1 = \begin{pmatrix} \overline{y}.(\cdot) \\ \cdots \\ \overline{y}.(\cdot) \\ \hline \vdots \\ \hline \overline{y}.(\cdot) \\ \cdots \\ \overline{y}.(\cdot) \end{pmatrix}, \mathbf{y}_2 = \begin{pmatrix} \overline{y}.(1) - \overline{y}.(\cdot) \\ \cdots \\ \overline{y}.(1) - \overline{y}.(\cdot) \\ \hline \vdots \\ \hline \overline{y}.(J) - \overline{y}.(\cdot) \\ \cdots \\ \overline{y}.(J) - \overline{y}.(\cdot) \end{pmatrix}, \mathbf{y}_3 = \begin{pmatrix} \tilde{y}_1 - \overline{y}.(1) \\ \cdots \\ \tilde{y}_{N_1} - \overline{y}.(1) \\ \hline \vdots \\ \hline \tilde{y}_{N_{J-1}} + 1 - \overline{y}.(J) \\ \cdots \\ \tilde{y}_{N_J} - \overline{y}.(J) \end{pmatrix},$$

are orthogonal to each other and reside in one-dimensional, $(J-1)$-dimensional and $(N-J)$-dimensional vector spaces, respectively. The degrees of freedom associated with each vector is the dimension of the corresponding subspace.

Arguably, the concept of degrees of freedom was first noted in the early nineteenth century by Carl Friedrich Gauss and expounded by William Sealy Gosset, popularly known by his pen name "Student" [90], although he did not specifically use the term "degrees of freedom". The term was popularized by Sir R. A. Fisher in the nineteen twenties.

The orthogonal decomposition of the vector of observed outcomes $\mathbf{y}$ (or its permuted version $\widetilde{\mathbf{y}}$), along with equation (5.23), leads to the following equation:

$$\|\mathbf{y}\|^2 = \|\widetilde{\mathbf{y}}\|^2 = \|\mathbf{y}_1\|^2 + \|\mathbf{y}_2\|^2 + \|\mathbf{y}_3\|^2, \tag{5.24}$$

where $\|\boldsymbol{x}\| = \sqrt{x_1^2 + \ldots + x_n^2}$ denotes the norm or length of the vector $\boldsymbol{x} = (x_1, \ldots, x_n)^{\mathrm{T}}$. The equivalence of equation (5.24) and the sum of squares decomposition given by (5.17) is immediate.

5.4.1.3 The F statistic

We now explain how the ANOVA decomposition (5.17) provides us with a useful tool for comparing the estimated effects of several treatments on the outcome variable. First, we define a mean square as the ratio of a sum of squares and the corresponding degree of freedom. Thus, the mean squares of treatment $\text{MSTre} = \text{SSTre}/(J-1)$ and the mean squares of residual $\text{MSRes} = \text{SSRes}/(N-J)$, where SSTre and SSRes are given by equations (5.19)-(5.20).The F statistic

$$F = \frac{\text{MSTre}}{\text{MSRes}} \tag{5.25}$$

is defined as the ratio of the mean squares of treatment MSTre to the mean squares of residual MSRes. Larger observed values of F correspond to stronger evidence against the sharp null hypothesis of no treatment effect.

Based on the geometric insight provided in Section 5.4.1.2, the F ratio can be geometrically interpreted as the ratio of the projection of the observed response onto the treatment space and the orthogonal residual space in terms of squared length per dimension. The following insight is particularly useful: if the treatments have absolutely no effect, then by symmetry the treatment subspace is no different from any other subspace of the same dimension orthogonal to the mean subspace, and so in expectation, F defined in (5.25) must be one when there is no treatment effect on any individual.

Returning to the example in Table 5.4, the observed value of the F-statistic is

$$F_{\text{obs}} = \text{MSTre}/\text{MSRes} = \frac{\text{SSTre}/2}{\text{SSRes}/6} = \frac{67.33/2}{28/6} = 3.714.$$

The results of the ANOVA can be summarized in a table of the form of Table 5.5, which is called the *one-way* ANOVA table, or Analysis of variance table proposed by Fisher (1925) [32].

TABLE 5.5
ANOVA Table from observed outcomes in Table 5.4

Source	Degrees of freedom	Sum of squares	Mean Squares	F-ratio
Treatments	$J-1=2$	SSTre = 69.33	MSTre = 34.67	$F=$
Residuals	$N-J=3$	SSRes = 28.00	MSRes = 9.33	3.714
Total	$N-1=5$	SSTot = 97.33		

An easy way to obtain the ANOVA table and compute the F-statistic using R is to use the anova(lm()) function, shown below in the context of the current example.

```
--------------------------------------------------------------
> y=c(8,18,12,10,9,5)
> w=c(1,2,2,1,3,3)
> anova(lm(y ~ factor(w)))

Analysis of Variance Table

Response: y
          Df Sum Sq Mean Sq F value Pr(>F)
factor(w)  2 69.333  34.667  3.7143 0.1543
Residuals  3 28.000   9.333
--------------------------------------------------------------
```

At this point, we ignore the last column with header Pr(> F) (p-value) in the ANOVA table generated above by R. The only thing we need to conduct a randomization test is the observed value of the F-statistic, which can be obtained without generating the entire ANOVA table as shown below:

```
--------------------------------------------------------------
> anova(lm(y ~ factor(w)))$F[[1]]

[1] 3.714286
--------------------------------------------------------------
```

Before proceeding to the randomization test using the F statistic, we note that when $J = 2$ (the case considered in Chapters 2 and 3), the ANOVA decomposition and F statistic can still be defined in the same way as done in this subsection. In this case, the F statistic is the square of the $T_{\text{avg-eqvar}}$ statistic defined in (2.18). See Problem 5.9.

5.4.1.4 Randomization test with the F statistic

Returning to the problem of assessing the sharp null hypothesis $H_{0\text{F}}$ of no treatment difference for *any* unit stated in (5.14), based on the argument in the previous section, the F-statistic is a natural choice for measuring how much the treatments matter. For this test, under $H_{0\text{F}}$, all J potential outcomes $Y_i(1), \ldots, Y_i(J)$ for unit $i = 1, \ldots, N$ are equal to the observed outcome y_i, and consequently any possible realization of the treatment assignment vector would generate an identical vector of observed outcomes, which implies that under $H_{0\text{F}}$ and for any realization $\mathbf{W} = \mathbf{w}$, the observed outcomes $\mathbf{y}$ are fixed, and hence SSTot is fixed. Therefore, different realizations of the treatment assignment vector simply lead to different decompositions of the same SSTot into different SSTre and SSRes. Because large values of F indicate departures from $H_{0\text{F}}$, larger values can be considered extreme under $H_{0\text{F}}$. Hence the p-value can be obtained as the proportion of values of the randomization distribution of F under $H_{0\text{F}}$ that are larger than or equal to its observed value.

We now illustrate the Fisher randomization test for the example in Table 5.4 for which the ANOVA decomposition is presented in Table 5.4, and the observed value of the F-statistic is 3.714. To obtain the randomization distribution of F under the sharp null hypothesis, we use the fact that the sharp null hypothesis of no treatment effect permits us to impute the missing values in Table 5.4, as shown in Table 5.6. In each sub-table, shown in Table 5.6, the observed outcomes are circled. The assignment vectors in the two tables are $(1, 2, 2, 1, 3, 3)^{\text{T}}$ (top) and $(1, 1, 2, 2, 3, 3)^{\text{T}}$ (bottom). Note that the former is the observed assignment vector; therefore the data from the table on the left yield an observed F value of 3.714, whereas the data on the right yield an observed F value of 0.933.

The histogram constructed from the 90 such possible F values under the sharp null hypothesis is shown in Figure 5.1, with the vertical line representing the actual observed value of 3.714. The p value is computed as 0.12, which can be considered small, resulting in some evidence against the sharp null hypothesis.

The R code used to conduct the Fisher randomization test is shown below. A simple and efficient computational algorithm to generate the randomization distribution of the F statistic is to randomly generate draws $\mathbf{w}_{\text{new}}$ from the distribution of the treatment assignment vectors $\mathbf{W}$, and calculate the test statistic F_{new} by changing only the treatment assignment vector, keeping the observed outcomes $\mathbf{y}$ unchanged in each iteration, i.e.,

TABLE 5.6
Computing F for two possible randomizations

Unit	Tr 1	Tr 2	Tr 3
1	(8)	8	8
2	18	(18)	18
3	12	(12)	12
4	(10)	10	10
5	9	9	(9)
6	5	5	(5)
Mean	$\overline{y}_{.1} = 9$	$\overline{y}_{.2} = 15$	$\overline{y}_{.3} = 7$

Unit	Tr 1	Tr 2	Tr 3
1	(8)	8	8
2	(18)	18	18
3	12	(12)	12
4	10	(10)	10
5	9	9	(9)
6	5	5	(5)
Mean	$\overline{y}_{.1} = 13$	$\overline{y}_{.2} = 11$	$\overline{y}_{.3} = 7$

```
new_F = anova(lm(y ~ factor(new_w)))$F[[1]]
```

```
---------------------------------------------------------------
# F-test Fisher

y=c(8,18,12,10,9,5)
w=c(1,2,2,1,3,3)
toy_anova <- anova(lm(y ~ factor(w)))
(Fobs = toy_anova$F[[1]])
[1] 3.714286

ITER = 5000
F_array = NULL
set.seed(9999)
for (i in 1:ITER) {
 new_w = sample(w)
 new_F = anova(lm(y ~ factor(new_w)))$F[[1]]
 F_array = c(F_array, new_F)
}
hist(F_array, nclass=5)
```

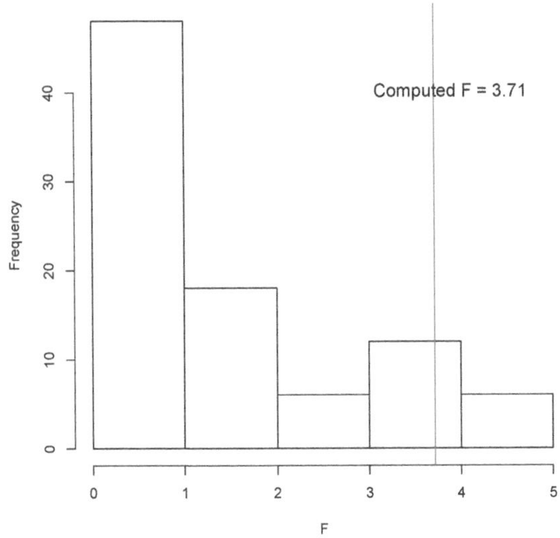

FIGURE 5.1
Histogram of computed F values from 90 randomizations under the sharp null hypothesis

```
abline(v=Fobs)
(pval = sum(F_array >= Fobs)/ITER)
[1] 0.1208
---------------------------------------------------
```

5.4.1.5 Approximate distribution of F under $H_{0\mathrm{F}}$

In most standard textbooks on design of experiments and linear models, e.g., [10, 93], an $F_{J-1,N-J}$ distribution with degrees of freedom $J-1$ and $N-J$ is often used in the analysis of variance table obtained from fitting a normal linear model. However, in our case, the potential outcomes are assumed fixed, and the probability distribution of F is a only a consequence of randomization. However, it is interesting to note that using purely randomization-based arguments, the randomization distribution of F under $H_{0\mathrm{F}}$ can be approximated by an $F_{J-1,N-J}$ distribution for reasonably large samples. The original F approximation for randomization inference for a finite population was derived by a method known as moment matching [91, 69, 50]. A simpler proof based on the finite population central limit theorem can be found in Ding and Dasgupta (2018) [28].

To see how well the distribution of the $F_{J-1,N-J}$ random variable approximates the randomization distribution of the F-statistic, we generate $N \times 4$ matrices of potential outcomes with strict additivity. The first column, i.e., the vector $(Y_1(1), \dots, Y_N(1))^{\mathrm{T}}$ is generated from a normal distribution with mean

10 and standard deviation 0.5. The next three columns are identical to the first column, ensuring that strict additivity holds. Potential outcomes, once generated, are kept fixed. Next, 10,000 completely randomized assignments of N units to 4 treatments are generated, and the F-statistic is computed from the 10,000 outcome vectors $\mathbf{y}$ generated from these assignments. Histograms of the randomization distribution of the F-statistic generated in this way are plotted for $N = 12$, $N = 20$ and $N = 80$ in Figure 5.2. The probability density functions of $F_{3,8}$, $F_{3,16}$ and $F_{3,76}$ are respectively superimposed on these histograms. It is clear that the approximation gets better as N increases.

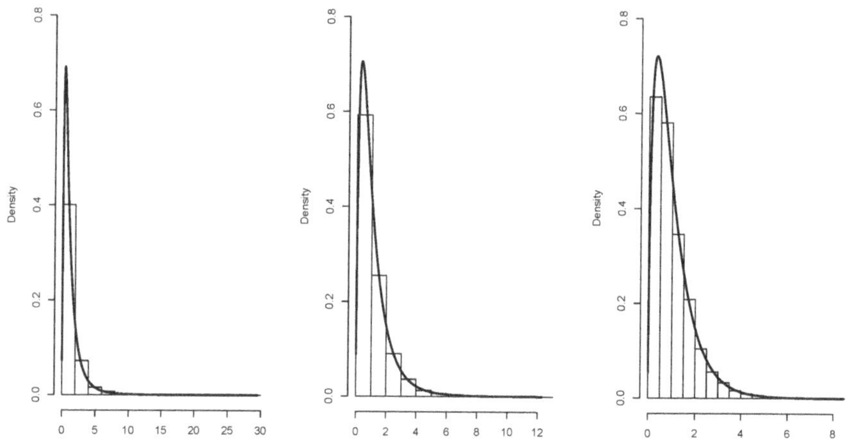

FIGURE 5.2
Approximating the randomization distribution of F by $F_{J-1,N-J}$ for $J = 4$. From left to right: $N = 12$, $N = 20$ and $N = 80$

Recall that in the toy example, the ANOVA table generated by the R function anova(lm()) had a p-value of 0.1543 associated with the F statistic:

```
------------------------------------------------------------------
Analysis of Variance Table

Response: y
          Df Sum Sq Mean Sq F value Pr(>F)
factor(w)  2 69.333  34.667  3.7143 0.1543
Residuals  3 28.000   9.333

------------------------------------------------------------------
```

This p-value is the area under the density curve of an $F_{2,3}$ distribution that lies to the right of the observed value 3.714 of the F statistic:

```
------------------------------------------------------------------
> 1-pf(3.714,2,3)
[1] 0.1543051

------------------------------------------------------------------
```

In this case, the randomization distribution of the F-statistic can be approximated by an $F_{2,3}$ distribution, and consequently, 0.1543 can be used as an approximate p-value obtained from the randomization test.

5.4.1.6 Paired Comparisons

If the randomization test described in Section 5.4.1.4 leads to evidence against the sharp null hypothesis, an obvious next step is to identify the similar and dissimilar treatments among the group of J treatments. Pairwise comparisons between treatments is a natural way to approach this problem. Therefore, we investigate which of the $\binom{J}{2}$ pairs of treatments appear similar or dissimilar in terms of their estimated effects on the unit-level responses. Formally, we define the sharp null hypothesis

$$H_{0\mathrm{F}}^{j\widetilde{j}} : \tau_i(j, \widetilde{j}) = 0, i = 1, \ldots, N, \tag{5.26}$$

where $\tau_i(j, \widetilde{j})$ is defined in (5.1).

The Fisher randomization test can again be applied to test $H_{0F}^{j\widetilde{j}}$, using the methodology described in Section 2.5 of Chapter 2, using the test statistic $T_{\text{avg-unegvar}}$, $T_{\text{avg-eqvar}}$ or T_{rank}. However, all N experimental units cannot be directly used to compare treatments j and $\widetilde{j}$. Such a comparison can be based only on the $N_j + N_{\widetilde{j}}$ units that are exposed either to treatment j or $\widetilde{j}$, i.e., satisfy $W_i \in \{j, \widetilde{j}\}$. Consider again the example in Table 5.4. Because the randomization test using the F-statistic provides evidence against the sharp null hypothesis of no treatment effect, we can compare treatments 1 and 2 using observed outcomes y_1, y_2, y_3 and y_4; treatments 1 and 3 using observed outcomes y_1, y_4, y_5 and y_6; and treatments 1 and 3 using observed outcomes y_2, y_3, y_5 and y_6.

Another important aspect to consider is the fact that simultaneous assessment of $\binom{J}{2}$ hypothesis tests leads to the problem of having a higher probability of rejecting some null hypothesis even when it is true. Suppose, every pair of treatments $j, \widetilde{j}$ is declared significantly different if the p-value $p_{j,\widetilde{j}}$ is smaller than a pre-defined level of significance p^*. Under this rule, even if all $\binom{J}{2}$ null hypotheses of no effects are true, the probability that at least one pair will be declared significantly different increases with J.

A standard simple but "conservative" (in the sense of not rejecting often enough) approach to tackle this problem was proposed by Dunn (1961) [30] and is called the *Bonferroni adjustment*, named after Italian mathematician Carlo Emilio Bonferroni. The method compensates for multiple comparisons by dividing the significance level by the number of comparisons. Specifically, the treatment pair $j, \widetilde{j}$ is declared significantly different if $p_{j,\widetilde{j}} < p^*/\binom{J}{2}$. This is equivalent to adjusting the p-value $p_{j,\widetilde{j}}$ by the "ensemble adjusted p-value," $\binom{J}{2}p_{j,\widetilde{j}}$ proposed by Rosenthal and Rubin (1983) [73]. If there are certain treatment pairs that are of more interest to the experimenter than the others, the level associated with such pairs can be adjusted by setting larger p^*-levels for these pairs in such that way that maintains the overall p^*-level [73].

To illustate this approach for the example in Table 5.4, following rejection of the sharp null hypothesis of no treatment effect on any unit, we assess the following three sharp null hypotheses:

$$H_{0F}^{1,2} : \tau_i(1,2) = 0, \;\; H_{0F}^{1,3} : \tau_i(1,3) = 0, \;\; H_{0F}^{2,3} : \tau_i(2,3) = 0.$$

The outcomes used to assess $H_{0F}^{1,2}$ are shown in Table 5.7, from which the observed value of the test statistic $T_{\text{avg-eqvar}}$ is computed as 1.9, and the p-value for the one-sided alternative is obtained as $p^{1,2} = 1/6$. Using the same procedure, the p-values for testing $H_{0F}^{1,3}$ and $H_{0F}^{2,3}$ against one-sided alternatives are obtained as $p^{1,3} = 2/6$ and $p^{2,3} = 1/6$ respectively. Because three hypotheses are being assessed simultaneously, the ensemble-adjusted p-values for pairs $(1,2)$, $(1,3)$ and $(2,3)$ are $1/2$, 1 and $1/2$ respectively.

TABLE 5.7
Comparison of treatments 1 and 2 from Table 5.4

Unit	Treatment 1	Treatment 2
1	8	?
2	?	18
3	?	12
4	10	?
Mean	$\overline{y}_.(1) = 9$	$\overline{y}_.(2) = 15$
Var	$s_1^2 = 2$	$s_2^2 = 18$

5.4.1.7 Case study: randomization test from the marketing experiment

We now demonstrate the analysis of an experiment with a completely randomized treatment assignment using data from the product marketing experiment described at the beginning of this chapter. To compare eleven "test" detergents with respect to customer satisfaction, the products were randomly assigned and mailed to 2,427 American households with identical instructions for using them. The distribution of the 2,427 households into the 11 treatment groups is shown in Table 5.8.

TABLE 5.8
Distribution of households to the eleven treatment groups

1	2	3	4	5	6	7	8	9	10	11
264	285	241	241	256	247	209	141	151	149	243

Each household returned a survey regarding the household view of the product's performance. Here we discuss the analysis of the primary outcome of interest, i.e., the household's response to the survey question "How would you rate the product overall?" on a three-point ordinal scale with 1 denoting poor and 3 denoting excellent. The first goal is to test whether the products made any difference with respect to the reported level of satisfaction of users. We therefore test Fisher's sharp null hypothesis of no treatment effect on any household. We demonstrate the analysis with the F-statistic.

The observed value of the F statistic obtained from the ratings of 11 products by 2427 households was $F^{\text{obs}} = 12.44$. Assuming that the treatment assignment was completely randomized with the distribution of 2427 units into 11 groups as shown in Table 5.8, 10,000 draws were obtained from the distribution of the assignment variable (implemented by permuting the 2427×1 vector $\mathbf{W}$ 10,000 times). Figure 5.3 shows the histogram of the 10,000 values of the F-statistic and the observed value superimposed on the histogram. Clearly, the observed value is very large compared to the randomization distribution of the statistic under the sharp null hypothesis, yielding a p-value smaller than 1/10,000, and we conclude that there exists a significant difference among the 11 products with respect to the rating received from the 2427 households.

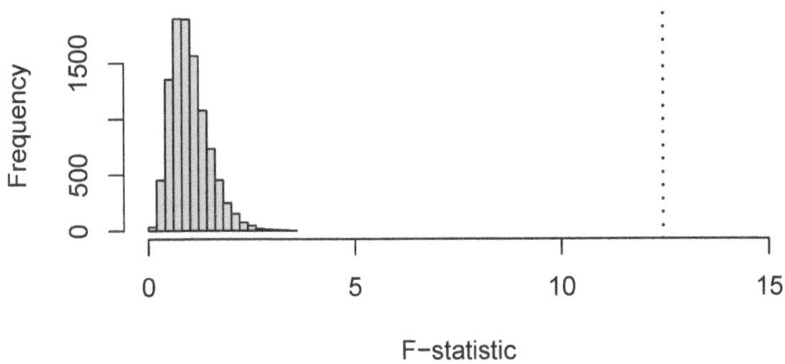

FIGURE 5.3
Randomization distribution of F for the marketing experiment

The R code used to generate Figure 5.3 is shown below:

```
---------------------------------------
data = read.csv("market_survey_data_mod.csv")
### get covariates and outcomes we want
var_name = c("Product.ID","ratprod")
data1 = data[,var_name]
y=data1$ratprod
T=data1$Product.ID
```

```
Fobs <- anova(lm(y ~ factor(T)))$F[1]

> Fobs
[1] 12.44322

Farray=NULL
ITER=10000
for (i in 1:ITER) {
  Tnew = sample(T)
  Fnew = anova(lm(y ~ factor(Tnew)))$F[1]
  Farray=c(Farray, Fnew)
}
pval=sum(Farray >= Fobs)/ITER

> pval
[1] 0

hist(Farray, main=NULL, xlab="F-statistic", xlim=c(0,15))
abline(v=Fobs, lty=3, lwd=2)
--------------------------------------
```

Having rejected the sharp null hypothesis of no treatment effect, we now proceed to identify specific pairwise differences among treatments. The mean outcome for each treatment is shown in Figure 5.4 in decreasing order. We find product 1 to have the highest average rating and product 7 to have the lowest.

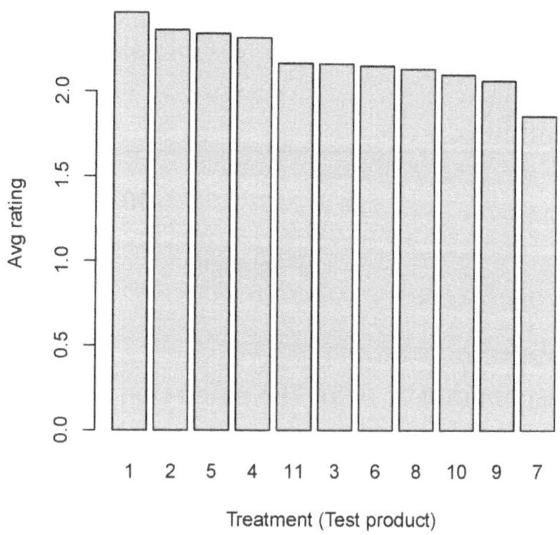

FIGURE 5.4
Average rating for each test product

To find out which pairs of treatments are different from each other, we test the sharp null hypothesis $H_{0F} : Y_i(j) = Y_i(\tilde{j})$ for $i = 1, \ldots, N$. As mentioned in Section 5.4.1.6, such a comparison can be based only on the $N_j + N_{\widetilde{j}}$ units that are exposed either to treatment j or $\widetilde{j}$, i.e., satisfy $W_i \in \{j, \widetilde{j}\}$. In this case, we use the test statistic $T_{\text{avg-eqvar}}$, which is a special case of the F statistic for comparing two treatments. Because 55 pairs of treatments will be compared, we should adjust for multiple testing using the ensemble-adjusted p-value approach described in Section 5.4.1.6. If we decide α to be the threshold one-sided p-value such that a p-value smaller than α will provide evidence against a particular $H_{0\text{F}}$, we will compare an ensemble-adjusted p-value $p* = p \times 55$ with α for testing the ensemble of 55 hypotheses. The two-sided ensemble-adjusted p-values for the 55 pairs is shown in Table 5.9. The R code used to obtain these ensemble-adjusted p-values is shown below:

```
--------------------------------------
# Pairwise with t

p=matrix(numeric(121), nrow=11)
for (i in 1:11){
  for (j in 1:11){
    if (i==j) {
      p[i,j]= NA
    }
    else {
      yi=y[T==i]
      Ti=T[T==i]
      yj=y[T==j]
      Tj=T[T==j]
      yij=c(yi,yj)
      Tij=c(Ti,Tj)
      Tobs=t.test(yi, yj, var.equal=TRUE)$statistic

      Tarray=NULL
      ITER=5000
      for (count in 1:ITER) {
        Tnew = sample(Tij)
        yi = yij[Tnew == i]
        yj = yij[Tnew == j]
        Tnew = t.test(yi, yj, var.equal=TRUE)$statistic
        Tarray=c(Tarray, Tnew)
      }
      pval=sum(abs(Tarray) >= abs(Tobs))/ITER
      p[i,j]=pval
    } # end else
  } # end j
} # end i

# Ensemble adjusted p-values
padj=p*55
for (i in 1:11) {
```

```
  for (j in 1:11){
padj[i,j] = min(padj[i,j],1)
  }
}

# Order them by treatment means
col.order <- sortedTrmeans[,1]
row.order <- sortedTrmeans[,1]
orderedpadj=padj[row.order,col.order]
library(rstatix)
pull_upper_triangle(orderedpadj)
--------------------------------------
```

TABLE 5.9
Two-sided p-values for 55 pairs of products

Treatment	Treatment									
	2	5	4	11	3	6	8	10	9	7
1	1.000	1.000	0.605	0.000	0.000	0.000	0.000	0.000	0.000	0.000
2		1.000	1.000	0.077	0.055	0.066	0.022	0.000	0.000	0.000
5			1.000	0.396	0.341	0.297	0.275	0.077	0.000	0.000
4				1.000	1.000	0.737	0.770	0.341	0.055	0.000
11					1.000	1.000	1.000	1.000	1.000	0.000
3						1.000	1.000	1.000	1.000	0.011
6							1.000	1.000	1.000	0.000
8								1.000	1.000	0.033
10									1.000	0.363
9										0.814

The ensemble-adjusted p-values smaller than .05 are underlined. Declaring these pairs as "significantly different" will ensure that the overall level of the test does not exceed .05. The results suggest that product 1 received the best rating among the eleven products, with ratings significantly (at level .05) higher than the ratings received by products 11, 3, 6, 8,10, 9 and 7. Product 2 received ratings signifcantly higher than products 8, 10, 9 and 7, whereas product 5 received ratings significantly higher than products 9 and 7. However, products 2 and 5 were not found to be significantly different. Also, products 4, 11, 3, and 6 received ratings significantly higher than product 7, but were not significantly different from products 8, 10 and 9. Products 8, 10, 9 and 7 received the lowest ratings and among these six pairs, only 8 and 7 were found significantly different.

5.4.2 Neymanian inference for treatment contrasts from CRD

We now return to randomization-based inference of some common population-level causal estimands $\overline{\tau}(j, \widetilde{j})$ defined in (5.4) and $\overline{\tau}(j)$ defined in (5.5) that were shown to be special cases of linear contrasts of treatment averages $\overline{\tau}(\lambda_1, \ldots, \lambda_J)$ in (5.6), and then discuss a general procedure for drawing inference on such contrasts using the Neymanian approach. This is a generalization of the Neymanian inference for the average treatment effect $\overline{\tau} = \overline{Y}(1) - \overline{Y}(0)$ discussed in Chapter 2.

Denoting the vector of contrast coefficients $(\lambda_1, \dots, \lambda_J)$ by $\boldsymbol{\lambda}$, note that under a completely randomized assignment of treatments $1, \dots, J$ respectively to $N_1, \dots, N_J$ units such that $\sum_{j=1}^{J} N$, an unbiased estimator of $\overline{\tau}(\boldsymbol{\lambda}) = \sum_{j=1}^{J} \lambda_j \overline{Y}(j)$ is

$$\widehat{\overline{\tau}}(\boldsymbol{\lambda}) = \sum_{j=1}^{J} \lambda_j \overline{y}.(j). \tag{5.27}$$

This follows from part (i) of Problem 5.16, because $\overline{y}.(j)$ is an unbiased estimator of $\overline{Y}(j)$. Before introducing the sampling variance of this estimator, we examine Table 5.10 of potential outcomes, which is almost the same as Table 5.1 with columns of covariates and average potential outcomes removed, and a column of contrasts introduced.

TABLE 5.10
Potential outcomes and contrasts for the N experimental units

Experimental Unit	Treatment 1	2	$\cdots$	J	Contrast
1	$Y_1(1)$	$Y_1(2)$	$\cdots$	$Y_1(J)$	$\tau_1(\boldsymbol{\lambda})$
2	$Y_2(1)$	$Y_2(2)$	$\cdots$	$Y_2(J)$	$\tau_2(\boldsymbol{\lambda})$
$\cdots$	$\cdots$	$\cdots$	$\cdots$	$\cdots$	$\cdots$
N	$Y_N(1)$	$Y_N(2)$	$\cdots$	$Y_N(J)$	$\tau_N(\boldsymbol{\lambda})$
Average	$\overline{Y}(1)$	$\overline{Y}(2)$	$\cdots$	$\overline{Y}(J)$	$\overline{\tau}(\boldsymbol{\lambda})$
Variance	S_1^2	S_2^2	$\cdots$	S_J^2	$S^2_{\overline{\tau}(\boldsymbol{\lambda})}$

The last row of Table 5.10 gives the variances (with divisor $N-1$) of the unit-level potential outcomes and contrasts. Specifically, for $j = 1, \dots, J$,

$$S_j^2 = \frac{1}{N-1} \sum_{i=1}^{N} \left(Y_i(j) - \overline{Y}(j)\right)^2, \tag{5.28}$$

denotes the variance of all N potential outcomes for treatment j with divisor $N-1$, and

$$S^2_{\overline{\tau}(\boldsymbol{\lambda})} = \frac{1}{N-1} \sum_{i=1}^{N} \left(\tau_i(\boldsymbol{\lambda}) - \overline{\tau}(\boldsymbol{\lambda})\right)^2, \tag{5.29}$$

denotes the variance of all unit-level contrasts. Then the sampling variance of the estimated contrast is

$$\text{Var}\left(\widehat{\overline{\tau}}(\lambda)\right) = \sum_{j=1}^{J} \lambda_j^2 \frac{S_j^2}{N_j} - \frac{1}{N} S^2_{\overline{\tau}(\boldsymbol{\lambda})}. \tag{5.30}$$

Note that (5.30) is a generalization of the result on the sampling variance of the estimated average treatment effect in treatment-control experiments given by (2.24) of Chapter 2. Recall that the average difference between two treatments j and $\tilde{j}$ is $\tau(j,\tilde{j})$ defined in (5.4), which is a special case of $\overline{\tau}(\boldsymbol{\lambda})$ with $\lambda_j = 1$ and $\lambda_{\tilde{j}} = -1$. An unbiased estimator of this estimand is $\widehat{\overline{\tau}}(j,\tilde{j}) = \overline{y}.(j) - \overline{y}.(\tilde{j})$. By (5.30), the sampling variance of this estimator is:

$$\mathrm{Var}\left(\widehat{\overline{\tau}}(j,\tilde{j})\right) = \frac{S_j^2}{N_j} + \frac{S_{\tilde{j}}^2}{N_{\tilde{j}}} - \frac{1}{N} S^2_{\overline{\tau}(j,\tilde{j})},$$

which is the same as (2.24).

To perform inference on a contrast $\overline{\tau}(\boldsymbol{\lambda})$, we estimate its sampling variance, i.e., $\mathrm{Var}\left(\widehat{\overline{\tau}}(\lambda)\right)$ in (5.30). Let s_j^2 denote the sample variances of observed values for treatment j, $j = 1, \ldots, J$, i.e.,

$$s_j^2 = \frac{1}{N_j - 1} \sum_{i:W_i=j}^{N} W_i \left(y_i - \overline{y}.(j)\right)^2. \tag{5.31}$$

Then, by standard theory of finite population sampling (e.g., see [18]), s_j^2 is an unbiased estimator of S_j^2, i.e., $E[s_j^2] = S_j^2$ for $j = 1, \ldots, J$. However the last term on the right side of (5.30), i.e., $S^2_{\overline{\tau}(\boldsymbol{\lambda})}$ does not admit an unbiased estimator from the observed data, because the unit-level treatment contrasts $\tau_i(\boldsymbol{\lambda})$ are not observable for *any* of the N units. However, under the assumption of strict additivity (as per Definition 5.1), all the unit level contrasts $\tau_i(\boldsymbol{\lambda})$ will be equal, and $S^2_{\overline{\tau}(\boldsymbol{\lambda})}$ will vanish, making $\sum_{j=1}^{J} \lambda_j^2 (s_j^2/N_j)$ an unbiased estimator of $\mathrm{Var}\left(\widehat{\overline{\tau}}(\lambda)\right)$. If strict additivity does not hold, then

$$E\left[\sum_{j=1}^{J} \lambda_j^2 (s_j^2/N_j)\right] - \mathrm{Var}\left(\widehat{\overline{\tau}}(\lambda)\right) = S^2_{\tau(\boldsymbol{\lambda})} > 0,$$

which means $\sum_{j=1}^{J} \lambda_j^2 (s_j^2/N_j)$ will have a positive bias, i.e., it will typically overestimate the sampling variance $\mathrm{Var}\left(\widehat{\overline{\tau}}(\lambda)\right)$.

We thus define the following estimator of the sampling variance of the estimated treatment contrast:

$$\widehat{\mathrm{Var}}_{\mathrm{unpooled}}\left(\widehat{\overline{\tau}}(\lambda)\right) = \sum_{j=1}^{J} \lambda_j^2 (s_j^2/N_j), \tag{5.32}$$

where s_j^2 is the observed sample variance of outcomes for treatment j. As just stated, this estimator is generally a positively biased estimator of $\mathrm{Var}\left(\widehat{\overline{\tau}}(\lambda)\right)$, and is unbiased if strict additivity holds.

If, however, there are reasons to believe that strict additivity holds, one can further improve estimator (5.32) by noting that under strict additivity, the variances $S_1^2, \ldots, S_J^2$ will all be equal. Thus, it makes sense to pool the J estimators $s_1^1, \ldots, s_J^2$ to obtain a more precise estimator of this common variance. Let S^2 denote the common value of $S_1^2, \ldots, S_J^2$ under strict additivity. Kempthorne (1955) [47] showed that under the assumption of strict additivity, this common variance S^2 can be unbiasedly estimated by MSRes, the mean squared residual obtained from the ANOVA decomposition. Consequently, the following variance estimator can be used:

$$\widehat{\text{Var}}_{\text{pooled}}\left(\widehat{\overline{\tau}}(\lambda)\right) = \text{MSRes}\sum_{j=1}^{J}\frac{\lambda_j^2}{N_j}, \tag{5.33}$$

which is an unbiased estimator of the sampling variance of the estimated contrast $\widehat{\overline{\tau}}(\lambda)$ under strict additivity.

To perform inference on a contrast $\overline{\tau}(\boldsymbol{\lambda})$, we can use a normal approximation of the estimator $\widehat{\overline{\tau}}(\lambda)$. Such approximation follows from the fact that $\widehat{\tau}(\lambda)$ is a linear combination of $\overline{y}.(j)$'s, which are approximately normal. Thus an approximate $100(1-\alpha)\%$ confidence interval for the contrast $\overline{\tau}(\boldsymbol{\lambda})$ can be obtained as:

$$\widehat{\overline{\tau}}(\lambda) \pm z_{\alpha/2}\sqrt{\widehat{\text{Var}}\left(\widehat{\overline{\tau}}(\lambda)\right)},$$

where $z_{\alpha/2}$ denotes the upper $\alpha/2$-point of a standard normal distribution and either $\widehat{\text{Var}}_{\text{unpooled}}\left(\widehat{\overline{\tau}}(\lambda)\right)$ or $\widehat{\text{Var}}_{\text{pooled}}\left(\widehat{\overline{\tau}}(\lambda)\right)$ can be substituted for $\widehat{\text{Var}}\left(\widehat{\overline{\tau}}(\lambda)\right)$.

Returning to the example in Table 5.4, suppose we are interested in the contrast $\overline{\tau}^* = \overline{Y}(3) - 2\overline{Y}(2) + \overline{Y}(1)$. A point estimate of $\overline{\tau}^*$ is

$$\widehat{\overline{\tau}^*} = \overline{y}.(3) - 2\overline{y}.(2) + \overline{y}.(1) = 7 - 2 \times 15 + 9 = -14,$$

obtained by substituting the values of $\overline{y}.(j)$ for $j = 1, 2, 3$ from Table 5.4. Calculating the values of the sample variances as $s_1^2 = 2$, $s_2^2 = 18$ and $s_3^2 = 8$, and substituting them into (5.32), we obtain $\widehat{\text{Var}}_{\text{unpooled}}(\widehat{\overline{\tau}^*}) = (2 \times 1 + 18 \times 4 + 8 \times 1)/2 = 41$. Thus a 95% confidence interval for $\overline{\tau}^*$ using this variance estimator is: $-14 \pm 1.96 \times \sqrt{41}$, i.e., $[-26.55, -1.45]$.

Substituting the value of MSRes from Table 5.5 into (5.33), we obtain $\widehat{\text{Var}}_{\text{pooled}}(\widehat{\overline{\tau}^*}) = 28/3 \times (1 + 4 + 1)/2 = 28$. A 95% confidence interval for τ^* using this variance estimator is: $-14 \pm 1.96 \times \sqrt{28}$, i.e., $[-24.37, -3.63]$. This interval is clearly shorter than the one obtained using $\widehat{\text{Var}}_{\text{unpooled}}(\widehat{\overline{\tau}^*})$, but may not contain the true value of $\overline{\tau}^*$ with the desired probability when the assumption of strict additivity is violated. Working out Exercise 5.6 will provide additional practice on how to obtain Neymanian confidence intervals for treatment contrasts.

Testing Neymanian hypothesis of equality of all average potential outcomes: is the F statistic appropriate?

Whereas the Fisher randomization test is a natural way to test the sharp null hypothesis $H_{0\text{F}}$ in (5.14), Neyman's null hypothesis of zero average treatment effect [65, 66], i.e.,

$$H_{0\text{N}} : \overline{Y}(1) = \ldots = \overline{Y}(J), \tag{5.34}$$

is often of interest. The null hypothesis $H_{0\text{N}}$ imposes weaker restrictions on the potential outcomes than $H_{0\text{F}}$, making it impossible to impute missing potential outcomes under that hypothesis and consequently create the randomization distribution of a test statistic like F. When $J = 2$, one can use the asymptotic normality of test statistic like $T_{\text{t-stat-uneqvar}}$ defined in Chapter 2 to test $H_{0\text{N}}$. We have already seen in Section 5.4.1.5 that the randomization distribution of the F statistic can be approximated by an $F_{J-1,\,N-J}$ distribution under Fisher's sharp null hypothesis $H_{0\text{F}}$. However, this approximation does not hold under $H_{0\text{N}}$ unless the potential outcomes are strictly additive (and $H_{0\text{N}}$ along with strict additivity implies that $H_{0\text{F}}$ holds).

Moreover, Ding and Dasgupta (2018) [28] argued that using the F statistic to test $H_{0\text{N}}$ leads to an invalid test in the sense that the test may reject a null hypothesis more or less often than the nominal level. An alternative test statistic, using which leads to a valid test for $H_{0\text{N}}$ was proposed. Interested readers are referred to the Appendix for details of this statistic and its use to assess $H_{0\text{N}}$.

5.5 Randomization-based inference from RBD: two-way decomposition

Section 5.2 introduced a randomized block design with J treatments, where the N units are grouped into B blocks indexed by $b = 1, \ldots, B$ with block sizes $M_1, \ldots, M_b$ such that $\sum_b M_b = N$. Also, we denoted the predetermined number of units in block b to be assigned to treatment j by N_{bj}, such that $\sum_{b=1}^{N} N_{bj} = N_j$. In the simplified stem cell experiment described in the introductory section of this chapter, $N = 24$ units, $B = 2$ blocks of sizes $M_1 = M_2 = 12$, $J = 3$ treatments, $N_{11} = N_{12} = N_{13} = 4$ units to receive each treatment within block 1, and $N_{21} = N_{22} = N_{23} = 4$ units to receive each treatment within block 2. In the example of Table 5.18, $N = 18$, $J = 3$, $B = 3$, $M_1 = M_2 = M_3 = 6$, $N_1 = N_2 = N_3 = 2$ and $N_{bj} = 2$ for $b = 1, 2, 3$ and $j = 1, 2, 3$. Both of these experiments are examples of Balanced RBDs defined below:

Definition 5.2 (Balanced RBD) *If $M_1 = M_2 = \cdots = M_B = N/B = r$, and N_{bj}'s are equal for all $b = 1, \dots, B$ and $j = 1, \dots, J$, i.e., all the B blocks are of equal size r, and each treatment is assigned to an equal number of units within and across blocks, then the design is called a Balanced RBD.*

This inferential procedures under a CRD described in Section 5.4 can be easily extended to the case of RBDs by adding another layer of decomposition of observed outcomes. Recall that a one-way ANOVA decomposition splits the total variance among the observed outcomes (measured by SSTot) into two components - the treatment sum of squares SSTre and the residual sum of squares SSRes (which captures the variation among units exposed to the same treatment). If it is possible to further decompose this residual variation into two components - one that is explained by the blocking structure, and the other part that is not explained by blocking and represents variation in outcomes among units exposed to the same treatment within each block - then the treatment effects can be compared typically with higher precision. Thus, if instead of a one-way decomposition, we are able to implement a two-way decomposition of the total variance in the data, departures from the sharp null hypothesis of no treatment effect on any unit can possibly be detected with greater "power", defined, as the probability of detecting a true difference when there actually exists one. We will now introduce such two-way decompositions in randomized block designs and the procedure for determining alternative F statistic for testing the sharp null hypothesis of no treatment effect on any unit.

As shown in classical experimental design textbooks [93, 10], for a balanced RBD, the total variation in the observed outcomes $\text{SSTot} = \sum_{i=1}^{N} \{y_i - \overline{y}.(\cdot)\}^2$ defined in (5.18) can be decomposed as follows:

$$\text{SSTot} = \text{SSTre} + \text{SSBlk} + \text{SSInt} + \text{SSRes}, \tag{5.35}$$

where as before, SSTot, SSTre and SSRes are acronyms for the Total sum of squares, Treatment sum of squares and Residual sum of squares respectively, and the new terms, SSBlk and SSInt, are acronyms for the Block sum of squares and Interaction (between treatment and block). Block effects and block-treatment interactions were defined in Chapter 3, and the interpretations of the sums of squares are analogous. A large value of SSBlk suggests differences among block-level average outcomes, whereas a large value of SSInt suggests that the treatment effects vary across blocks. See Exercise 5.12 to see how (5.35) follows from a decomposition of the observed outcome y_i similar to (5.15), but including estimators of mean-centered block effects and block-treatment interactions.

As in Section 5.4.1.1, the decomposition in (5.35) can be summarized in Table 5.11, which outlines the ANOVA table for a balanced RBD with $N = MB$ units in B blocks of size M each, J treatments, with each treatment replicated $r = M/J$ times within each block.

TABLE 5.11
ANOVA Table and F-statistic for RBD

Source	DF	SS	MS	F statistic
Treatments	$J-1$	SSTre	MSTre	$F_{\text{Tre}} = \text{MSTre}/\text{MSRes}$
Block	$B-1$	SSBlk	MSBlk	$F_{\text{Blk}} = \text{MSBlk}/\text{MSRes}$
Blk-Tre interaction	$(J-1)(B-1)$	SSInt	MSInt	$F_{\text{Int}} = \text{MSInt}/\text{MSRes}$
Residuals	$JB(r-1)$	SSRes	MSRes	
Total	$N-1$	SSTot		

Note that unlike Table 5.4 that has a single F-statistic used for comparison of treatments by the Fisher randomization test, Table 5.11 has three F statistics - F_{Tre}, F_{Blk} and F_{Int}.

Arguing along the lines of Sections 5.4.1.1 and 5.4.1.4, the F-statistic defined in (5.25) as the ratio of MSTre and MSRes would be an appropriate test statistic for testing $H_{0\text{F}}$, the sharp null hypothesis of no treatment effect on any unit. We refer to this F statistic as F_{Tre}.

We now introduce two other F-statistics that can be defined on the basis of the ANOVA decomposition for balanced RBD as shown in Table 5.11. First, define

$$F_{\text{Blk}} = \frac{\text{MSBlk}}{\text{MSRes}}, \tag{5.36}$$

as the F-ratio that measures block-to-block variation relative to residual variation. Note that $\text{MSBlk} = \text{SSBlk}/(B-1)$, where SSBlk measures the observed difference among block averages. In some standard textbooks on design of experiments, this F-statistic is used to "test" for the presence of block effects. However, in the present context, all the randomness in the observed data, and hence in any test statistic, comes from the randomness of the treatment assignment mechanism. Whereas repeated randomizations change the treatment assignment within blocks, and consequently change SSTre and SSInt, the quantity SSBlk remains unchanged across randomizations. To see this, consider the observed outcomes in Table 5.18, and note that the observed averages in block 1, block 2 and block 3 are 8, 10 and 6 respectively. Given the observed outcomes, these numbers will remain unchanged, irrespective of how the treatment assignment of the six units within each block changes. Consequently, SSBlk and MSBlk will not be affected by the treatment assignment, and will not have any non-trivial randomization distribution, and therefore F_{Blk} is constant in the randomization test.

Next, we introduce another F-ratio:

$$F_{\text{Int}} = \frac{\text{MSInt}}{\text{MSRes}}, \tag{5.37}$$

which is a measure of the interaction between block and treatment. This F-ratio can be interpreted as a measure of the block-treatment interaction, and gets larger as this interaction increases. Unlike F_{Blk}, given the observed data,

F_{Int} will vary with the treatment assignments and thus will have a randomization distribution. Note that, under $H_{0\text{F}}$, there will be no block treatment interaction, and the computed value of F_{Int} should be small. However, departure from $H_{0\text{F}}$ will not necessarily result in a larger value of F_{Int}. For example, if the alternative hypothesis is a sharp null hypothesis

$$\tau_i(j, \widetilde{j}) = \tau(j, \widetilde{j})$$

for all $i = 1, \ldots, N$ and all $j \neq \widetilde{j}$, $j, \widetilde{j} \in \{1, \ldots, J\}$, then strict additivity holds, and there is still no block-treatment interaction. Under this alternative, F_{Int} will still be small, and its randomization distributions under the null and alternative hypotheses should be similar.

However, if the alternative hypothesis is of the form:

$$\tau_i(j, \widetilde{j}) = \overline{\tau}_b(j, \widetilde{j})$$

for all i in block b, $b = 1, \ldots, B$ and all $j \neq \widetilde{j}$, $j, \widetilde{j} \in \{1, \ldots, J\}$, then F_{Int} can indicate departures from $H_{0\text{F}}$ in favor of such an alternative, and become larger as $\overline{\tau}_b(j, \widetilde{j})$ varies with b. Thus, it can be used to test sharp null hypothesis of constant treatment effect for all units against the alternative that the treatment effect varies across blocks.

5.5.1 Example: simplified stem cell experiment

Consider another hypothetical version of the simplified stem cell experiment described in 5.1 where three modulators, labeled 1,2 and 3, are compared on 24 experimental units (wells) in two blocks (plates). The three treatments are assigned to the 12 units in each plate by a completely randomized assignment mechanism. The observed outcomes (proportions of stem cells in each well that were converted to β cells) are shown in Table 5.12.

We obtain the observed values of F_{Tre} and F_{Int} as shown in the two-way ANOVA decomposition in Table 5.11 using the following R functions:

```
-------------------------------------------------------------
y = c(0.32,0.07,0.13,0.05,0.82,0.75,0.03,0.34,0.29,0.23,0.41,0.50,0.31,0.22,
0.04,0.12,0.12,0.38,0.09,0.12,0.30,0.15,0.04,0.01)
W = c(1,3,2,1,2,2,3,2,1,3,3,1,2,1,3,2,1,2,3,1,1,2,3,3)
B = c(rep(1,12),rep(2,12))

> Ftreobs=anova(lm(y~ factor(W) + factor(B) + factor(W):factor(B)))$F[1]
> Ftreobs
[1] 4.108674

> Fintobs=anova(lm(y~ factor(W) + factor(B) + factor(W):factor(B)))$F[3]
> Fintobs
[1] 0.479919
-------------------------------------------------------------
```

Comparing the observed values 4.10 and 0.48 of F_{Tre} and F_{Int} respectively with their randomization distributions under the sharp null hypothesis of no

TABLE 5.12
Observed outcomes (y) from the hypothetical simplified stem cell experiment with three treatments

Unit	Block(plate)	x_1 (row)	x_2 (column)	W	y
1	1	1	1	1	0.32
2	1	1	2	3	0.07
3	1	1	3	2	0.13
4	1	1	4	1	0.05
5	1	2	1	2	0.82
6	1	2	2	2	0.75
7	1	2	3	3	0.03
8	1	2	4	2	0.34
9	1	3	1	1	0.29
10	1	3	2	3	0.23
11	1	3	3	3	0.41
12	1	3	4	1	0.50
13	2	1	1	2	0.31
14	2	1	2	1	0.22
15	2	1	3	3	0.04
16	2	1	4	2	0.12
17	2	2	1	1	0.12
18	2	2	2	2	0.38
19	2	2	3	3	0.09
20	2	2	4	1	0.12
21	2	3	1	1	0.30
22	2	3	2	2	0.15
23	2	3	3	3	0.04
24	2	3	4	3	0.01

treatment effect on any unit, we obtain p-values of 0.047 and 0.64 (see Figure 5.5), suggesting the possible presence of treatment effects, but no evidence of its variation across blocks (i.e., no block-treatment interaction). The R code is shown below:

```
---------------------------------------------------------------
# Fisher randomization test

Ftreobs=anova(lm(y~ factor(W) + factor(B) + factor(W):factor(B)))$F[1]
Fintobs=anova(lm(y~ factor(W) + factor(B) + factor(W):factor(B)))$F[3]
ITER=5000
Ftre=NULL
Fint=NULL
set.seed(10)
for (i in 1:ITER) {
Wnew=numeric(24)
Wnew[1:12] = sample(c(1,1,1,1,2,2,2,2,3,3,3,3))
Wnew[13:24] = sample(c(1,1,1,1,2,2,2,2,3,3,3,3))
Ftre = c(Ftre, anova(lm(y~ factor(Wnew) + factor(B) + factor(Wnew):
factor(B))
```

```
)$F[1])
Fint = c(Fint, anova(lm(y~ factor(Wnew) + factor(B) + factor(Wnew):
factor(B)))
$F[3])
}

pvalTre=sum(Ftre >= Ftreobs)/5000
hist(Ftre, freq=FALSE, ylim=c(0,.8),xlim=c(0,20), main=NULL,
xlab=expression(F[Tre]))
abline(v=Ftreobs, lty=3, lwd=1.5)
text(10,0.6,paste("p-value=",pvalTre))

pvalint=sum(Fint >= Fintobs)/5000
hist(Fint, freq=FALSE, ylim=c(0,.8),xlim=c(0,20), main=NULL,
xlab=expression(F[Int]))
abline(v=Fintobs, lty=3, lwd=1.5)
text(10,0.6,paste("p-value=",pvalint))
------------------------------------------------------------
```

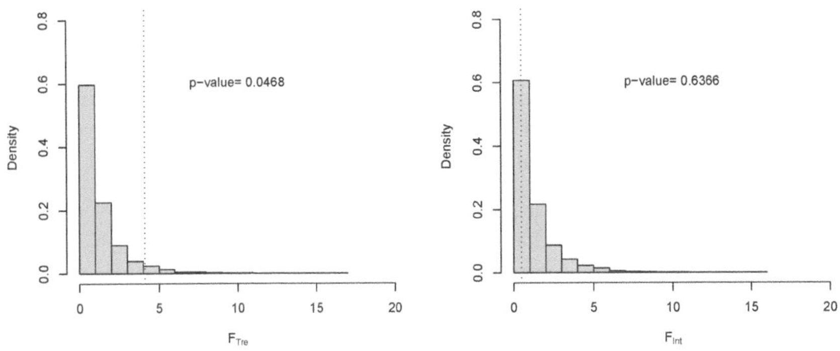

FIGURE 5.5
Randomization test with F_{Tre} (left) and F_{Int} (right)

5.5.2 Unreplicated RBD

Box, Hunter and Hunter [10] describe an experiment in which four types of paints used for marking lanes on highways are compared with respect to their durability. Six highway sites, each with four locations, e.g., lanes, to apply paint, are available. Each of the 24 locations is an experimental unit in this study. In a completely randomized design, the 24 units (locations) would be randomly assigned to the four treatments (paints). Clearly such a strategy would not be advisable if the factors affecting durability of paints (e.g., volume and nature of traffic, weather conditions) vary from one site to the other. Therefore an RBD with the four locations within each highway site comprising a block is a more appropriate design in this situation. In this experiment, there are $B = 6$ blocks, $M = 4$ units (locations) within each block, and $J = 4$ treatments (paints) to be compared. This is an example of an *unreplicated* RBD in which each treatment is assigned to one and only one unit within each block, so that $M = J$. A formal definition is given below.

Definition 5.3 (Unreplicated RBD) *If $M_1 = M_2 = \cdots = M_B = J$, and $N_{bj} = 1$ for all $b = 1, \ldots, B$ and $j = 1, \ldots, J$, i.e., all the B blocks are of the same size as the number of treatments, so that exactly one unit is assigned to each treatment within every block, then the design is called an unreplicated RBD.*

Unreplicated RBDs typically arise due to resource constraints, when the experimenter does not have enough experimental units to ensure that each treatment is assigned to at least two units within every block. The total number of possible assignments in an unreplicated RBD is $\{(N/B)!\}^B = (J!)^B$. Unreplicated RBD being a special case of balanced RBDs, most of the ideas, methods and results discussed in this section so far apply to unreplicated RBDs. However, some are some subtle differences that we will point out below.

In the ANOVA decomposition for an unreplicated RBD, the degrees of freedom for treatments and blocks are respectively $J-1$ and $B-1$, as before. Recalling that the degrees of freedom for the block-treatment interaction is $(J-1)(B-1)$, the degrees of freedom from these three sources add up to

$$(J-1) + (B-1) + (J-1)(B-1) = JB - 1 = N - 1,$$

which is the total degrees of freedom available from the observed outcomes of the experiment, leaving no degrees of freedom for the residuals. This phenomenon occurs because there is only one observation corresponding to every (block, treatment) combination in an unreplicated RBD, not allowing us to distinguish the residual from the block-treatment interaction. Consequently, it is possible to draw inference from the ANOVA decomposition of an unreplicated RBD if and only if it is assumed that there is no block-treatment interaction, i.e., the treatment effects within each block are the same.

Thus, for an unreplicated RBD, we have following decomposition:

$$\text{SSTot} = \text{SSTre} + \text{SSBlk} + \text{SSRes}, \tag{5.38}$$

where SSTot, SSTre and SSBlk are the same as in (5.35), with degrees of freedom $BJ-1$, $J-1$ and $B-1$ respectively. The residual sum of squares SSRes is the same as SSInt, having degrees of freedom $(BJ-1)-(B-1)-(J-1) = (B-1)(J-1)$, and can be obtained by subtraction of SSTre + SSBlk from SSTot. The above decomposition can be summarized in Table 5.13, which is the ANOVA table for an unreplicated balanced RBD with $N = MB = MJ$ units in B blocks of size M each, J treatments, so that each treatment is sassigned to one and only one unit within each block.

TABLE 5.13
ANOVA Table for Unreplicated RBD

Source	DF	SS	MS	F
Treatments	$J-1$	SSTre	MSTre	$F_{\text{Tre}} = \text{MSTre}/\text{MSRes}$
Block	$B-1$	SSBlk	MSBlk	
Residuals	$(J-1)(B-1)$	SSRes	MSRes	
Total	$JB-1$	SSTot		

The randomization test to assess Fisher's sharp null hypothesis of no treatment effect for the unreplicatd RBD can be performed in exactly the same way as in the case of a balanced replicated RBD. The observed value of the test statistic F_{Tre} needs to be computed and compared with its randomization distribution obtained by repeated draws from the distribution of the assignment vector $\mathbf{W}$, leading to computation of the p-value. Analysis of the data obtained from the paint experiment using this approach is left as an exercise (5.17).

5.6 Classical and modern strategies to avoid undesirable allocations

We now discuss general strategies to find an acceptable allocation of N units to $J(>2)$ treatments in the presence of multiple premeasured covariates. When the number of such covariates is small, blocking can be a reasonable strategy. As mentioned in 5.2, the problem of assigning $J(>2)$ treatments to units when there are two blocking variables with levels $B_1(>2)$ and $B_2(>2)$ has been studied under the name "row-column designs" [87, 88, 94], a name that originates from agricultural experiments where experimental units were plots of land, and undesirable allocations were likely to make effects of treatments (e.g., types of fertilizers) confounded with the effects of fertility gradients that could exist along two perpendicular directions (rows and columns). The problem of finding a row-column design can be formulated as assigning J Latin letters (each representing a treatment factor) on a $B_1 \times B_2$ rectangular array of cells in such a way that there exists "some balance" among the number of alphabets along the rows and columns. Arguably, the simplest and most commonly used row-column designs are Latin square designs, which can be used only when $J = B_1 = B_2$.

Consider another hypothetical version of the simplified stem cell experiment described in Section 5.1 where five modulators, labeled A, B, C, D, E, are compared on 25 experimental units (wells) arranged in a 5×5 array of units (plates). Assignment of the five treatments to the 25 units in each plate using a completely randomized assignment mechanism can lead to undesirable allocations as shown in Figure 5.6.

A	B	C	D	E
A	B	C	D	E
A	B	C	D	E
A	B	C	D	E
A	B	C	D	E

A	A	A	A	A
B	B	B	B	B
C	C	C	C	C
D	D	D	D	D
E	E	E	E	E

FIGURE 5.6
Two undesirable allocations labeled w_{bad1} (left) and w_{bad2} (right) of five treatments on a 5×5 square

As in Chapter 4, the units, covariates (row and column) and the potential outcomes for the 25 units is displayed in Table 5.14, where the mapping between i and the cell rows and columns ($x_1, x_2 = 1, \ldots, 5$) is lexicographic

TABLE 5.14
Table of potential outcomes in a 5×5 Square

Unit	Row	Column	Potential Outcomes				
(i)	(x_1)	(x_2)	$Y(A)$	$Y(B)$	$Y(C)$	$Y(D)$	$Y(E)$
1	1	1	$Y_1(A)$	$Y_1(B)$	$Y_1(C)$	$Y_1(D)$	$Y_1(E)$
2	1	2	$Y_2(A)$	$Y_2(B)$	$Y_2(C)$	$Y_2(D)$	$Y_2(E)$
3	1	3	$Y_3(A)$	$Y_3(B)$	$Y_3(C)$	$Y_3(D)$	$Y_3(E)$
...		...			...		
24	5	4	$Y_{24}(A)$	$Y_{24}(B)$	$Y_{24}(C)$	$Y_{24}(D)$	$Y_{24}(E)$
25	5	5	$Y_{25}(A)$	$Y_{25}(B)$	$Y_{25}(C)$	$Y_{25}(D)$	$Y_{25}(E)$

The general solution to the design problem here is similar to the one described in Chapter 4: ensure that the five treatment groups are "acceptably similar" in terms of covariate differences. Let $D_{\text{M}}(j, j')$ denote the Mahalanobis distance between the two-dimensional vectors of covariate averages $\overline{\mathbf{x}}(j)$ and $\overline{\mathbf{x}}(j')$. Then a criterion for acceptance of a specific assignment vector is

$$\max_{j,j'} D_{\text{M}}(j, j') \leq a, \tag{5.39}$$

where $a > 0$ is a threhold criterion.

Let us now compare the twoundesirable assignments in Figure 5.6 with the two in Figure 5.7. Labeling the five treatments $A, \ldots, B$ as $1, \ldots, 5$, the assignment vectors corresponding to these four assignments are:

$$\begin{aligned}
w_{\text{bad1}} &= (1,2,3,4,5,1,2,3,4,5,1,2,3,4,5,1,2,3,4,5,1,2,3,4,5), \\
w_{\text{bad2}} &= (1,1,1,1,1,2,2,2,2,2,3,3,3,3,3,4,4,4,4,4,5,5,5,5,5), \\
w_{\text{LSD}} &= (1,2,3,4,5,2,3,4,5,1,3,4,5,1,2,4,5,1,2,3,5,1,2,3,4). \\
w_{\text{test}} &= (1,1,5,3,4,3,3,5,5,2,2,2,1,5,1,4,2,3,4,5,3,2,4,4,1).
\end{aligned}$$

A	B	C	D	E
B	C	D	E	A
C	D	E	A	B
D	E	A	B	C
E	A	B	C	D

A	A	E	C	D
C	C	E	E	B
B	B	A	E	A
D	B	C	D	E
C	B	D	D	A

FIGURE 5.7
Two acceptable allocations of five treatments labeled w_{LSD} (left) and w_{test} (right) on a 5×5 square

The first two are unacceptable allocations, making the treatment effects completely confounded with the effects of the covariates x_1 (row) and x_2 (column) respectively. The third allocation is special in the sense that (i) the five units within *each* row receive five *distinct* treatments and (ii) the five units within *each* column receive five *distinct* treatments. The fourth one is a randomly chosen assignment from the $\binom{25}{5}$ possible allocations generated by a completely randomized mechanism.

The Mahalanobis distances between vectors of covariate means for pairs of treatment groups as shown as 5×5 symmetric matrices in Table 5.15. Both the undesirable assignment vectors yield a constant (and hence maximum) Mahalanobis distance of 9 between pairs of treatment groups. The third (and the best) assignment vector results in perfect covariate mean balance between all pairs, resulting in a maximum of zero Mahalanobis distance. The randomly chosen assignment vector yields a maximum Mahalanobis distance of 3.7 between treatment groups 3 and 5.

TABLE 5.15
Matrices of D_{M} for different $\mathbf{w}$'s - Left: w_{bad1} and w_{bad2} (both lead to identical matrices), Middle: w_{LSD}, Right: w_{test}

$$\begin{pmatrix} 0 & 9 & 9 & 9 & 9 \\ 9 & 0 & 9 & 9 & 9 \\ 9 & 9 & 0 & 9 & 9 \\ 9 & 9 & 9 & 0 & 9 \\ 9 & 9 & 9 & 9 & 0 \end{pmatrix}, \begin{pmatrix} 0 & 0 & 0 & 0 & 0 \\ 0 & 0 & 0 & 0 & 0 \\ 0 & 0 & 0 & 0 & 0 \\ 0 & 0 & 0 & 0 & 0 \\ 0 & 0 & 0 & 0 & 0 \end{pmatrix}, \begin{pmatrix} 0.0 & 1.9 & 1.3 & 1.3 & 2.5 \\ 1.9 & 0.0 & 0.9 & 2.0 & 3.5 \\ 1.3 & 0.9 & 0.0 & 3.4 & \underline{3.7} \\ 1.3 & 2.0 & 3.4 & 0.0 & 2.2 \\ 2.5 & 3.5 & 3.7 & 2.2 & 0.0 \end{pmatrix}$$

Figure 5.8 shows the empirical distribution of $\max_{j,j'} D_{\text{M}}(j, j')$ for $j, j' \in \{1, \ldots, 5\}$ obtained from 5000 random permutations of the treatment assignment vactor. Choosing $a = 1.095$ in the acceptance criterion (5.39) accepts 1% of the allocations. One can choose other percentiles of the distribution of the maximum Mahalanobis distance to determine the threshold.

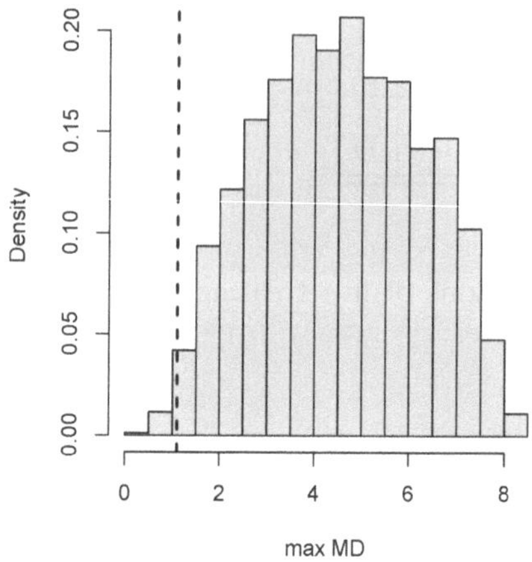

FIGURE 5.8
Histogram of the maximum Mahalanobis distance between covariate mean vectors for pairs of treatments obtained from 5000 randomly selected treatment assignment vectors to allocate 25 units to five treatments. The dotted vertical line (at 1.095) indicates the 1st percentile of the empirical distribution

If we want to restrict acceptable randomizations only to the type of allocations as in w_{LSD} where we have a "perfect" balance, then a naive algorithm to implement such a treatment assignment mechanism is to generate repeatedly assignment vectors using a completely randomized assignment mechanism, and then reject those vectors that do not satisfy the condition of each level of each blocking factor appearing once for each treatment level. Such a strategy, however, is computationally relatively inefficient. A Latin Square is a useful tool that gives a mathematically precise definition of a perfectly balanced treatment assignment with two blocking factors, and helps the experimenter adopt a computationally efficient strategy to choose assignment vectors that satisfy those requirements. The feature of this square that results in a perfectly balanced treatment assignment is that, in the square, each letter appears ONLY once in each row and each column.

5.6.1 Latin square designs and Fisherian inference

A Latin square is defined as a square consisting of Latin letters, in which each Latin letter appears ONLY once in each row and each column. Of course, the elements of the square do not necessarily have to be Latin letters, and can be any symbols, e.g., Greek letters or integers. It is known that an $n \times n$ Latin Square exists for any integer n.

A Latin square in which the first row and column are arranged in alphabetical order is called a *Reduced Latin square* or a Latin square in *standard* form. The assignment on the left of Figure 5.7 is an example of a 5×5 Latin square. There are different methods for constructing Latin Squares and the reader is referred to [5] for details. One of the straightforward methods is the cyclic method, in which the letters in the first row can be written in any order. In the second row, all the letters should be shifted by one place to the right so that the letter occupying the last position in the first row moves to the first position in the second row. Continuing in the same manner for each row, i.e., by shifting the letters of the nth row by one place to the right in the $(n+1)$th row and moving the last letter of the nth row to the first position in the $(n+1)$th row, a Latin square can be generated. An interesting property of Latin Squares is, if the rows or columns of a Latin square are permuted, the resulting square still remains a Latin square. This property is often used to randomly assign units to treatments in a Latin square design. The R package "magic" provides a simple R function, called rlatin(n), to generate a random Latin square of order n. A demonstration is given below:

```
---------------------------------------------------------------
> library(magic)
> rlatin(5)
     [,1] [,2] [,3] [,4] [,5]
[1,]    1    4    5    2    3
[2,]    3    1    4    5    2
[3,]    4    5    2    3    1
[4,]    2    3    1    4    5
[5,]    5    2    3    1    4
---------------------------------------------------------------
```

Latin squares were first used for designing experiments by Sir R. A. Fisher, while he was working at the Rothamsted agricutural experimental station in England. During 1919–1925, Fisher proposed several fundamental principles of experimental design that included the application of Latin Squares in designing randomized experiments with two blocking factors. In many of the agricultural experiments conducted at Rothamstead, a large plot of land would be divided into an array of sub-plots that served as experimental units for comparison of treatments such as different types of fertilizers. Therefore balancing treatments across two blocking factors and the idea of Latin Squares was natural in such experiments. Fisher's pioneering work on experimental design during these years eventually led to publication of the book "The Design of Experiments" [33] in 1935.

We now discuss Fisherian inference using data from a Latin square design. Consider the hypothetical data from the hypothetical stem cell experiment with five treatments on a 5×5 array of wells shown in Table 5.16. Note that this data can also be alternatively represented in the form of observed and missing potential outcomes for each unit, analogous to Table 5.14.

TABLE 5.16
Latin square design and observed data for the hypothetical stem cell experiment with five treatments.

Row (x_1)	Columns (x_2) 1	2	3	4	5
1	$A = 0.43$	$B = 0.38$	$C = 0.38$	$D = 0.46$	$E = 0.41$
2	$B = 0.40$	$C = 0.43$	$D = 0.58$	$E = 0.50$	$A = 0.52$
3	$C = 0.39$	$D = 0.51$	$E = 0.46$	$A = 0.47$	$B = 0.41$
4	$D = 0.46$	$E = 0.45$	$A = 0.48$	$B = 0.41$	$C = 0.51$
5	$E = 0.44$	$A = 0.55$	$B = 0.41$	$C = 0.47$	$D = 0.49$

As in the case of CRD and RBD, the total sum of squares in the 25 observed outcomes shown in Table 5.16 can be decomposed as

$$\text{SSTot} = \text{SSRow} + \text{SSCol} + \text{SSTre} + \text{SSRes}, \tag{5.40}$$

where SSRow and SSCol respectively represent the sums of squares for the two blocking factors represented by the row and column. Decomposition (5.40) assumes that there is no interaction among the two blocking factors (row and column) and treatment factors. The ANOVA decomposition of the observed outcomes in Table 5.16 is shown below. The observed value of the F-statistic for treatment is seen to be equal to 7.53.

```
------------------------------------------------------------
#Hypothetical stem cell (5x5 Latin square)
y = c(0.43,0.40,0.39,0.46,0.44,0.38,0.43,0.51,0.45,0.55,0.38,0.58,0.
46,0.48,
0.41,0.46,0.50,0.47,0.41,0.47,0.41,0.52,0.41,0.51,0.49)
row = c(rep(1,5), rep(2,5), rep(3,5), rep(4,5), rep(5,5))
col = c(rep(c(1,2,3,4,5),5))
w= c(1,2,3,4,5,2,3,4,5,1,3,4,5,1,2,4,5,1,2,3,5,1,2,3,4)
anova(lm(y ~ factor(row) + factor(col) + factor(w)))

Analysis of Variance Table

Response: y
            Df  Sum Sq   Mean Sq F value   Pr(>F)
factor(row)  4 0.00652 0.0016300  1.5281 0.255812
factor(col)  4 0.01596 0.0039900  3.7406 0.033649 *
factor(w)    4 0.03212 0.0080300  7.5281 0.002832 **
Residuals   12 0.01280 0.0010667
------------------------------------------------------------
```

Suppose we want to test a sharp null hypothesis of zero treatment effect, i.e., the hypothesis that the five modulators have the same effect on “cell differentiation” (i.e., proportion of cells in each well converted to beta-cells).

A natural test statistic is the F-statistic F_{Tre} associated with the treatment in the ANOVA decomposition, the ratio of how much variability in the data results from the levels of the treatment factor relative to the variability remaining after fitting the main effects of the treatment, row and column factors. The observed value of this statistic was obtained as 7.53.

TABLE 5.17
Completed table of potential outcomes for the hypothetical stem cell (Latin square) example, with missing potential outcomes imputed under the null hypothesis of zero treatment effect. Observed potential outcomes are in bold. Imputed potential outcomes are in parentheses.

				Potential Outcomes				
Unit	Row	Column	Assignment	$Y(A)$	$Y(B)$	$Y(C)$	$Y(D)$	$Y(E)$
1	1	1	A	**0.43**	(0.43)	(0.43)	(0.43)	(0.43)
2	1	2	B	(0.40)	**0.40**	(0.40)	(0.40)	(0.40)
3	1	3	C	(0.39)	(0.39)	**0.39**	(0.39)	(0.39)
...		...				...		
24	5	4	C	(0.47)	(0.47)	**0.47**	(0.47)	(0.47)
25	5	5	D	(0.49)	(0.49)	(0.49)	**0.49**	(0.49)

Under this sharp null hypothesis of zero treatment effects, the missing potential outcomes for the ith unit are imputed exactly as observed for that unit (see Table 5.17). As before, after imputing the missing potential outcomes under the sharp null hypothesis of zero treatment effect, we repeatedly re-assign the treatments to the units such that each treatment appears once in each row and once in each column, according to the way the experiment was actually conducted. A simple and efficient computational algorithm to do this is to randomly generate draws $\mathbf{w}_{\text{new}}$ from the distribution of the "allowable" treatment assignment vectors satisfying the restriction imposed by a Latin square, and calculating the test statistic by changing only the treatment assignment vector, keeping the observed outcomes, row and column vectors unchanged in each iteration.

Based on 2,000 randomly drawn treatment assignments satisfying the Latin square restrictions, we obtain a randomization-based p-value of 0.0055 — the F-statistics from the hypothetical "observed" datasets were as large as or larger than the observed value of 7.53 in just 11 of the 2,000 permutations. That is, even when the observed F-statistic is combined with more extreme possible F-statistics that could have been observed under the sharp null hypothesis, a rare event(=11/2000) has occurred. The randomization distribution and observed value of the test statistic are shown in Figure 5.9. The low p-value for the F-statistic from the observed dataset suggests that the sharp null hypothesis is unlikely to be true, and different modulators *do* have a non-null effect on the cell differentiation. The R code for the Fisher randomization test is shown below:

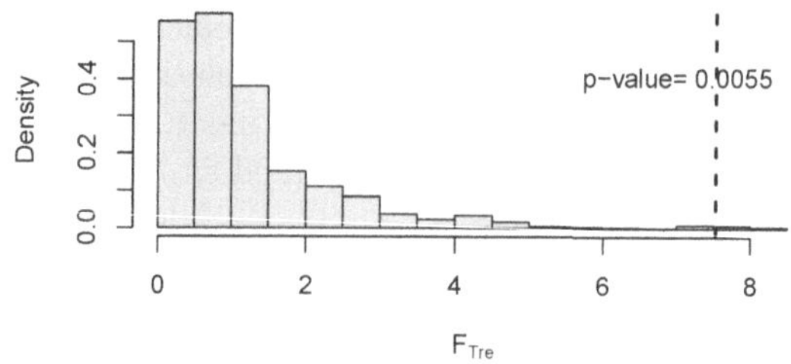

FIGURE 5.9
Randomization distribution and actual observed value of the F-statistic in the hypothetical stem cell (LSD) example, based on 2,000 random treatment assignments satisfying the Latin square restriction. The bold vertical line indicates the actual observed F-statistic value

```
------------------------------------------------------------
# Randomization test from hypothetical stem cell expt (5x5 Latin square)
Fobs <- anova(lm(y ~ factor(row) + factor(col) + factor(w)))$F[3]

library(magic)
F <- NULL
set.seed(20)
for (i in 1:2000) {
  new_w <- as.vector(rlatin(5))
  Fnew <- anova(lm(y ~ factor(row) + factor(col) + factor(new_w)))$F[3]
  F <- c(F, Fnew)
}

pval <- sum(round(F,2) >= round(Fobs,2))/2000
hist(F, main=NULL, freq=FALSE, xlab=expression(F[Tre]))
abline(v = Fobs, lty=2, lwd=2)
text(7,0.4,paste("p-value=",pval))
------------------------------------------------------------
```

Although the Latin square design permits an orthogonal ANOVA decomposition of the observed outcomes, it is clearly very restrictive in terms of possible allocations, and is only applicable when (i) there are two blocking factors (ii) the number of levels of the two blocking factors are the same, and are also equal to the number of levels of the treatment factor. Thus, if in the stem cell experiment, we need to compare five modulators on an 12×8 array of wells, a pure Latin square design cannot be used. Similarly, if there are 100 units, with 10 pre-measured covariates, and four treatments to be compared, a classical Latin square strategy will not work. The ideas are important but need to be embedded in a more general framework to be more applicable.

5.7 Problems

5.1 Describe a situation where SUTVA may be violated in the PTSD experiment. Describe how such a situation can be avoided by better design of the study.

5.2 Show that the unit-level mean-centered treatment effect $\tau_i(j)$ defined in (5.2) are unit-level treatment contrasts and derive the coefficients $\lambda_1, \ldots, \lambda_J$ for each j. Similarly, argue that the population-level mean-centered treatment averages $\overline{\tau}(j)$ defined in (5.5) are population-level treatment contrasts.

5.3 Consider the potential outcomes in Table 5.2.

(a) Find the unit-level adjusted treatment effects $\tau_i(j)$ defined in (5.2) for $i = 1, \ldots, 6$ and $j = 1, \ldots, J$, and the corresponding population-level mean-centered treatment averages $\overline{\tau}(j)$ defined in (5.5).

(b) Find the unit-level treatment contrasts $Y_i(1) - (Y_i(2) + Y_i(3))/2$ and their population-level counterpart.

5.4 In the simplified stem cell experiment for which the units, covariates and potential outcomes are shown in Table 5.3, determine the total number of possible allocations in the following situations, and identify one allocation that you consider "worst" among these, explaining under what assumption such an allocation will be worst.

(i) $N_1 = 12$, $N_2 = N_3 = 6$ and a CRD is used for assigning the three treatments to the 24 units.

(ii) Plate is used as a blocking variable, and within each plate a CRD is used for assigning the three treatments to the 12 units, so that four units receive each treatment.

(iii) Both plate and column within plate are used as blocking variables. Within each plate and each column, three treatments are randomly assigned to the three units.

5.5 In each of the three designs in Problem 5.4, write down the probability distribution $\Pr[\mathbf{W} = \mathbf{w}]$ of the assignment vector $\mathbf{W} = (W_1, \ldots, W_N)$, where, as usual, W_i denotes the assignment variable for unit i.

5.6 From the observed outcomes in Table 5.4, find *point estimates* and 95% approximate confidence intervals of (a) the adjusted treatment effects $\overline{\tau}(j)$ defined in (5.5) for $j = 1, 2, 3$, (b) the paired differences $\tau(j, \tilde{j}) = \overline{\tau}(\tilde{j}) - \overline{\tau}(j)$ defined in (5.4) for $j \neq \tilde{j}$, $j, \tilde{j} \in \{1, 2, 3\}$, and (c) the contrast $\tau(1, -1/2, -1/2) = \overline{Y}(1) - \left(\overline{Y}(2) + \overline{Y}(3)\right)/2$.

5.7 Table 5.18 shows observed outcomes from an experiment with 18 units in three blocks of equal size $M_1 = M_2 = M_3 = J = 6$. As in problem 5.6 above, find *point estimates* of (a) the mean-centered treatment effects $\overline{\tau}(j)$ defined in (5.5) for $j = 1, 2, 3$, (b) the paired differences $\tau(j, \widetilde{j}) = \overline{\tau}(\widetilde{j}) - \overline{\tau}(j)$ defined in (5.4) for $j \neq \widetilde{j}$, $j, \widetilde{j} \in \{1, 2, 3\}$, and (c) the contrast $\tau(1, -1/2, -1/2) = \overline{Y}(1) - \left(\overline{Y}(2) + \overline{Y}(3)\right)/2$ first for each block, and then for the entire population.

TABLE 5.18
Examples of observed and unobserved (missing) outcomes with 18 units in 3 blocks

Block	Experimental Unit	Treatment Tr 1	Tr 2	Tr 3
Block 1	1	?	9	?
	2	?	?	8
	3	7	?	?
	4	6	?	?
	5	?	?	9
	6	?	9	?
Block Mean		$\overline{y}_1(1) = 6.5$	$\overline{y}_1(2) = 9$	$\overline{y}_1(3) = 8.5$
Block 2	7	7	?	?
	8	?	?	10
	9	?	13	?
	10	?	11	?
	11	?	?	10
	12	9	?	?
Block Mean		$\overline{y}_2(1) = 8$	$\overline{y}_2(2) = 12$	$\overline{y}_2(3) = 10$
Block 3	13	?	?	5
	14	4	?	?
	15	5	?	?
	16	?	?	7
	17	?	8	?
	18	?	7	?
Block Mean		$\overline{y}_3(1) = 4.5$	$\overline{y}_3(2) = 7.5$	$\overline{y}_3(3) = 6$
Population Mean		$\overline{y}.(1) = 6.33$	$\overline{y}.(2) = 9.50$	$\overline{y}.(3) = 8.17$

5.8 Starting from (5.15), prove the ANOVA decomposition (5.17).

5.9 Show that for $J = 2$ treatments indexed by 0 and 1, where N_0 and N_1 units are randomly assigned to treatments 0 and 1,

$$\begin{aligned} \text{SSTre} &= \frac{N_0 N_1}{N} \left(\overline{y}.(1) - \overline{y}.(0)\right)^2, \\ \text{SSRes} &= (N_0 - 1)s_0^2 + (N_1 - 1)s_1^2, \end{aligned}$$

where

$$s_j^2 = \frac{1}{N_j - 1} \sum_{i:W_i=j} \left(y_i - \overline{y}.(j)\right)^2$$

is the sample variance of the observed outcomes for treatment $j = 0, 1$. Hence show that the F-statistic is equal to the square of the $T_{\text{avg-eqvar}}$ statistic defined in (2.18).

5.10 Consider the example in Table 5.18 where 18 units in three blocks are assigned to three treatments using a balanced completely randomized assignment mechanism within each each block. Calculate the F statistic using the *one-way* ANOVA decomposition described in Section 5.4.1.1. Obtain the randomization distribution of F, under the sharp-null hypothesis of no treatment effect for any unit, by generating repeated draws from the assignment variable $\mathbf{W}$ based on the actual assignment mechanism, i.e., a randomized block design, and compare the distribution with its observed value to assess the sharp null hypothesis of no treatment effect on any unit.

5.11 You are interested in three treatments proposed to reduce low-density lipoprotein (LDL) or "bad" cholesterol for females. One of the three treatments is the standard treatment ($T0$) and the other two are new ($T1$ and $T2$). You want to decide whether either of the new ones is better than the old one and if so, which one. Nine female participants with similar pre-treatment LDL cholesterol levels (around 180 mg/dL, considered high) are selected for the study. These nine participants are labeled $1, \ldots, 9$ in increasing order of their ages, which are 29, 32, 35, 42, 45, 45, 54, 57 and 60 years respectively.

(a) Create a blank table of potential outcomes for this experiment. Make sure to create the right number of rows and columns in the table.

(b) Suppose you wish to assess whether there is *any* differential effect of *any* of the three treatments on *any* of the nine participants whatsoever. Formulate this scientific question using a statistical test of hypothesis. Write down the hypothesis clearly using appropriate notations, explaining the meaning of each notation you use.

(c) Suppose you wish to compare the average (over all participants) effect of the old treatment with the average of the new ones. Write down a suitable estimand (explaining the notations involved) that measures such an effect.

For all the remaining parts of the problem, let W_i be a random variable taking possible values 0, 1 or 2 depending on the whether treatment $T0$, $T1$ or $T2$ is assigned to unit i using some randomized assignment mechanism.

(d) Suppose the vector $(W_1, \ldots, W_9)$ is observed to be $(0, 2, 2, 1, 0, 2, 1, 0, 1)$. Indicate the appropriate cells (with a tick or a "x" mark) in the table of potential outcomes created in part (a) that you will observe.

For parts (e) to (f), consider a randomized assignment mechanism without any restrictions on the number of units assigned to a particular treatment.

(e) How many realizations of $(W_1, \ldots, W_9)$ are possible under this assignment mechanism?

(f) Write down a realization of $(W_1, \ldots, W_9)$ generated by this assignment that you feel is the "worst" possible one, keeping in mind the objective of the experiment. If you think there are multiple worst realizations (i.e., vectors $(W_1, \ldots, W_9)$ that are equally bad), write down any one of them. Explain *in a single sentence* why you consider such a realization to be worst.

For the following parts, consider a completely randomized assignment mechanism where exactly three units will be assigned to each treatment

(g) How many realizations of $(W_1, \ldots, W_9)$ are possible under the assignment mechanism described above?

(h) Write down a blank ANOVA table that can be constructed from the observed data, specifying the sources of variation and corresponding degrees of freedom.

5.12 *Note: This problem is created to provide the algebraic details of the two-way ANOVA decomposition for a balanced RBD and can be skipped by readers not interested in doing mundane and tedious algebraic derivations. However, just reading through the problem and solving the next one (5.13) should provide a good understanding of the decomposition.*

Consider a balanced RBD with B blocks of size M each. Let $r = M/J$ units be assigned to each of the J treatments within each block. Let $W_i \in \{1, \ldots, J\}$ denote the treatment assigned to unit i, and $b_i \in \{1, \ldots, B\}$ denote the block to which unit i belongs. We have the following mathematical identity among the observed values:

$$\begin{aligned} y_i &= \overline{y}. + (\overline{y}.(W_i) - \overline{y}.(\cdot)) + (\overline{y}_{b_i}(\cdot) - \overline{y}.(\cdot)) \\ &+ (\overline{y}_{b_i}(W_i) - \overline{y}.(W_i) - \overline{y}_{b_i}(\cdot) + \overline{y}.(\cdot)) + (y_i - \overline{y}_{b_i}(W_i)) \\ &= \overline{y}.(\cdot) + \widehat{\overline{\tau}}(W_i) + \widehat{\overline{\beta}}_{b_i} + \widehat{\overline{\gamma}}_{b_i}(W_i) + e_i, \end{aligned} \quad (5.41)$$

where

(a) The term $\widehat{\overline{\tau}}(W_i) = \overline{y}.(W_i) - \overline{y}.(\cdot)$ is an unbiased estimator of the mean-centered treatment effect $\overline{\tau}(j)$ (as in the one-way decomposition).

(b) The term $\widehat{\overline{\beta}}_{b_i} = \overline{y}_{b_i}(\cdot) - \overline{y}.(\cdot)$ unbiasedly estimates the mean-centered block effect.

(c) The term $\widehat{\overline{\gamma}}_{b_i}(W_i)$ unbiasedly estimates the mean-centered block-treatment interaction effect.

(d) $e_i = y_i - \overline{y}_{b_i}(W_i)$ denotes the *observed residual* for y_i, after accounting for the treatment, block and block-treatment interaction effects.

Show that, transferring $\overline{y}.(\cdot)$ on the left hand side of (5.41), followed by squaring and summing over i, leads to decomposition (5.35), i.e.,

$$\text{SSTot} = \text{SSTre} + \text{SSBlk} + \text{SSInt} + \text{SSRes},$$

where

$$\begin{aligned}
\text{SSTot} &= \sum_{i=1}^{N} \{y_i - \overline{y}.(\cdot)\}^2, \\
\text{SSTre} &= \sum_{i=1}^{N} \left\{\widehat{\overline{\tau}}(W_i)\right\}^2, \\
\text{SSBlk} &= \sum_{i=1}^{N} \left\{\widehat{\overline{\beta}}_{b_i}\right\}^2, \\
\text{SSInt} &= \sum_{i=1}^{N} \left\{\overline{\gamma}_{b_i}(j)\right\}^2, \\
\text{SSRes} &= \sum_{i=1}^{N} e_i^2.
\end{aligned}$$

5.13 Using the observed outcomes in Table 5.18 of problem 5.7, complete Table 5.19 to decompose the vector of observed outcomes into different components defined in (a)-(e) of problem 5.12. Calculate the different sums of squares by squaring and summing columns 6–9. Compute these sums of squares directly using suitable R functions, and compare them with the results obtained through your step-wise calculations.

5.14 For the hypothetical stem cell example with five treatments in Section 5.6, assess the sharp null hypothesis of no treatment effect using a randomization test, assuming that the design was based on the acceptance criterion $\max_{j,j'} D_{\text{M}}(j, j') \leq 1.095$. Note that in this case, orthogonal decomposition of the observed outcomes is not possible and hence the F statistic cannot be computed. An alternative statistic that can be used is the R-statistic

$$R = \max_{j,j'} |\widehat{\overline{\tau}}(j, j')| = \max_{j,j'} |\overline{y}_.(j') - \overline{y}_.(j)| \,.$$

5.15 Consider a completely randomized design with N units randomly assigned to J treatments, with N_j units assigned to treatment j, $j = 1, \ldots, J$. For $i = 1, \ldots, N$ and $j = 1, \ldots, J$, define a binary treatment indicator variable:

$$W_i(j) = \begin{cases} 1 & \text{if the } i\text{th unit is assigned to treatment } j \\ 0 & \text{otherwise,} \end{cases}$$

TABLE 5.19
Decomposition of observed outcomes in the toy example

Unit	b_i	w_i	y_i	$\overline{y}_{\cdot}(\cdot)$	$\widehat{\overline{\tau}}(w_i)$	$\widehat{\overline{\beta}}_{b_i}$	$\widehat{\overline{\gamma}}_b(j)$	e_i
1								
2								
3								
4								
5								
6								
7								
8								
9								
10								
11								
12								
13								
14								
15								
16								
17								
18								
Sum of squares					SSTre =	SSBlk =	SSInt =	SSRes =

(i) Express the assignment variable W_i in terms of the $W_i(j)$'s.

(ii) Show that

$$\begin{aligned}
\mathbb{E}[W_i(j)] &= \frac{N_j}{N}, \quad j = 1, \ldots, J, \\
\mathrm{Var}[W_i(j)] &= \frac{N_j}{N}(1 - \frac{N_j}{N}), \quad j = 1, \ldots, J, \\
\mathrm{Cov}[W_i(j), W_i(\widetilde{j})] &= -\frac{N_j N_{\widetilde{j}}}{N^2}, \quad j \neq \widetilde{j}, \\
\mathrm{Cov}[W_i(j), W_{\widetilde{i}}(j)] &= -\frac{N_j(N - N_j)}{N^2(N-1)}, \quad i \neq \widetilde{i}. \\
\mathrm{Cov}[W_i(j), W_{\widetilde{i}}(\widetilde{j})] &= \frac{N_j N_{\widetilde{j}}}{N^2(N-1)}, \quad i \neq \widetilde{i}, \quad j \neq \widetilde{j}.
\end{aligned}$$

5.16 Consider a completely randomized design with N units randomly assigned to J treatments such that N_j units receive treatment $j \in \{1, \ldots, J\}$. Let $\overline{y}_{\cdot}(j)$ denote the average of the observed outcomes corresponding to treatment $j = 1, \ldots, J$.

(i) Show that $\overline{y}_{\cdot}(j)$ is an unbiased estimator of $\overline{Y}(j)$, the mean of the potential outcomes corresponding to treatment j.

(ii) Show that $\overline{y}_{\cdot}(\cdot)$, the grand average of the N observed outcomes, is an unbiased estimator of $\overline{Y}$, the grand mean of all potential outcomes, if the N_j's are all equal.

(iii) Prove that $\overline{y}.(j) - \overline{y}.(\widetilde{j})$ is an unbiased estimator of $\overline{\tau}(j) - \overline{\tau}(\widetilde{j})$ for $j \neq \widetilde{j}$.

(iv) Calculate $\text{Var}\left(\overline{y}.(j) - \overline{y}.(\widetilde{j})\right)$ for $j \neq \widetilde{j}$.

Hint: Use the results of Problem 5.15.

5.17 Box, Hunter and Hunter (2005) [10] describe an RBD experiment in which four types of paints (treatments) used for marking lanes (units) on highways are compared with respect to their durability. Six highway sites (blocks), each with four locations, e.g., lanes, to apply paint, are available. Within each block (highway site), the four units (lanes) are assigned to the four treatments (types of paint) using a completely randomized assignment mechanism. The observed outcomes (durability in months) are shown below.

TABLE 5.20
Paint experiment data from Box, Hunter and Hunter (2005)

Site	Paint 1	Paint 2	Paint 3	Paint 4	Total
1	69	59	55	70	253
2	83	65	65	75	288
3	74	64	59	74	271
4	61	52	59	62	234
5	78	71	67	74	290
6	69	64	58	74	265
Total	434	375	363	429	1601
Average	72.33	62.50	60.50	71.50	66.71

Table 5.20 can be further condensed in a 24×3 matrix, each row representing a unit and columns representing the block $(1 \ldots, 4)$, treatment assigned $(1 \ldots, 4)$ and observed outcome, to make it amenable to analysis in R using the lm() function.

```
---------------------------------------------------------
> paintdata
      [,1] [,2] [,3]
 [1,]    1    1   69
 [2,]    1    2   59
 [3,]    1    3   55
 [4,]    1    4   70
 [5,]    2    1   83
 [6,]    2    2   65
 [7,]    2    3   65
 [8,]    2    4   75
```

```
 [9,]   3   1   74
[10,]   3   2   64
[11,]   3   3   59
[12,]   3   4   74
[13,]   4   1   61
[14,]   4   2   52
[15,]   4   3   59
[16,]   4   4   62
[17,]   5   1   78
[18,]   5   2   71
[19,]   5   3   67
[20,]   5   4   74
[21,]   6   1   69
[22,]   6   2   64
[23,]   6   3   58
[24,]   6   4   74
--------------------------------------------------------------
```

Assess whether there is any differential effect of any of the four paints on any lane. If you find evidence of such an effect, find out which paint(s) are different from the others using a pairwise comparison approach.

6

The 2^2 factorial experiment

Angrist et al. (2009) [1] reported an experimental evaluation of two strategies (factors) designed to improve academic performance among college freshmen at one of the satellite campuses of a Canadian university. One factor was the Student Support Program (SSP), which provided students with some extra services for study. The other factor was Student Fellowship Program (SFP), which offered financial incentives for good grades. Each factor had a treatment (1) and a control (0) level. In this experiment, also analyzed in [52], a total of 1656 freshmen were assigned randomly to the four treatment groups – 1006 students received neither SSP nor SFP (notationally level combination 00), 250 students were offered only SFP (notationally level combination 01), 250 were offered only SSP (level combination 10) and 150 students were offered both SSP and SFP (level combination 11). This is an example of a 2^2 factorial experiment, defined as an experiment conducted with the objective of simultaneously assessing the causal effects of *two* intervention factors on an outcome of interest, where each intervention factor has two levels, denoted by 0 and 1. A 2^2 factorial experiment has four different possible treatments, each of which is assigned to the experimental units using a randomized mechanism.

One may wonder why such an experiment cannot be treated as a randomized design with a single factor with $J = 2^2$ levels discussed in Chapter 5. The answer is that, it can be, but a factorial structure permits us to assess, in precise senses, the effects of each individual factor as well their combinated effects (interactions) on the outcome of interest. In the following subsection, we will explicitly define the estimands that permit such assessments.

6.1 Factorial effects: Causal estimands in a 2^2 experiment

As in the educational experiment just described, a 2^2 factorial experiment has four treatment combinations labeled 00, 01, 10, and 11, or notationally $z_1 z_2$ where $z_1 \in \{0, 1\}$ and $z_2 \in \{0, 1\}$. The order of writing down the four treatment combinations, starting with 00 and ending with 11 is referred to as

DOI: 10.1201/9781003048794-6

lexicographic order. The jth treatment combination is thus

$$j = 2z_1 + z_2 + 1. \tag{6.1}$$

Note that $(z_1 z_2)$ is the binary representation of the integer $j - 1$. Thus the four treatment combinations 00, 01, 10, and 11 can be indexed as $j = 1, 2, 3$ and 4 respectively.

Let N be the number of experimental units. Under SUTVA, the ith unit has four potential outcomes denoted by $Y_i(z_1 z_2)$, where $z_1 \in \{0, 1\}$ and $z_2 \in \{0, 1\}$. More explicitly, unit i has four potential outcomes: $Y_i(00)$, $Y_i(01)$, $Y_i(10)$, and $Y_i(11)$. Thus the matrix of all potential outcomes has dimension $N \times 4$. Columns 2–5 of Table 6.1 comprise the 5×4 matrix of potential outcomes with five units.

TABLE 6.1
The potential outcomes (left portion), conditional effects (middle portion: $A : \tau_i(F_1|F_2 = 0)$, $B : \tau_i(F_1|F_2 = 1)$, $C : \tau_i(F_2|F_1 = 0)$, $D : \tau_i(F_2|F_1 = 1)$) and factorial effects (right portion) for the 2^2 experiment.

Unit	Potential outcome for				Conditional effects				Factorial effects		
(i)	(00)	(01)	(10)	(11)	A	B	C	D	$\tau_i(F_1)$	$\tau_i(F_2)$	$\tau_i(F_1F_2)$
1	4	4	4	4	0	0			0	0	0
2	4	5	6	7	2	2			2	1	0
3	4	5	6	8	2	3			2.5	1.5	0.5
4	4	5	7	6	3	1			2	0	–1
5	4	6	6	8	2	2			2	2	0
Avg	4.00	4.75	5.75	6.25	1.75	1.50			1.625	0.625	0.125
Var	0	0.25	1.58	2.92	1.58	1.67			1.230	0.560	0.400

For each unit, any comparison of the four potential outcomes is a causal effect of possible interest. In particular, we can consider linear contrasts defined in (5.3) of the form

$$\begin{aligned} \tau_i(\boldsymbol{\lambda}) &= \lambda(00)Y_i(00) + \lambda(01)Y_i(01) + \lambda(10)Y_i(10) + \lambda(11)Y_i(11) \\ &= \sum_{z_1=0}^{1}\sum_{z_2=0}^{1} \lambda(z_1 z_2)Y_i(z_1 z_2), \end{aligned} \tag{6.2}$$

where in (6.2), $\boldsymbol{\lambda}$ is the contrast vector $(\lambda(00), \lambda(01), \lambda(10), \lambda(11)^{\mathrm{T}}$, satisfying $\lambda(00) + \lambda(01) + \lambda(10) + \lambda(11) = 0$ where not all λs are zero.

We now define some treatment contrasts of common interest in the experiment. First, we define, for unit i, the *conditional main effect* (CME) of factor F_1 on Y when factor F_2 is held fixed at level 0:

$$\tau_i(F_1|F_2 = 0) = Y_i(10) - Y_i(00). \tag{6.3}$$

A comparison with (6.2) shows that the contrast vector $\boldsymbol{\lambda}$ for $\tau_i(F_1|F_2 = 0)$ is (–1,0,1,0). Analogously, the conditional main effect of factor F_1 given F_2 is at level 1 is

$$\tau_i(F_1|F_2 = 1) = Y_i(11) - Y_i(01), \tag{6.4}$$

for which the contrast vector is (0,–1,0,1).

For example, the potential outcomes for unit 2 in Table 6.1 under the four treatment combinations 00, 01, 10, and 11 are 4, 5, 6 and 7 respectively. Thus, for unit 2, the effect of changing factor F_1 from 0 to 1 when factor F_2 is held fixed at 0 is $\tau_2(F_1|F_2 = 0) = Y_2(10) - Y_2(00) = 6 - 4 = 2$. Similarly we have $\tau_2(F_1|F_2 = 1) = Y_2(11) - Y_2(01) = 7 - 5 = 2$. The conditional main effects $\tau_i(F_1|F_2 = 0)$ and $\tau_i(F_1|F_2 = 1)$ for the four units are shown in columns 6 and 7 of Table 6.1 where they are denoted by A and B respectively.

Similar to the CMEs defined in (6.3) and (6.4), we define the CME of factor F_2 on the response when factor F_1 is held fixed at level 0 as

$$\tau_i(F_2|F_1 = 0) = Y_i(01) - Y_i(00), \tag{6.5}$$

and the CME of factor F_2 given F_1 is at level 1 as

$$\tau_i(F_2|F_1 = 1) = Y_i(11) - Y_i(10). \tag{6.6}$$

Calculating the contrast vectors for these two contrasts for the potential outcomes in Table 6.1 is left as an exercise (see problem 6.1).

We now define the unit-level *unconditional* (or average) main effect of each factor as the average of the two CMEs for that factor. Thus, the unconditional main effect, or simply the main effect (ME) of factor F_1 is:

$$\begin{aligned}
\tau_i(F_1) &= \frac{\tau_i(F_1|F_2 = 1) + \tau_i(F_1|F_2 = 0)}{2} \\
&= \frac{1}{2}\left(-Y_i(00) - Y_i(01) + Y_i(10) + Y_i(11)\right) && (6.7) \\
&= \frac{Y_i(11) + Y_i(10)}{2} - \frac{Y_i(01) + Y_i(00)}{2}. && (6.8)
\end{aligned}$$

Comparison of representation (6.7) of the ME of F_1 with (6.2) establishes it as a contrast with contrast vector (–1/2, –1/2,1/2, 1/2). Equation (6.8) shows another representation of this ME that some may consider a more natural way of defining an unconditional ME: the difference between the average potential outcomes of unit i when F_1 is at its high level (1) and when at its low level (0).

Similarly, the unit-level ME of factor F_2 is:

$$\begin{aligned}
\tau_i(F_2) &= \frac{\tau_i(F_2|F_1 = 1) + \tau_i(F_2|F_1 = 0)}{2} \\
&= \frac{1}{2}\left(-Y_i(00) + Y_i(01) - Y_i(10) + Y_i(11)\right) && (6.9) \\
&= \frac{Y_i(11) + Y_i(01)}{2} - \frac{Y_i(10) + Y_i(00)}{2}. && (6.10)
\end{aligned}$$

The difference between the two CMEs $\tau_i(F_1|F_2 = 1)$ and $\tau_i(F_1|F_2 = 0)$ measures the unit-level interaction between factors F_1 and F_2. To scale the interaction effect in the same way as main effects, we divide this difference by 2 to obtain the unit-level interaction effect as

$$\begin{aligned}\tau_i(F_1F_2) &= \frac{\tau_i(F_1|F_2=1) - \tau_i(F_1|F_2=0)}{2} \\ &= \frac{1}{2}\left(Y_i(00) - Y_i(01) - Y_i(10) + Y_i(11)\right). \qquad (6.11)\end{aligned}$$

This interaction effect measures how the effect of factor F_1 on the potential outcome of unit i changes as the level of factor F_2 changes. Note that the contrast vector associated with this effect is (1/2, −1/2, −1/2, 1/2).

The unit-level interaction effect $\tau_i(F_1F_2)$ is symmetric with respect to the order of the factors F_1 and F_2, i.e., $\tau_i(F_1F_2)$ defined in (6.11) equals $\tau_i(F_2F_1)$, that can be defined by taking the difference between the CMEs $\tau_i(F_2|F_1=1)$ and $\tau_i(F_2|F_1=0)$. Proving this symmetry is left as an exercise (Problem 6.2).

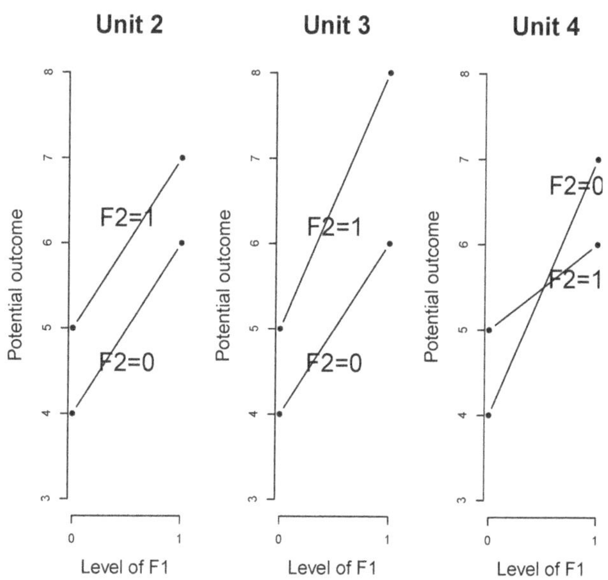

FIGURE 6.1
Visual representation of unit-level factorial effects in Table 6.1

The unit-level factorial effects are shown in columns 9–11 of Table 6.1. Note that because all the four potential outcomes for unit 1 are the same, each of the three unit-level factorial effects equals zero, meaning neither of the two factors, nor their interaction, has any causal effect on unit 1. Figure 6.1 gives a visual representation of the three unit-level factorial effects (MEs of factors F_1 and F_2, and the interaction between F_1 and F_2) for units 2, 3 and 4 by plotting the potential outcomes against the levels 0 and 1 of factor F_1 separately for the two levels of factor F_2. For all the three units, a positive

TABLE 6.2
Contrast coefficients $\lambda(z_1 z_2)$ for the three factorial effects (without divisor 2)

Level combination	Factorial effect		
$z_1 z_2$	ME of F_1	ME of F_2	Interaction
00	−1	−1	+1
01	−1	+1	−1
10	+1	−1	−1
11	+1	+1	+1

effect of changing the level of F_1 from 0 to 1 is observed, irrespective of the level of factor F_2. For unit 2, this positive effect remains the same as the level of F_2 changes from 0 to 1 (reflected in the two lines being parallel. However, this is not the case for units 3 and 4, indicating the presence of interaction effects for units 3 and 4, but none for unit 2. Note that this interaction is particularly severe for unit 4 where the lines cross each other – a consequence of the fact that the effect of changing F_2 from 0 to 1 changes from positive to negative as F_1 is changed from level 0 to level 1.

Table 6.2 shows the contrast coefficient of each potential outcome for the three factorial effects without the common divisor 2. Note that the contrast vectors for the ME of F_1, ME of F_2 and the interaction between F_1 and F_2 shown in the columns of Table 6.2 are orthogonal to each other (i.e., their inner, that is, dot products are zero). Further, the contrast vector for the interaction can be obtained by element-wise multiplication of the two ME contrasts, which is a nice feature that will help us to define an interaction among multiple factors in general 2^K designs. Lastly, along with these three contrast vectors shown in Table 6.2, for completeness, define unit i's average potential outcome as

$$\tau_i(0) = \frac{Y_i(00) + Y_i(01) + Y_i(10) + Y_i(11)}{4}.$$

The vector of coefficients of the potential outcomes in the expression for $\tau_i(0)$ is (1,1,1,1), which is orthogonal to each of the three contrast vectors in Table 6.2. However, note that although $\tau_i(0)$ is a linear combination of the potential outcomes for unit i, it is not a contrast, and does not have a causal interpretation.

The contrast matrix in Table 6.2 is useful because it provides a direct way for computing the three factorial effects in the last three columns of Table 6.1 from the potential outcomes, bypassing the computation of the conditional effects $A - D$. For example, the three unit-level factorial effects $\tau_2(F_1), \tau_2(F_2)$ and $\tau_2(F_1 F_2)$ for unit 2 can be calculated as a 1×3 row vector as

$$\begin{aligned}
& (\tau_2(F_1) \;\; \tau_2(F_2) \;\; \tau_2(F_1 F_2)) \\
= \; & \frac{1}{2}(4, 5, 6, 7) \begin{pmatrix} -1 & -1 & +1 \\ -1 & +1 & -1 \\ +1 & -1 & -1 \\ +1 & +1 & +1 \end{pmatrix} \qquad (6.12) \\
= \; & (2, 1, 0).
\end{aligned}$$

On the other hand, the unit-level factorial effects can be obtained as a column vector by transposing the contrast matrix and changing the order the multiplication on the right hand side of (6.17), i.e.,

$$\begin{pmatrix} \tau_2(F_1) \\ \tau_2(F_2) \\ \tau_2(F_1F_2) \end{pmatrix} = \frac{1}{2}\begin{pmatrix} -1 & -1 & +1 & +1 \\ -1 & +1 & -1 & +1 \\ +1 & -1 & -1 & +1 \end{pmatrix}\begin{pmatrix} 4 \\ 5 \\ 6 \\ 7 \end{pmatrix} = \begin{pmatrix} 2 \\ 1 \\ 0 \end{pmatrix}.$$

Consistent with the traditional definition of causal effects in the previous chapters, and in the factorial design literature, the causal estimands at the *population level* are the averages of all unit level factorial effects over all N units in the population. To define these quantities, we first define the averages of all N potential outcomes in the finite population of N units for each treatment z_1z_2, $z_1 \in \{0,1\}$, $z_2 \in \{0,1\}$ as

$$\overline{Y}(z_1z_2) = \frac{1}{N}Y_i(z_1z_2), \quad i = 1, \ldots, N. \tag{6.13}$$

These averages are shown in the penultimate row in Table 6.1.

The population level main effects of each treatment factor and their interaction, denoted by $\overline{\tau}(F_1)$, $\overline{\tau}(F_2)$ and $\overline{\tau}(F_1, F_2)$ respectively, are averages of the corresponding unit-level factorial effects. These effects can also be expressed as contrasts of the average potential outcomes $\overline{Y}(z_1z_2)$ with the same contrast vectors shown in Table 6.2:

$$\begin{aligned} \overline{\tau}(F_1) &= \frac{-\overline{Y}(00) - \overline{Y}(01) + \overline{Y}(10) + \overline{Y}(11)}{2} \\ &= \frac{1}{N}\sum_{i=1}^{N} \tau_i(F_1), \end{aligned} \tag{6.14}$$

$$\begin{aligned} \overline{\tau}(F_2) &= \frac{-\overline{Y}(00) + \overline{Y}(01) - \overline{Y}(10) + \overline{Y}(11)}{2} \\ &= \frac{1}{N}\sum_{i=1}^{N} \tau_i(F_2), \end{aligned} \tag{6.15}$$

$$\begin{aligned} \overline{\tau}(F_1F_2) &= \frac{\overline{Y}(00) - \overline{Y}(01) - \overline{Y}(10) + \overline{Y}(11)}{2} \\ &= \frac{1}{N}\sum_{i=1}^{N} \tau_i(F_1F_2). \end{aligned} \tag{6.16}$$

The population-level factorial effects are shown in the last three columns of the penultimate row of Table 6.1 and can be computed from the columns of conditional effects, e.g.,

$$\overline{\tau}(F_1) = \frac{B+A}{2} = \frac{1.75+1.50}{2} = 1.625,$$
$$\overline{\tau}(F_1) = \frac{B-A}{2} = \frac{1.75-1.50}{2} = 0.125.$$

They can also be directly from the columns of potential outcomes by multiplying the vector of average potential outcomes by half of the contrast matrix in Table 6.1, i.e., by averaging both sides of (6.17) over all units:

$$\begin{aligned}
&(\overline{\tau}(F_1)\ \ \overline{\tau}(F_2)\ \ \overline{\tau}(F_1F_2)) \\
&= \frac{1}{2}(4.00, 4.75, 5.75, 6.25)\begin{pmatrix} -1 & -1 & +1 \\ -1 & +1 & -1 \\ +1 & -1 & -1 \\ +1 & +1 & +1 \end{pmatrix} \qquad (6.17) \\
&= (1.625, 0.625, 0.125).
\end{aligned}$$

As before, for the sake of completeness, we denote the average of all $4N$ potential outcomes by

$$\overline{\tau}_0 = \frac{\overline{Y}(00)+\overline{Y}(01)+\overline{Y}(10)+\overline{Y}(11)}{4} = \frac{1}{N}\sum_{i=1}^{N}\tau_i(0), \qquad (6.18)$$

which is a linear combination (not a contrast) of the average potential outcomes $\overline{Y}(z_1z_2)$ with coefficient vector $(1/4, 1/4, 1/4, 1/4)^{\mathrm{T}}$.

We conclude this Section with the following two remarks: (a) Henceforth, instead of representing treatment combinations by z_1z_2, we will represent them by their more compact lexicographic order j, so that $j = 1, 2, 3, 4$ correspond to 00, 01, 10, and 11 respectively, (b) only one potential outcome in each row of Table 6.1 is actually observed from an experiment, and the remaining three are missing.

6.2 Assignment mechanisms, observed outcomes and unbiased estimators

The treatment assignment mechanism selects the subset of potential outcomes that will be revealed and observed; the other potential outcomes are missing. In the education experiment described at the beginning of this Chapter, 1656 units received the four treatment combinations through a completely randomized assignment mechanism with $N_1 = 1006$, $N_2 = N_3 = 250$ and $N_4 = 150$. However, other assignment mechanisms could have been possible. The actual distribution of the 1656 subjects into the four treatment groups by sex is shown in Table 6.3. It appears that the completely randomized assignment mechanism resulted in an allocation that is acceptable with respect to the covariate (sex) distribution in the four treatment groups. However, bad randomizations were also possible - for example, all the 948 female students could have been allocated to the treatment group 00. To prevent such possibilities, the experimenters could have allocated the 948 females and 708 males to the four treatment groups proportionately to the total number of units allocated to the four treatment groups, and then used a completely randomized assignment mechanism within each block (defined by sex). Such a design would be a randomized block design with predetermined N_{bj}'s (number of units in block b allocated to treatment j) as per Table 6.3, where $b = 1, 2$ and $j = 1, \ldots, 4$.

TABLE 6.3
Allocation of 1656 units into the four treatment groups by sex

Sex	00	01	10	11	Total
Female	574	150	142	82	948
Male	432	100	108	68	708
Total	1006	250	250	150	1656

A third possibility would have been to assign 1656 units to the four treatment combinations in a manner such that each unit receives treatment j with fixed probability π_j *independent* of other units for $j = 1, \ldots, 4$ such that $\pi_1 + \pi_2 + \pi_3 + \pi_4 = 1$; in that case $N_1, \ldots, N_4$ would have been random variables.

In addition to sex, the study also had several premeasured covariates for each subject, e.g., native language, parents' education. As discussed in Chapters 4 and 5, these covariates could also have been used to construct blocks or more general criteria for acceptable randomized allocations.

Let W_i denote the random variable denoting the treatment combination j assigned to unit i and let $\mathbf{W} = (W_1, \ldots, W_N)^{\mathrm{T}}$ denote the N-dimensional column vector of random variables W_i's for $i \in 1, \ldots, N$, and let $\mathbf{w}$ be a specific realization, i.e., a vector comprising all the individual treatment assignments *after* randomization. Hence, the joint probability distribution of $(W_1, \ldots, W_N)$ defines the design, or treatment assignment mechanism for $\mathbf{W}$. The vector of actual treatment assignments, $\mathbf{w}$, is a draw from this distribution.

For a completely randomized 2^2 factorial experiment with N_j units assigned to the jth treatment combination, $j = 1, \ldots, 4$, the total number of possible assignments is $N_1!N_2!N_3!N_4!$. Thus, similar to Section 5.2 of Chapter 5, we can write

$$Pr(\mathbf{W} = \mathbf{w}) = \begin{cases} \frac{1}{N!/(N_1!N_2!N_3!N_4!)} & \text{if } \sum_{i=1}^{N} \mathbb{I}_{\{W_i = j\}} = N_j \\ 0 & \text{otherwise,} \end{cases},$$

where, as earlier, $\mathbb{I}_A$ denotes the indicator random variable that takes value one of the random event A occurs, and zero otherwise.

Denote the observed outcome corresponding to the ith experimental unit by $y_i, i = 1, \ldots, N$, so that

$$y_i = Y_i(W_i) = \sum_{i=1}^{N} \mathbb{I}_{\{W_i = j\}} Y_i(j),, \tag{6.19}$$

and let $\mathbf{y}$ be the $N \times 1$ column vector of observed outcomes. This means unit i receives the treatment j for which $W_i = j$.

Suppose the observed treatment assignment vector $\mathbf{w}$ in the toy example of Table 6.1 is (1,4,3,2,3). The *observed* potential outcomes are displayed in Table 6.4; the missing potential outcomes are represented by question marks.

TABLE 6.4
Observed Outcomes for the 2^2 experiment with $\mathbf{w} = (1, 4, 3, 2, 3)$.

Unit (i)	Assignment (w_i)	observed outcome for treatment combination 1(00)	2(01)	3(10)	4(11)
1	1	4	?	?	?
2	4	?	?	?	7
3	3	?	?	6	?
4	2	?	5	?	?
5	3	?	?	6	?
Average		$\overline{y}.(1) = 4$	$\overline{y}.(2) = 5$	$\overline{y}.(3) = 6$	$\overline{y}.(4) = 7$

Let

$$\overline{y}.(j) = \frac{1}{N_j} \sum_{i:W_i=j} y_i$$

denote the average of all observed outcomes for treatment combination j, where N_j is the number of observations for treatment j. Then, it is easy to see (exercise) that $\overline{y}.(j)$ is an unbiased estimator (with respect to its randomization distribution) of $\overline{Y}(j)$, the average potential outcome for treatment combination j. Consequently, unbiased estimators of the population-level causal estimands $\overline{\tau}(F_1)$, $\overline{\tau}(F_2)$ and $\overline{\tau}(F_1F_2)$ can be obtained by substituting $\overline{y}(j)$ for $\overline{Y}(z_1z_2)$ in equations (6.14) - (6.16), and noting that $j = 1, 2, 3, 4$ correspond to $z_1z_2 = 00$, 01, 10, and 11 respectively.

$$\widehat{\overline{\tau}}(F_1) = \frac{-\overline{y}.(1) - \overline{y}.(2) + \overline{y}.(3) + \overline{y}.(4)}{2} \tag{6.20}$$

$$\widehat{\overline{\tau}}(F_2) = \frac{-\overline{y}.(1) + \overline{y}.(2) - \overline{y}.(3) + \overline{y}.(4)}{2} \tag{6.21}$$

$$\widehat{\overline{\tau}}(F_1F_2) = \frac{\overline{y}.(1) - \overline{y}.(2) - \overline{y}.(3) + \overline{y}.(4)}{2}. \tag{6.22}$$

Finally, an unbiased estimator of $\overline{\tau}_0$ is

$$\widehat{\overline{\tau}}_0 = \frac{\overline{y}.(1) + \overline{y}.(2) + \overline{y}.(3) + \overline{y}.(4)}{4}. \tag{6.23}$$

Note that $\widehat{\overline{\tau}}_0$ equals $\overline{y}_.(\cdot)$, the average of N observed outcomes if and only if $N_j = N/4 = r$ for each $j = 1, \ldots 4$.

From the observed data in Table 6.4, unbiased estimates of the three factorial effects can be calculated as: $\widehat{\overline{\tau}}(F_1) = 2$, $\widehat{\overline{\tau}}(F_2) = 1$, and $\widehat{\overline{\tau}}(F_1F_2) = 0$. An unbiased estimator of $\overline{\tau}_0$ is $\widehat{\overline{\tau}}_0 = 5.5$, which in this case is *not* equal to $\overline{y}_.(\cdot) = 5.6$. As in the case of the true (population-level) factorial effects, the estimated effects can be computed from the $\overline{y}_.(j)$ using the following transformation which is equivalent to (6.20)-(6.22) and the observed counterpart of (6.17).

$$\begin{aligned} & (\overline{\tau}(F_1) \;\; \overline{\tau}(F_2) \;\; \overline{\tau}(F_1F_2)) \\ = \;& \frac{1}{2}(4,5,6,7)\begin{pmatrix} -1 & -1 & +1 \\ -1 & +1 & -1 \\ +1 & -1 & -1 \\ +1 & +1 & +1 \end{pmatrix} \\ = \;& (2,1,0). \end{aligned} \tag{6.24}$$

6.3 Computing estimated factorial effects using R

Let $\widehat{\overline{\boldsymbol{\tau}}} = \left(\widehat{\overline{\tau}}_0, \widehat{\overline{\tau}}(F_1), \widehat{\overline{\tau}}(F_2), \widehat{\overline{\tau}}(F_1F_2)\right)^{\mathrm{T}}$ denote the 4×1 vector of estimators of the factorial effects (including the population average). From equations (6.20)–(6.23), we can write

$$\widehat{\overline{\boldsymbol{\tau}}} = \begin{pmatrix} \widehat{\overline{\tau}}_0 \\ \widehat{\overline{\tau}}(F_1) \\ \widehat{\overline{\tau}}(F_2) \\ \widehat{\overline{\tau}}(F_1F_2) \end{pmatrix} = \begin{pmatrix} 1/4 & 1/4 & 1/4 & 1/4 \\ -1/2 & -1/2 & 1/2 & 1/2 \\ -1/2 & 1/2 & -1/2 & 1/2 \\ 1/2 & -1/2 & -1/2 & 1/2 \end{pmatrix} \begin{pmatrix} \overline{y}_.(1) \\ \overline{y}_.(2) \\ \overline{y}_.(3) \\ \overline{y}_.(4) \end{pmatrix}.$$

To convert the 4×4 matrix on the right hand side into a nicer form, we introduce $\widehat{\overline{\boldsymbol{\tau}}}_{\mathrm{adj}}$, an adjusted version of the vector of estimators of the factorial effects $\widehat{\overline{\boldsymbol{\tau}}}$ as follows:

$$\begin{aligned} \widehat{\overline{\boldsymbol{\tau}}}_{\mathrm{adj}} = \begin{pmatrix} \widehat{\overline{\tau}}_0 \\ \widehat{\overline{\tau}}(F_1)/2 \\ \widehat{\overline{\tau}}(F_2)/2 \\ \widehat{\overline{\tau}}(F_1F_2)/2 \end{pmatrix} &= \frac{1}{4}\begin{pmatrix} 1 & 1 & 1 & 1 \\ -1 & -1 & 1 & 1 \\ -1 & 1 & -1 & 1 \\ 1 & -1 & -1 & 1 \end{pmatrix} \begin{pmatrix} \overline{y}_.(1) \\ \overline{y}_.(2) \\ \overline{y}_.(3) \\ \overline{y}_.(4) \end{pmatrix} \\ &= \frac{1}{4}\mathbf{L}^{\mathrm{T}}\overline{\mathbf{y}}, \end{aligned} \tag{6.25}$$

where $\overline{\mathbf{y}} = (\overline{y}_.(1), \overline{y}_.(2), \overline{y}_.(3), \overline{y}_.(4))^{\mathrm{T}}$ denotes the 4×1 vector of observed treatment averages, and

$$\mathbf{L} = \begin{pmatrix} 1 & -1 & -1 & 1 \\ 1 & -1 & 1 & -1 \\ 1 & 1 & -1 & -1 \\ 1 & 1 & 1 & 1 \end{pmatrix} \tag{6.26}$$

is a coefficient matrix that augments a column of ones to the three contrast vectors for the three factorial effects shown in Table 6.2. The matrix $\mathbf{L}$, referred to as the *model matrix* in some experimental design textbooks, is easy to generate in R. The command gen.factorial() available in the package "AlgDesign" [92] generates the contrast vectors for the M.E.'s in reverse lexicographic order and can be used to generate $\mathbf{L}$ with minor manipulations as follows:

```
-----------------------------------------------------------
> library(AlgDesign)
> D = gen.factorial(2,2) ## ME contrast columns in reverse order
>
> D[, c(1:2)] = D[,c(2:1)] ## Reverse the order of the columns
> colnames(D) = c("F1","F2")
> L = model.matrix(~ F1 + F2 + F1*F2,D)
> L
  (Intercept) F1 F2 F1:F2
1           1 -1 -1     1
2           1 -1  1    -1
3           1  1 -1    -1
4           1  1  1     1
-----------------------------------------------------------
```

The vector $\overline{\mathbf{y}}$ can be generated from the observed data $(\mathbf{y}, \mathbf{w})$ and pre-multiplied by $(1/4)\mathbf{L}^{\mathrm{T}}$ to obtain the vector $\widehat{\overline{\boldsymbol{\tau}}}_{\mathrm{adj}}$ defined in (6.25). For the data in Table 6.4, these steps yield the estimated factorial effects as 2, 1 and 0, as seen earlier.

```
---------------------------------------
>y=c(4,7,6,5,6)
>w=c(1,4,3,2,3)
>ybarvector = NULL
>for (l in 1:4) {
>ybarvector = c(ybarvector,mean(y[w==l]))
}

> ybarvector
[1] 4 5 6 7

> taubarhatadj = (t(L) %*% ybarvector)/4  # adjusted tau bar hat
> taubarhat = 2*taubarhatadj[2:4] # adjust factorial effects by 2
> taubarhat
> [1] 2 1 0
---------------------------------------
```

Note that in the last step of the code, the estimated adjusted factorial effects are multiplied by two to obtain estimates of the original effects.

6.4 Orthogonal decomposition of data in 2^2 experiments

We have seen in Chapter 5, how decomposing the observed data into different components facilitate in finding some natural statistics for testing of Fisherian and Neymanian hypotheses in randomized experients with multiple treatments. Because a 2^2 experiment can be viewed as one with 4 treatment arms, the observed data can be projected onto three orthogonal spaces - the overall average of dimension 1, the treatment space of dimension $4-1=3$ and residual space of dimension $N-4$, as extensively discussed and explained in Section 5.4.1.1.

Assume that for $j = 1, \ldots, 4$, N_j units receive treatment combination j. Then, each observed outcome can be written as

$$y_i = \overline{y}.(\cdot) + \{\overline{y}.(W_i) - \overline{y}.(\cdot)\} + \{y_i - \overline{y}.(W_i)\},$$

exactly as in (5.15), leading to the decomposition (5.17), i.e.,

$$\text{SSTot} = \text{SSTre} + \text{SSRes},$$

where SSTot, SSTre and SSRes are defined in (5.18), (5.19) and (5.20) respectively. Following the argument in Section 5.4.1.1, SSTre and SSRes have degrees of freedom 3 and $N-4$ respectively, and we can construct an ANOVA table like the one in Table 5.5. Thus, the F statistic

$$F_{\text{Tre}} = \text{MSTre}/\text{MSRes} = \frac{\text{SSTre}/3}{\text{SSRes}/(N-4)}$$

measures the contributions of the treatment factors relative to the unexplained (residual) variation in the observed data.

Table 6.5 shows observed and missing outcomes from a hypothetical balanced replicated 2^2 experiment in which each treatment combination is randomly assigned to $r = 5$ units.

Applying the computational method described in Section 6.3, the row vector of treatment group averages $\overline{\mathbf{y}}$ is obtained as $(4.62, 6.38, 6.18, 7.56)$, as shown in the code below and in the penultimate row of Table 6.5. The row vector of sample variances $(s_1^2, s_2^2, s_3^2, s_4^2)$ of the observed outcomes in the four treatment groups is computed to be $(0.432, 0.622, 0.267, 0.523)$ and is shown in the last row of Table 6.5.

Also, as shown below, the estimated main effects and interaction effect are obtained as:

$$\widehat{\overline{\tau}}(F_1) = 1.37, \quad \widehat{\overline{\tau}}(F_2) = 1.57, \quad \widehat{\overline{\tau}}(F_1, F_2) = -0.19.$$

TABLE 6.5
Observed and missing outcomes in a hypothetical replicated 2^2 experiment

Unit	Treatment combination ($z_1 z_2$)				Assignment
(i)	(00)	(01)	(10)	(11)	w_i
1		7.4			2
2	5.0				1
3			6.6		3
4				7.6	4
5	4.8				1
6			6.3		3
7		5.8			2
8				6.5	4
9				7.3	4
10			5.3		3
11	3.6				1
12		6.6			2
13	4.4				1
14			6.5		3
15		5.4			2
16				8.4	4
17				8.0	4
18	5.3				1
19		6.7			2
20			6.2		3
$\bar{y}.(j)$	4.62	6.38	6.18	7.56	
s_j^2	0.432	0.622	0.267	0.523	

```
--------------------------------------
# Enter data unit-wise
y=c(7.4,5.0,6.6,7.6,4.8,6.3,5.8,6.5,7.3,5.3,3.6,6.6,
4.4,6.5,5.4,8.4,8.0,5.3,6.7,6.2)
N = length(y)

# Enter assignment vector
w=c(2,1,3,4,1,3,2,4,4,3,1,2,1,3,2,4,4,1,2,3)

# Calculate trt group averages and variances
ybarvector = varyvector = NULL
for (l in 1:4) {
ybarvector = c(ybarvector,mean(y[w==l]))
varyvector = c(varyvector,var(y[w==l]))
}

> ybarvector
[1] 4.62 6.38 6.18 7.56

> varyvector
[1] 0.432 0.622 0.267 0.523

taubarhatadj = (t(L) %*% ybarvector)/4
taubarhat = 2*taubarhatadj[2:4] # adjust factorial effects by 2
```

```
> taubarhat
[1]  1.37  1.57 -0.19
```

Recall from (5.18)-(5.19) that

$$\begin{aligned} \text{SSTot} &= \sum_{i=1}^{N}\{y_i - \overline{y}.(\cdot)\}^2, \\ \text{SSTre} &= \sum_{j=1}^{J} N_j\{\overline{y}.(j) - \overline{y}.(\cdot)\}^2, \end{aligned}$$

and SSRes = SSTot − SSTre. Thus the ANOVA table is created using the following R code:

```
 # ANOVA from definition
> (Nvec = rep(5,4)) # vector of N_j's
[1] 5 5 5 5
> (N = sum(Nvec)) # Total population size
[1] 20
> (ybar = mean(y))
[1] 6.185

# Calculation of SSTot
> (SSTot = (N-1)*var(y))
[1] 29.2655

# Calculation of SSTre
SSTre = 0
for (j in 1:4) {
+   SSTre = SSTre + Nvec[j]*(ybarvector[j] - ybar)^2
+ }
> SSTre
[1] 21.8895

# Calculation of SSRes
> (SSRes = SSTot - SSTre)
[1] 7.376

 # Mean squared errors
> (MSTre = SSTre/3)
[1] 7.2965
> (MSRes = SSRes/16)
[1] 0.461

 # Calculation of F
 (FTre = MSTre/MSRes)
[1] 15.82755
```

The ANOVA Table obtained from the above calculations is shown in Table 6.6.

TABLE 6.6
ANOVA Table from observed outcomes in Table 6.5

Source	Degrees of freedom	Sum of squares	Mean Squares	F-ratio
Treatments	$J-1=3$	SSTre = 21.89	MSTre = 7.30	$F=$
Residuals	$N-J=16$	SSRes = 7.38	MSRes = 0.46	15.83
Total	$N-1=19$	SSTot = 29.27		

6.4.1 Special case of balanced 2^2 (or 2^K experiments in general

In the case of balanced 2^2 (or 2^K experiments in general) as in the example of Table 6.5, the treatment sum of squares, SSTre, can further be decomposed into sums of squares of individual factorial effects. Specifically, for a balanced and replicated experiment with $r = N/4$ units receiving each treatment, it can be shown that (Section 6.4.2 and Exercise problem 6.5 - readers not interested can skip these materials)

$$\text{SSTre} = r\left(\{\widehat{\tau}(F_1)\}^2 + \{\widehat{\tau}(F_2)\}^2 + \{\widehat{\tau}(F_1F_2)\}^2\right). \qquad (6.27)$$

Thus, the treatment sum of squares, which has three degrees of freedom for a 2^2 factorial experiment, can further be decomposed into three components if the experiment is balanced – the main effect of F_1, the main effect of F_2 and the interaction between F_1 and F_2 – each component having one degree of freedom. The ANOVA based on this decomposition is shown in Table 6.7.

TABLE 6.7
ANOVA Table for a balanced 2^2 experiment with $N = 4r$

Source	DF	SS	MS	F-ratio
M.E. of F_1	1	$r\widehat{\tau}^2(F_1)$	$r\widehat{\tau}^2(F_1)$	$r\widehat{\tau}^2(F_1)$/MSRes
M.E. of F_2	1	$r\widehat{\tau}^2(F_2)$	$r\widehat{\tau}^2(F_2)$	$r\widehat{\tau}^2(F_2)$/MSRes
Int. $F_1 \times F_2$	1	$r\widehat{\tau}^2(F_1F_2)$	$r\widehat{\tau}^2(F_1F_2)$	$r\widehat{\tau}^2(F_1F_2)$/MSRes
Residuals	$4(r-1)$	SSRes	MSRes	
Total	$4r-1$	SSTot		

Each F-statistic defined in Table 6.7 measures the relative contribution of a particular factorial effect (main effect or two-factor interaction) to the total variation. The F statistic that can be considered an estimator of an overall measure of treatment effects is the sum of the three individual F-statistics and equals

$$F_{\text{Tre}} = \frac{r\left(\{\widehat{\overline{\tau}}(F_1)\}^2 + \{\widehat{\overline{\tau}}(F_2)\}^2 + \{\widehat{\overline{\tau}}(F_1F_2)\}^2\right)}{\text{MSRes}}. \tag{6.28}$$

It can be seen (Exercise 6.6) that for the observed data in Table 6.5, the observed value of F ratios for the MEs of F_1, ME of F_2 and the interaction between F_1 and F_2 are 20.35, 26.73, and 0.39 respectively. The readers are also asked to expand the ANOVA table 6.6 into an ANOVA table like the one in Table 6.7 that shows the contribution of each factorial effect.

6.4.2 Decomposition for an unreplicated 2^2 experiment (readers not interested may skip)

For an easy exposition of the proof of decomposition (6.27), we consider an unreplicated 2^2 experiment with $N = 4$ units, where each $\overline{y}.(j)$ is the observation y_i for which $W_i = j$. Note that in this case, SSTot equals SSTre, and the degrees of freedom for residuals is zero. Therefore the statistic F_{Tre} cannot be computed, making this decomposition useless.

Because $\overline{y}.(j)$ equals the observation y_i, the vector $\mathbf{y}$ of observed outcomes can be written as a permutation of the vector $\overline{\mathbf{y}}$. Now, from (6.25), we have that

$$\overline{\mathbf{y}} = 4(\mathbf{L}^{\text{T}})^{-1}\widehat{\overline{\boldsymbol{\tau}}}_{\text{adj}}, \tag{6.29}$$

where $\mathbf{A}^{-1}$ denotes the inverse of an invertible matrix $\mathbf{A}$.

From (6.26), we see that multiplication of $\mathbf{L}$ by its transpose $\mathbf{L}^{\text{T}}$ yields $\mathbf{L}\mathbf{L}^{\text{T}} = 4\mathbf{I}_4$, where $\mathbf{I}_4$ is the identity matrix of order 4. The mutual orthognality of the three contrast vectors and the vector of ones makes the off-diagonal elements zero. Therefore $\mathbf{L} = 4(\mathbf{L}^{\text{T}})^{-1}$, and consequently from (6.29), it follows that

$$\overline{\mathbf{y}} = \mathbf{L}\widehat{\overline{\boldsymbol{\tau}}}_{\text{adj}},$$

which implies,

$$\overline{\mathbf{y}} = \begin{pmatrix} \overline{y}.(1) \\ \overline{y}.(2) \\ \overline{y}.(3) \\ \overline{y}.(4) \end{pmatrix} = \begin{pmatrix} 1 & -1 & -1 & 1 \\ 1 & -1 & 1 & -1 \\ 1 & 1 & -1 & -1 \\ 1 & 1 & 1 & 1 \end{pmatrix} \begin{pmatrix} \widehat{\overline{\tau}}_0 \\ \widehat{\overline{\tau}}(F_1)/2 \\ \widehat{\overline{\tau}}(F_2)/2 \\ \widehat{\overline{\tau}}(F_1F_2)/2 \end{pmatrix}$$

$$= \begin{pmatrix} \widehat{\overline{\tau}}_0 \\ \widehat{\overline{\tau}}_0 \\ \widehat{\overline{\tau}}_0 \\ \widehat{\overline{\tau}}_0 \end{pmatrix} + \begin{pmatrix} -\widehat{\overline{\tau}}(F_1)/2 \\ -\widehat{\overline{\tau}}(F_1)/2 \\ \widehat{\overline{\tau}}(F_1)/2 \\ \widehat{\overline{\tau}}(F_1)/2 \end{pmatrix} + \begin{pmatrix} -\widehat{\overline{\tau}}(F_2)/2 \\ \widehat{\overline{\tau}}(F_2)/2 \\ -\widehat{\overline{\tau}}(F_2)/2 \\ \widehat{\overline{\tau}}(F_2)/2 \end{pmatrix} + \begin{pmatrix} -\widehat{\overline{\tau}}(F_1F_2)/2 \\ \widehat{\overline{\tau}}(F_1F_2)/2 \\ -\widehat{\overline{\tau}}(F_1F_2)/2 \\ \widehat{\overline{\tau}}(F_1F_2)/2 \end{pmatrix} \tag{6.30}$$

Equation (6.30) shows that the four-dimensional vector of observed treatment averages $\overline{\mathbf{y}}$ can be decomposed into four orthogonal vectors, each belonging to a space of dimension one. Consequently, the squared norm of $\overline{\mathbf{y}}$ can be expressed as the sum of the squared norms of the four vectors on the final term of (6.30). Noting that the norm of the observed data vector $||\mathbf{y}|| = \sum_{i=1}^4 y_i^2$ equals $||\overline{\mathbf{y}}||$, we can write from (6.30)

$$\sum_{i=1}^{4} y_i^2 = 4\widehat{\overline{\tau}}_0^2 + \{\widehat{\overline{\tau}}(F_1)\}^2 + \{\widehat{\overline{\tau}}(F_2)\}^2 + \{\widehat{\overline{\tau}}(F_1F_2)\}^2.$$

Transferring $4\widehat{\overline{\tau}}_0^2 = 4\{\overline{y}.(\cdot)\}^2$ to the LHS, and noting that $\sum_{i=1}^4 y_i^2 - 4\{\overline{y}.(\cdot)\}^2 = \sum_{i=1}^4 \{y_i - \overline{y}.(\cdot)\}^2$, we can write

$$\sum_{i=1}^{4} \{y_i - \overline{y}.(\cdot)\}^2 = \{\widehat{\overline{\tau}}(F_1)\}^2 + \{\widehat{\overline{\tau}}(F_2)\}^2 + \{\widehat{\overline{\tau}}(F_1F_2)\}^2. \quad (6.31)$$

Noting that $\sum_{i=1}^4 \{y_i - \overline{y}.(\cdot)\}^2$ is SSTot, the above decomposition can be easily extended to the case of a balanced and replicated experiment with $r = N/4$ units receiving each treatment, because from (6.30), it is straightforward (exercise problem 6.5) to show that (6.27) holds for such an experiment.

6.5 Fisherian inference from 2^2 experiments

Fisher's sharp null hypothesis of no treatment effect on any unit can be written as:

$$H_0 : Y_i(1) = Y_i(2) = Y_i(3) = Y_i(4), \quad i = 1, \ldots, N,$$

or equivalently as

$$H_0 : \tau_i(F_1) = \tau_i(F_2) = \tau_i(F_1, F_2) = 0,, \quad i = 1, \ldots, N.$$

Many possible test statistics could be used to test the H_0. A detailed discussion can be found in Espinosa et al. (2015) [31]. Each estimated factorial effect $\widehat{\overline{\tau}}(\cdot)$ can be used as a test statistic because each defines a specific type of deviation from the sharp null hypothesis. Therefore, a Fisher randomization test can be obtained by comparing each estimated factorial effect against its randomization distribution under the sharp null hypothesis of no treatment effects. Under this sharp null hypothesis, the missing potential outcomes in each row of Table 6.4 or Table 6.5 can be imputed by the single observed outcome in that row. The randomization distribution of an estimated factorial effect can be obtained by taking draws from the distribution of $\mathbf{W}$, and calculating the estimated factorial effect for each draw. It is interesting to note that, due to the symmetry of the definition of all factorial effects, the reference distribution under the sharp null hypothesis of no effects is the same for all three factorial effects. In this approach, one can adjust the p-value for multiple comparisons, for example, by using the ensemble-adjusted p-value described in Section 5.4.1.6 of Chapter 5, which simply modifies the raw p-value for each test to $\max\{3p, 1\}$. Scaled versions of the estimated factorial effects obtained by dividing them by their respective standard errors, such as

$$T_{\text{scaled}}(\overline{\tau}(F_1)) = \frac{\widehat{\overline{\tau}}(F_1)}{\sqrt{\frac{\text{MSRes}}{4} \sum_{j=1}^{4} \frac{1}{N_j}}} \tag{6.32}$$

can also be used. Derivation of the standard error, and hence a justification for the denominator in (6.32), will be presented in Section 6.6.

However, if interest is in assessing if there is a deviation from the null hypothesis without a particular direction in mind, a commonly used test statistic in the presence of replicates ($r > 1$) is the F_{Tre} ratio associated with the decomposition of the total sum of squares of the observed outcomes and defined in (6.28). Using the square root of this F-statistic, denoted by $T_{\text{Tre}} = \sqrt{F_{\text{Tre}}}$ yields identical results.

Another reasonable option is to use the estimated factorial effect

$$\widehat{\overline{\tau}}_{\max} = \max\left[|\widehat{\overline{\tau}}(F_1)|, |\widehat{\overline{\tau}}(F_2)|, |\widehat{\overline{\tau}}(F_1, F_2)|\right]$$

that has the largest absolute value among all the three estimated effects. A scaled version of $\widehat{\overline{\tau}}_{\max}$ obtained by dividing it by its standard error, can also be used.

The R code used to implement the randomization test to assess Fisher's sharp null hypothesis using the absolute values of the estimates of the factorial effects $\widehat{\overline{\tau}}(F_1)$, $\widehat{\overline{\tau}}(F_2)$ and $\widehat{\overline{\tau}}(F_1F_2)$, is shown below. As mentioned earlier, it suffices to compare the observed values of the estimated factorial effects with the randomization distribution of *any one* of these estimators under the sharp null hypothesis of no treatment effect. Here, we consider the randomization distribution of $\widehat{\overline{\tau}}(F_1)$. In each iteration, a new assignment vector $\mathbf{w}_{\text{rep}}$ is drawn, the vector $\overline{\mathbf{y}}_{\text{rep}}$ of observed treatment averages is computed, and eventually $\widehat{\overline{\tau}}_{\text{rep}}(F_1)$ is obtained using (6.25). The p-values for the main effects of F_1, F_2 and interaction F_1F_2 are obtained as the proportion of values of the randomization distribution of $|\widehat{\overline{\tau}}_{\text{rep}}(F_1)|$ that are greater than or equal to the observed absolute values of the three factorial effects (1.37, 1.57 and 0.19 respectively). The p values of 0.01, 0.0024 and 0.74 obtained from 5000 draws of $\mathbf{w}$ suggest possible existence of main effects of factors F_1 and F_2, but essentially no evidence for an interaction effect.

```
--------------------------------------
y=c(7.4,5.0,6.6,7.6,4.8,6.3,5.8,6.5,7.3,
    5.3,3.6,6.6,4.4,6.5,5.4,8.4,8.0,5.3,6.7,6.2)
```

```
N = length(y)
r = N/4
library(AlgDesign)
# L matrix
D = gen.factorial(2,2) ## ME contrast columns in reverse order
D[, c(1:2)] = D[,c(2:1)] ## Reverse the order of the columns
colnames(D) = c("F1","F2")
L = model.matrix(~ F1 + F2 + F1*F2,D)

ITER=5000 # Number of iterations

tau1hatobs= 1.37
tau2hatobs=1.57
tau3hatobs=-0.19

tau1hatarray=NULL # Generated values of tau1barhat will be stored
in this array
# for any one factorial effect

for (count in 1:ITER) {
  wrep = sample(c(rep(1,r),rep(2,r),rep(3,r),rep(4,r)))
  ybarvectorrep = NULL
  for (l in 1:4) {
    ybarvectorrep = c(ybarvectorrep,mean(y[wrep==l]))
      }
  taubarhatadjrep = (t(L) %*% ybarvectorrep)/4
  tau1barhatrep = 2*taubarhatadjrep[2]

  tau1hatarray = c(tau1hatarray,abs(tau1barhatrep))
}

# Plotting the histogram
hist(tau1hatarray, main=NULL,xlim=c(0,2.2),xlab=expression(hat(bar(tau))
[1]))
abline(v=abs(tau1hatobs), lty=3, lwd=2)
abline(v=abs(tau2hatobs), lty=3, lwd=2)
abline(v=abs(tau3hatobs), lty=3, lwd=2)
> (pval1 = sum(tau1hatarray >= abs(tau1hatobs))/ITER)
[1] 0.01
> (pval2 = sum(tau1hatarray >= abs(tau2hatobs))/ITER)
[1] 0.0024
> (pval3 = sum(tau1hatarray >= abs(tau3hatobs))/ITER)
[1] 0.7366

text(1.6,700,round(pval2,3),cex=0.8)
text(1.3,700,round(pval1,3),cex=0.8)
text(0.3,700,round(pval3,3),cex=0.8)
--------------------------------------
```

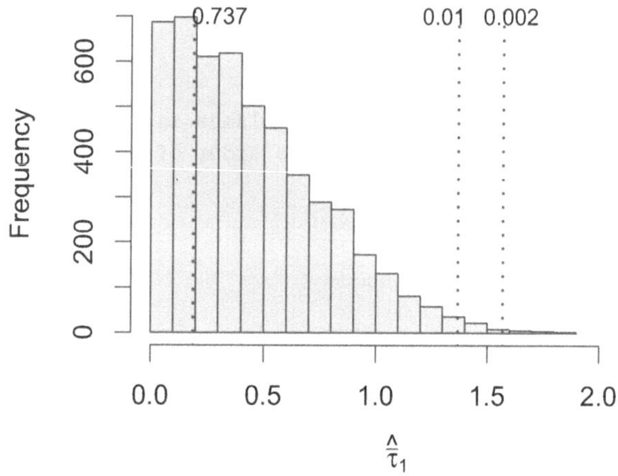

FIGURE 6.2
Randomization test for the data in Table 6.5 using $|\widehat{\overline{\tau}}(F_1)|$, $|\widehat{\overline{\tau}}(F_2)|$ and $|\widehat{\overline{\tau}}(F_1F_2)|$ as the test statistics, where the dashed vertical lines represent the observed values of the test statistics, and the numbers next to them the corresponding p-values

TABLE 6.8
Observed T_{scaled} statistic

Factorial effect	Estimated effect	S.E.	$T^{\text{obs}}_{\text{scaled}}$	p-value
M.E. of F_1	1.37	.303	4.52	0.0098
M.E. of F_2	1.57	.303	5.18	0.0020
INT F_1F_2	- 0.19	.303	-0.63	0.75

Next, to perform the randomization test to assess the sharp null hypothesis of no treatment effect using the scaled versions of absolute estimated effects as test statistics $|T_{\text{scaled}}(\overline{\tau}(F_1))|$, $|T_{\text{scaled}}(\overline{\tau}(F_2))|$ and $|T_{\text{scaled}}(\overline{\tau}(F_1F_2))|$, we first compute the observed values of these statistics. These values are shown in Table 6.8. Note that the standard error is computed as $\sqrt{\frac{\text{MSRes}}{4}\sum_{j=1}^{4}\frac{1}{N_j}} = \sqrt{0.46/5} = 0.303$, after substituting the value of MSRes from Table 6.6. Next, we impute the missing outcomes in Table 6.5, generate hypothetical draws of $\mathbf{W}$, and compute the test statistic for each set of observed outcomes. As mentioned earlier, each of the three statistic $T_{\text{scaled}}(F_1)$, $T_{\text{scaled}}(F_2)$ and $T_{\text{scaled}}(F_1, F_2)$ will have the same randomization distribution due to symmetry. The results of the randomization test from 20,000 hypothetical randomizations is shown in Figure 6.3. The p values for the main effects of F_1, F_2 and the interaction F_1F_2 are obtained as 0.0098, 0.0020 and 0.75 respectively, leading to the same conclusion as before.

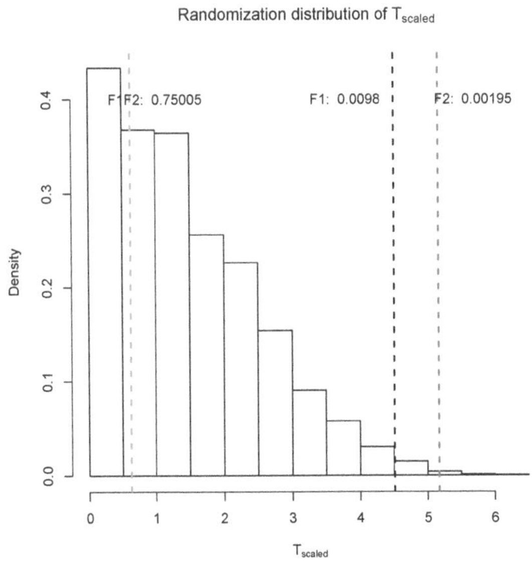

FIGURE 6.3
Randomization test for the data in Table 6.5 using the T_{scaled} statistics, where the dashed vertical lines represent the observed values, and the numbers next to them the corresponding p-values

6.5.1 Randomization test for sharp null hypothesis of non-zero treatment effects

The randomization test for sharp null hypothesis of zero treatment effect is straightforward to implement because the imputation of the three missing potential outcomes for each unit under this hypothesis simply involves imputing all the missing outcomes by the observed outcome for that unit. However, imputation of missing potential outcomes under a sharp null hypothesis of non-zero treatment effect is non-trivial. Such a sharp null hypothesis is of the form:

$$H_0 : \tau_i(F_1) = \tau_1^*, \ \tau_i(F_2) = \tau_2^*, \ \tau_i(F_1, F_2) = \tau_{12}^*, \ \ i = 1, \ldots, N. \tag{6.33}$$

or

$$H_0 : \boldsymbol{\tau}_i = \boldsymbol{\tau}^*, \ \ i = 1, \ldots, N \tag{6.34}$$

where $\boldsymbol{\tau}_i = (\tau_i(F_1), \tau_i(F_2), \tau_i(F_1F_2))^{\mathrm{T}}$ is the vector of unit-level factorial effects (not including the average of the four potential outcomes $\tau_i(0)$.)

Dasgupta, Pillai and Rubin [20] proposed a procedure for imputing missing outcomes in a 2^K factorial experiment, which is presented below as Procedure 6.1 for $K = 2$. We need to introduce some notation to facilitate understanding of this procedure. Consider partitioning the 4×4 matrix $\mathbf{L}$ defined in (6.26) into two parts - (i) row w_i, the treatment unit i is assigned to and (ii) three rows other than row w_i, i.e., rows corresponding to the treatment labels unit i is *not* assigned to. For example, if $W_i = 2$ for unit i, then the matrix

$$\mathbf{L} = \begin{pmatrix} 1 & -1 & -1 & 1 \\ 1 & -1 & 1 & -1 \\ 1 & 1 & -1 & -1 \\ 1 & 1 & 1 & 1 \end{pmatrix}$$

can be partitioned into the row vector $\ell_i = (1, -1, 1, -1)$ (second row of the matrix), the 3×4 matrix $\mathbf{L}_{-i}$ consisting of rows 1, 3 and 4 of $\mathbf{L}$. Let $\widetilde{\ell}_i = (-1, -1, 1)$ denote the vector ℓ_i without its first element 1.

Procedure 6.1 ***Steps for imputing missing potential outcomes under a sharp null hypothesis of non-zero treatment effects in a 2^2 factorial experiment:***

Step 1: Estimate the mean $\tau_i(0)$ of the four potential outcomes for unit i as

$$\widehat{\tau}_i(0) = y_i - \frac{1}{2}\tilde{\ell}_i \boldsymbol{\tau}^*$$

Step 2: Impute the 3×1 vector of missing outcomes y_i^{mis} as

$$y_i^{mis} = \mathbf{L}_{-i} \begin{pmatrix} \widehat{\tau}_i(0) \\ \boldsymbol{\tau}^*/2 \end{pmatrix}.$$

To illustrate the Procedure to impute the missing potential outcomes for unit 1 in Table 6.5 under the sharp null hypothesis (6.34) with $\boldsymbol{\tau}^* = (2, 0, 0)$, we note that $w_1 = 2$, $\ell_1 = (1, -1, 1, -1)$, the second row of $\mathbf{L}$, and $\widetilde{\ell}_i = (-1, 1, -1)$. Thus, in step 1, we estimate

$$\widehat{\tau}_i(0) = y_i - \frac{1}{2}\tilde{\ell}_i \boldsymbol{\tau}^* = 7.4 - (1/2) \times (-1) \times (2) = 8.4.$$

In step 2, we impute the missing potential outcomes for treatment combinations 1,3 and 4 as

$$y_1^{\text{mis}} = \mathbf{L}_{-1} \begin{pmatrix} \widehat{\tau}_i(0) \\ \boldsymbol{\tau}^*/2 \end{pmatrix} = \begin{pmatrix} 1 & -1 & -1 & 1 \\ 1 & 1 & -1 & -1 \\ 1 & 1 & 1 & 1 \end{pmatrix} \begin{pmatrix} 8.4 \\ 1 \\ 0 \\ 0 \end{pmatrix} = \begin{pmatrix} 7.4 \\ 9.4 \\ 9.4 \end{pmatrix}.$$

Thus the imputed values of the missing potential outcomes for unit 1, corresponding to treatment combinations $00, 10$ and 11 in Table 6.5, are 7.4, 9.4, 9.4, which is intuitive for $\boldsymbol{\tau}^* = (2, 0, 0)$.

The function twosqimpute() shown below takes $\mathbf{y}$, $\mathbf{w}$, and the three elements of τ^* as arguments, and returns an $N \times 4$ matrix of imputed potential outcomes.

```
----------------------------------------
twosqimpute = function(y,w,tau1=0,tau2=0,tau3=0){
  N = length(y)
  library(tidyverse)
  Data = tibble(y=y,w=w)
  N_j =  Data %>% count(w) %>% .$n
  beta = c(tau1/2, tau2/2, tau3/2)

  genPOT = matrix(c(rep(0,N*4)), ncol=4)

  library(AlgDesign)
  # L matrix
  D = gen.factorial(2,2) ## ME contrast columns in reverse order
  D[, c(1:2)] = D[,c(2:1)] ## Reverse the order of the columns
  colnames(D) = c("F1","F2")
  L = model.matrix(~ F1 + F2 + F1*F2,D)

  L_indexed = matrix(cbind(c(1:4),L), ncol=5)

  for (i in 1:N) {
    genPOT[i,w[i]] = y[i]
    Lobs = L_indexed[w[i],-c(1:2)] # 1x3 row vector
    tau0 = y[i] - t(Lobs) %*% beta # estimate average of the unit
    Lmis = L_indexed[-w[i],]
    trmis = L_indexed[-w[i],1]
    Ymis = Lmis[,-1] %*% c(tau0, beta) # impute missing outcomes

    for (j in 1:3) {
      genPOT[i,trmis[j]] = Ymis[j]
    }  # end j
  }  # end i
return(genPOT)
}
----------------------------------------
```

Below, we show application of the function in imputing the missing potential outcomes of Table 6.5 under the sharp null hypothesis (6.34) with $\boldsymbol{\tau}^* = (2, 0, 0)$.

```
----------------------------------------
y=c(7.4,5.0,6.6,7.6,4.8,6.3,5.8,6.5,7.3,
    5.3,3.6,6.6,4.4,6.5,5.4,8.4,8.0,5.3,6.7,6.2)
w=c(2,1,3,4,1,3,2,4,4,3,1,2,1,3,2,4,4,1,2,3)

> twosqimpute(y,w,2,0,0)
      [,1] [,2] [,3] [,4]
 [1,]  7.4  7.4  9.4  9.4
 [2,]  5.0  5.0  7.0  7.0
 [3,]  4.6  4.6  6.6  6.6
```

```
 [4,]  5.6  5.6  7.6  7.6
 [5,]  4.8  4.8  6.8  6.8
 [6,]  4.3  4.3  6.3  6.3
 [7,]  5.8  5.8  7.8  7.8
 [8,]  4.5  4.5  6.5  6.5
 [9,]  5.3  5.3  7.3  7.3
[10,]  3.3  3.3  5.3  5.3
[11,]  3.6  3.6  5.6  5.6
[12,]  6.6  6.6  8.6  8.6
[13,]  4.4  4.4  6.4  6.4
[14,]  4.5  4.5  6.5  6.5
[15,]  5.4  5.4  7.4  7.4
[16,]  6.4  6.4  8.4  8.4
[17,]  6.0  6.0  8.0  8.0
[18,]  5.3  5.3  7.3  7.3
[19,]  6.7  6.7  8.7  8.7
[20,]  4.2  4.2  6.2  6.2
---------------------------------------
```

The remaining steps of the randomization test are the same as before, but with a minor difference. Unlike the case of sharp null hypothesis of no treatment effect, where the randomization distribution of all three factorial effects are the same due to symmetry, here we need to consider separate randomization distributions for the three effects, allowing the three elements of $\boldsymbol{\tau}^*$ in (6.34) to differ. The one-sided and two-sided p-values of the randomization test to assess the sharp null hypothesis (6.33) with $\boldsymbol{\tau}^* = (1.2, 1.5, 0)$, from the data in Table 6.5 with the estimated factorial effects $\widehat{\overline{\tau}}(F_1)$, $\widehat{\overline{\tau}}(F_2)$ and $\widehat{\overline{\tau}}(F_1F_2)$ as test statistics are shown below. These p-values are derived by using the function Fishertesttwosquare(), which takes $\mathbf{w}$, $\mathbf{y}$, $\boldsymbol{\tau}^*$ and ITER, the number of draws from the distribution of $\mathbf{W}$ as arguments. The two-sided p-values are obtained as 0.29, 0.41, and 0.52, indicating no evidence against the null hypothesis. Note that this function can also be used to test the sharp null hypothesis of no treatment effect by setting $\boldsymbol{\tau}^* = (0, 0, 0)$.

```
---------------------------------------
Fishertesttwosquare = function(y,w,tau1,tau2,tau3,ITER) {
  library(tidyverse)
  N = length(y)
  Data = tibble(y=y,w=w)
  N_j =  Data %>% count(w) %>% .$n
  ybarvector = Data %>% group_by(w) %>%
    summarize(ybar=mean(y)) %>% .$ybar

  library(AlgDesign)
  # L matrix
  D = gen.factorial(2,2) ## ME contrast columns in reverse order
```

```
D[, c(1:2)] = D[,c(2:1)] ## Reverse the order of the columns
colnames(D) = c("F1","F2")
L = model.matrix(~ F1 + F2 + F1*F2,D)

Yimp = twosqimpute(y,w,tau1, tau2, tau3) # Imputed matrix

taubarhatadj = (t(L) %*% ybarvector)/4
taubarhat = 2*taubarhatadj[2:4] # adjust factorial effects by 2
taubarhat

tau1hatobs= taubarhat[1]
tau2hatobs= taubarhat[2]
tau3hatobs= taubarhat[3]

tau1hatarray=tau2hatarray = tau3hatarray = NULL
# for any one factorial effect

for (count in 1:ITER) {
  wnew = sample(c(rep(1,N_j[1]),rep(2,N_j[2]),rep(3,N_j[3]),
  rep(4,N_j[4])))

  ynew = numeric(N)
  for (i in 1:N) {
    ynew[i]=Yimp[i,wnew[i]]

  }
  ybarvectornew = NULL
  for (j in 1:4) {
  ybarvectornew = c(ybarvectornew,mean(ynew[wnew==j]))
  }
  taubarhatadjnew = (t(L) %*% ybarvectornew)/4
  tau1barhatnew = 2*taubarhatadjnew[2]
  tau2barhatnew = 2*taubarhatadjnew[3]
  tau3barhatnew = 2*taubarhatadjnew[4]
  tau1hatarray = c(tau1hatarray,tau1barhatnew)
  tau2hatarray = c(tau2hatarray,tau2barhatnew)
  tau3hatarray = c(tau3hatarray,tau3barhatnew)
}

pval1plus = sum(tau1hatarray >= tau1hatobs)/ITER
pval2plus = sum(tau2hatarray >= tau2hatobs)/ITER
pval3plus = sum(tau3hatarray >= tau3hatobs)/ITER

pval1minus = sum(tau1hatarray <= tau1hatobs)/ITER
pval2minus = sum(tau2hatarray <= tau2hatobs)/ITER
pval3minus = sum(tau3hatarray <= tau3hatobs)/ITER

pval1both = sum(abs(tau1hatarray) >= abs(tau1hatobs))/ITER
pval2both = sum(abs(tau2hatarray) >= abs(tau2hatobs))/ITER
```

```
  pval3both = sum(abs(tau3hatarray) >= abs(tau3hatobs))/ITER
return(list(pvalplus = c(pval1plus, pval2plus, pval3plus),
            pvalminus = c(pval1minus, pval2minus, pval3minus),
            pvalboth = c(pval1both, pval2both, pval3both)))

}

# Application
y=c(7.4,5.0,6.6,7.6,4.8,6.3,5.8,6.5,7.3,
    5.3,3.6,6.6,4.4,6.5,5.4,8.4,8.0,5.3,6.7,6.2)
w=c(2,1,3,4,1,3,2,4,4,3,1,2,1,3,2,4,4,1,2,3)
> Fishertesttwosquare(y,w,1.2,1.5,0,10000)
$pvalplus
[1] 0.2911 0.4125 0.7430

$pvalminus
[1] 0.7264 0.5949 0.2643

$pvalboth
[1] 0.2911 0.4125 0.5242
---------------------------------------
```

6.5.2 Fiducial intervals for 2^2 factorial experiments

Dasgupta, Pillai and Rubin (2015) [20] developed a procedure for obtaining Fiducial intervals for the factorial effects in a 2^2 factorial experiment. This procedure is an extension of the one outlined in Section 2.5.6, where two p-value functions $p^+(\cdot)$ and $p^-(\cdot)$ are obtained for each factorial effect, and the lower and upper limits of a $100(1-\alpha)\%$ interval for the true value of the factorial effect are obtained by separately solving $p^+(\tau) = \alpha/2$ and $p^-(\tau) = \alpha/2$ for τ.

The function INTERVALFISHERTWOSQ() shown below computes $100(1-\alpha)\%$ intervals for the three factorial effects. The function takes the following inputs: observed outcomes $\mathbf{y}$, assignment vector $\mathbf{w}$, number of draws (ITER) from the distribution of $\mathbf{W}$, the "search space", i.e., the minimum (taumin) and maximum (taumax) for each factorial effect and the number (m) of three-dimensional points within the search space at which at which the p-values are to be computed; and a simulation seed point that can be used to reproduce the results obtained. The function chooses m points within the search space using a Latin Hypercube Design (LHD) (Mackay et al. 1979 [59]).

The function INTERVALFISHERTWOSQ() used a function findinterval() shown below that takes an array of τ's, the corresponding p^+ and p^- values, confidence coefficient α and returns the lower and upper $\alpha/2$ limits of τ. Although we show this function separately, it can also be embedded within the INTERVALFISHERTWOSQ() function.

```
-------------------------------------
findinterval <- function(taustarvec,pvalarray_plus,
pvalarray_minus, alpha) {
  # Calculating lower limit
  if (sum(pvalarray_plus == alpha/2) == 1) {
    tau_l = taustarvec[pvalarray_plus==alpha/2]
  } else {
    plbelow=max(0, max(pvalarray_plus[pvalarray_plus<alpha/2]))
    plabove=min(1, min(pvalarray_plus[pvalarray_plus>alpha/2]))
    tau_l_below = taustarvec[pvalarray_plus==plbelow]
    tau_l_above = taustarvec[pvalarray_plus==plabove]
    tau_l = tau_l_below +
   (tau_l_above - tau_l_below)* (alpha/2 - plbelow)/(plabove-plbelow)
  }
  # Calculating upper limit
  if (sum(pvalarray_minus == alpha/2) == 1) {
    tau_u = taustarvec[pvalarray_minus==alpha/2]
  } else {
    pubelow=max(pvalarray_minus[pvalarray_minus<alpha/2])
    puabove=min(pvalarray_minus[pvalarray_minus>alpha/2])
    tau_u_below = taustarvec[pvalarray_minus==pubelow]
    tau_u_above = taustarvec[pvalarray_minus==puabove]
    tau_u = tau_u_below +
   (tau_u_above - tau_u_below)* (alpha/2 - pubelow)/(puabove-pubelow)
  }
  return(c(tau_l, tau_u))
} # end function
-------------------------------------
```

As seen from the application of the function INTERVALFISHERTWOSQ() below, 90% intervals for the three factorial effects - the main effect of F_1, main effect of F_2, and the interaction effect of F_1 and F_2 - are obtained as $(0.115, 2.808)$, $(0.602, 2.483)$, $(-1.935, 1.274)$.

```
-------------------------------------
INTERVALFISHERTWOSQ = function(y,w,ITER,taumin, taumax, m, alpha,
 seed) {

  library(tgp)
  rect = rbind(c(taumin,taumax),c(taumin,taumax),c(taumin,taumax))

  set.seed(seed)
  des = lhs(m, rect)

  pmat_plus = NULL
  pmat_minus = NULL

  for (i in 1:m) {
```

```
    inp = des[i,]

    pvec_plus = Fishertesttwosquare(y,w,inp[1],inp[2],inp[3],ITER)[[1]]
    pvec_minus = Fishertesttwosquare(y,w,inp[1],inp[2],inp[3],ITER)[[2]]
    pmat_plus = rbind(pmat_plus, c(inp, pvec_plus))
    pmat_minus = rbind(pmat_minus, c(inp, pvec_minus))

  } # end

pmat_plus <- data.frame(pmat_plus)
colnames(pmat_plus) <- c("tau1", "tau2", "tau3", "pplus1", "pplus2",
 "pplus3")

pmat_minus <- data.frame(pmat_minus)
colnames(pmat_minus) <- c("tau1", "tau2", "tau3", "pminus1", "pminus2",
 "pminus3")

pmat1 <- pmat_plus %>% select(tau1, pplus1) %>%
  mutate(pminus1 = pmat_minus$pminus1) %>%
  arrange(tau1)

pmat2 <- pmat_plus %>% select(tau2, pplus2) %>%
  mutate(pminus2 = pmat_minus$pminus2) %>%
  arrange(tau2)

pmat3 <- pmat_plus %>% select(tau3, pplus3) %>%
  mutate(pminus3 = pmat_minus$pminus3) %>%
  arrange(tau3)

interval1 <- findinterval(pmat1$tau1, pmat1$pplus1, pmat1$pminus1, alpha)
interval2 <- findinterval(pmat2$tau2, pmat2$pplus2, pmat2$pminus2, alpha)
interval3 <- findinterval(pmat3$tau3, pmat3$pplus3, pmat3$pminus3, alpha)

return(list(F1=interval1,F2=interval2,F3=interval3))

}


> INTERVALFISHERTWOSQ(y,w,ITER=5000,taumin=-3, taumax=3, m=30, alpha=0.1,
  seed=200)
$F1
[1] 0.115121 2.808408

$F2
[1] 0.6021476 2.4834192

$F3
[1] -1.934806  1.274177
-------------------------------------
```

6.6 Neymanian inference from completely randomized 2^2 experiments

Neymanian inference in a 2^2 factorial design involves drawing inference (by obtaining confidence intervals or testing hypothesis) on the three population-level factorial effects $\overline{\tau}(F_1), \overline{\tau}(F_2)$ and $\overline{\tau}(F_1F_2)$ using their unbiased estimators, $\widehat{\overline{\tau}}(F_1)$, $\widehat{\overline{\tau}}(F_2)$ and $\widehat{\overline{\tau}}(F_1F_2)$. To perform such inference, one needs to obtain the sampling properties of these estimators over repeated randomizations. Such properties were derived by Dasgupta, Pillai and Rubin (2015) [20] for balanced 2^K designs and extended to the case of unbalanced designs by Lu (2016), [54], who showed that in a 2^2 experiment in which N_j units receive treatment combination $j = 1, 2, 3, 4$ through a completely randomized assignment mechanism, the sampling variance of the estimated factorial effects are

$$\begin{aligned} \text{Var}(\widehat{\overline{\tau}}(F_1)) &= \frac{1}{4}\sum_{j=1}^{4}\frac{S_j^2}{N_j} - \frac{S^2(\overline{\tau}(F_1))}{N}, \\ \text{Var}(\widehat{\overline{\tau}}(F_2)) &= \frac{1}{4}\sum_{j=1}^{4}\frac{S_j^2}{N_j} - \frac{S^2(\overline{\tau}(F_2)}{N}, \\ \text{Var}(\widehat{\overline{\tau}}(F_1F_2)) &= \frac{1}{4}\sum_{j=1}^{4}\frac{S_j^2}{N_j} - \frac{S^2(\overline{\tau}(F_1F_2)}{N}, \end{aligned} \tag{6.35}$$

where S_j^2 denotes the variance (with divisor $N-1$) of all potential outcomes corresponding to treatment combination $j = 1, 2, 3, 4$, as defined in (2.25)-(2.26). The values of S_1^2, S_2^2, S_3^2 and S_4^2 for the matrix of potential outcomes in Table 6.1 are shown in the last row and columns 2–5 to be 0, 0.25, 1.58 and 2.92 respectively. The last term in each variance expressions of (6.35) represents the variances (with divisor $N-1$) of the relevant unit level factorial effects, i.e.,

$$\begin{aligned} S^2(\overline{\tau}(F_1)) &= \frac{1}{N-1}\sum_{i=1}^{N}\left(\tau_i(F_1) - \overline{\tau}(F_1)\right)^2, \\ S^2(\overline{\tau}(F_2)) &= \frac{1}{N-1}\sum_{i=1}^{N}\left(\tau_i(F_2) - \overline{\tau}(F_2)\right)^2, \\ S^2(\overline{\tau}(F_1F_2)) &= \frac{1}{N-1}\sum_{i=1}^{N}\left(\tau_i(F_1F_2) - \overline{\tau}(F_1F_2)\right)^2. \end{aligned} \tag{6.36}$$

These variances are shown in the last three columns in the last row of Table 6.1, where their values are 1.230, 0.560 and 0.400 respectively.

To estimate the variances in (6.35) unbiasedly, from the observed data, one needs to estimate S_j^2 for all four treatment combinations $j = 1, 2, 3, 4$, and also the three variance terms $S^2(\overline{\tau}(F_1))$, $S^2(\overline{\tau}(F_2))$ and $S^2(\overline{\tau}(F_1F_2))$ defined in (6.36). However, none of the variances $S^2(\overline{\tau}(F_1))$, $S^2(\overline{\tau}(F_2))$ and $S^2(\overline{\tau}(F_1F_2))$ is directly estimable, because none of the unit-level factorial effects can be computed from observed data. However, if *strict additivity* defined in Definition 5.1 of Chapter 5 holds, i.e., the potential outcomes for any treatment can be expressed as the sum of the corresponding potential outcomes for any other treatment and a constant, then all the unit-level factorial effects are the same, leading to the following three results:

1. The variances of the potential outcomes corresponding to all four treatment combinations (in columns 2–5 of Table 6.1) are identical, i.e.,
$$S_1^2 = S_2^2 = S_3^2 = S_4^2 = S^2.$$
2. The variances of the unit level factorial effects (in columns 7–9 Table 6.1) are zero, i.e., all the variance terms in (6.36) vanish.

Therefore, under strict additivity, each sampling variance in (6.35) reduces to
$$\frac{S^2}{4}\sum_{j=1}^{4}\frac{1}{N_j},$$
where S^2 is the common variance of all four groups of potential outcomes, and can be unbiasedly estimated by MSRes obtained from the ANOVA decomposition. Consequently, the three sampling variances in (6.35) can be unbiasedly estimated by
$$\widehat{V}_{\text{Ney}} = \frac{\text{MSRes}}{4}\sum_{j=1}^{4}\frac{1}{N_j}. \tag{6.37}$$

To test the Neymanian null hypothesis that a particular population-level factorial effect is zero, e.g.,
$$H_{0\text{N}} : \overline{\tau}(F_1) = 0,$$
we can use the scaled version of the estimator $\widehat{\overline{\tau}}(F_1)$, i.e. the estimator divided by its standard error:
$$\frac{\widehat{\overline{\tau}}(F_1)}{S.E.\left(\widehat{\overline{\tau}}(F_1)\right)} = \frac{\widehat{\overline{\tau}}(F_1)}{\sqrt{\widehat{V}_{\text{Ney}}}} = \frac{\widehat{\overline{\tau}}(F_1)}{\sqrt{\frac{\text{MSRes}}{4}\sum_{j=1}^{4}\frac{1}{N_j}}},$$
which is exactly the same as $T_{\text{scaled}}(\overline{\tau}(F_1))$ defined in (6.32). The denominator of the statistic $T_{\text{scaled}}(\overline{\tau}(F_1))$ is thus justified by (6.37). The large sample distribution of this statistic can be approximated by a standard normal distribution, and a two-sided p-value can be obtained as $Pr\left(|Z| \geq |T^{\text{obs}}_{\text{scaled}}(F_1)|\right)$, where $T^{\text{obs}}_{\text{scaled}}(F_1)$ is the observed value of the test statistic (6.32). The smaller the p-value, the stronger is the evidence against $H_{0\text{N}}$.

An approximate $100(1-\alpha)\%$ confidence interval for $\overline{\tau}(F_1)$ is:

$$\widehat{\overline{\tau}}(F_1) \pm z_{\alpha/2}\sqrt{\widehat{V}_{\text{Ney}}}. \qquad (6.38)$$

Creating the tests for the three individual Neymanian hypotheses, $\overline{\tau}(F_1) = 0$, $\overline{\tau}(F_2) = 0$ and $\overline{\tau}(F_1F_2) = 0$, from the observed data in Table 6.5 and constructing of 95% large sample confidence intervals for the three factorial effects, are left as exercises (exercise 6.7).

6.7 Analysis of the education experiment

We now present analyses of a 2^2 experiment using the data obtained from the education experiment reported in Angrist et al. (2009) [1] and described at the beginning of this Chapter, where the effects of two educational incentive strategies were investigated on academic performance of 1656 freshmen. Data from the experiment is available in the MIT data archive https://economics.mit.edu/people/josh-angrist/angrist-data-archive. Each freshman in the study received one of the four treatment combinations 00, 01, 10, and 11 in year one through a completely randomized assignment. The two responses of interest were the GPA of the freshmen after year 1 (Y_1) and the GPA after year 2 (Y_2). Out of the 1656 freshmen, the observed value of Y_1 was not reported for 119, and that for Y_2 was not reported for 309 freshmen. Here, we will infer the causal effects of the two new incentives on Y_1 using the data obtained from 1537 units, assuming that these units received one of the four treatments through a completely randomized assignment. Inference for causal effects of the two incentives on Y_2 will be left to the readers. The pre-processed data for outcome Y_1 is stored as a data frame named StardataY1Ch6 {available on the first author's web page} that consists of 1537 rows and 2 columns. The first 20 rows of this data frame are shown below. Note that $w = 1, 2, 3, 4$ correspond to the treatment combinations 00, 01, 10, and 11 respectively.

```
-------------------------------------------------------------------
> dim(StardataY1Ch6)
[1] 1537    2
> StardataY1Ch6[1:20,]
   w   y1
1  1 2.58
2  2 3.55
3  1 3.75
4  1 2.37
5  1 1.68
6  2 1.79
```

```
7  4 2.03
8  2 2.11
9  1 2.05
10 2 1.14
11 1 0.43
12 1 1.25
13 3 2.13
14 4 1.93
15 1 3.00
16 1 1.29
17 3 2.25
18 3 1.35
19 4 3.01
20 1 1.88
-------------------------------------------------------------------
```

Table 6.9, generated using the code below, shows the number of units N_j, observed average $\overline{y}(j)$ and observed variance s_j^2 for $j = 1, 2, 3, 4$ for response Y_1.

```
-------------------------------------------------------------------
library(tidyverse)
N_j <- StardataY1Ch6 %>% count(w) %>% .$n %>% as.integer()
meanvec = c(mean(y1[w==1]),mean(y1[w==2]),mean(y1[w==3]),
mean(y1[w==4]))
varvec = c(var(y1[w==1]),var(y1[w==2]),var(y1[w==3]),var(y1[w==4]))
rbind(N_j, meanvec, varvec)
-------------------------------------------------------------------
```

TABLE 6.9
Sample sizes, observed averages and observed variances of Y_1 (GPA after year 1)

Sample size; observed mean and variance	Treatment combination $j(z_1 z_2)$			
	1 (00)	2 (01)	3 (10)	4 (11)
N_j	935	237	231	134
$\overline{y}_j$	1.784	1.772	1.779	1.842
s_j^2	0.841	0.691	0.844	0.879

Before computing the estimates of the factorial effects and assessing their significance, a quick visual summary of the observed responses is often helpful. Figure 6.4 shows a boxplot of the observed outcomes for the four treatment combinations, generated using the following code:

```
-------------------------------------------------------------------
boxplot(y1 ~ w, xlab="Treatment combination", ylab=expression(Y[1]))
-------------------------------------------------------------------
```

Figure 6.5 displays a visual representation of the two main effects and the interaction, generated using the following code:

```
-----------------------------------------------------------------
dev.new(width=15)
par(mfrow=c(1,3))
## ME SSP plot
ssp0=mean(c(y1[w==1],y1[w==2]))
ssp1=mean(c(y1[w==3],y1[w==4]))
plot(c(0,1),c(ssp0,ssp1),type="b",main=NULL,axes=FALSE,xlab="SSP",
ylab="AVG GPA",ylim=c(1.75,1.90),pch=16, cex.lab=1.5, cex.axis=1.2)
axis(side=1, at=c(0:1))
axis(side=2, at=seq(1.75, 1.85, by=.01))
## ME SFP plot
sfp0=mean(c(y1[w==1],y1[w==3]))
sfp1=mean(c(y1[w==2],y1[w==4]))
plot(c(0,1),c(sfp0,sfp1),type="b",main=NULL,axes=FALSE,xlab="SFP",ylab=
"AVG GPA",ylim=c(1.75,1.90),pch=16, cex.lab=1.5, cex.axis=1.2)
axis(side=1, at=c(0:1))
axis(side=2, at=seq(1.75, 1.85, by=.01))
## Interaction plot
plot(c(0,1),c(meanvec[1],meanvec[2]),type="b",main=NULL,axes=FALSE,
xlab="SFP",ylab="AVG GPA",ylim=c(1.75,1.90),pch=16, cex.lab=1.5,
cex.axis=1.2)
points(c(0,1),c(meanvec[3],meanvec[4]),type="b",pch=16)
axis(side=1, at=c(0:1))
axis(side=2, at=seq(1.75, 1.85, by=.01))
text(.5,1.77,"SSP=0", cex=1.5)
text(.5,1.82,"SSP=1", cex=1.5)
-----------------------------------------------------------------
```

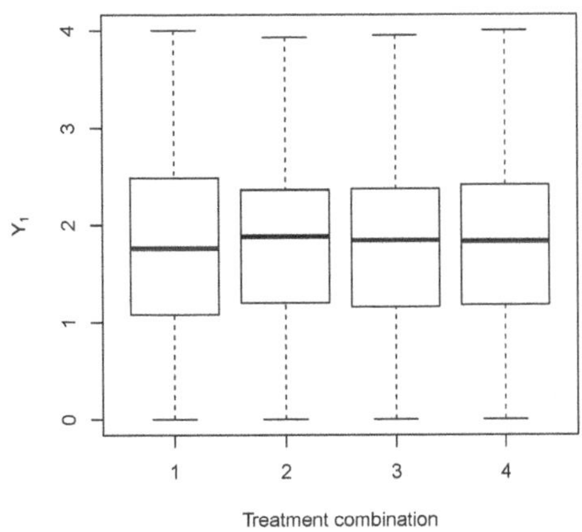

FIGURE 6.4
Boxplots of disributions of observed first year GPA for four treatment groups

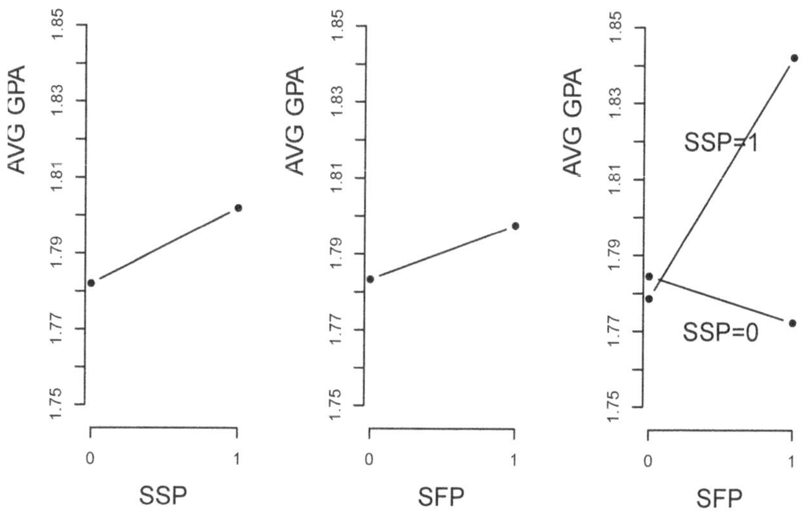

FIGURE 6.5
From the left: Plots of main effect of SSP, main effect of SFP and their interaction

Figure 6.4 does not reveal any striking difference between the distributions of the observed outcome in the four treatment groups. In particular, the variances in the four groups also appear similar, suggesting that the assumption of strict additivity may be reasonable. The main effects and interaction plots suggests small increases in average GPA at the end of the first year, after application of each treatment, and the possible presence of an interaction effect. However, the significance of these visual effects should also be assessed through formal statistical analyses.

Table 6.10 shows the ANOVA decomposition of the observed outcomes, obtained as follows:

```
------------------------------------------------------------------------
> anova(lm(y1 ~ factor(w)))
Analysis of Variance Table

Response: y1
            Df  Sum Sq Mean Sq F value Pr(>F)
factor(w)    3    0.48 0.16052  0.1954 0.8996
Residuals 1533 1259.66 0.82169
---------------------------------------------
```

The overall F-ratio (F_{Tre}) is small, and a randomization test with 20,000 draws from the distribution of $\mathbf{W}$ yields a p-value of 0.90, providing no evidence that any of the treatments is effective. Figure 6.6 shows the results of

TABLE 6.10
ANOVA Table for education experiment

Source	Degrees of freedom	Sum of squares	Mean Squares	F-ratio
Treatments	3	0.48	0.1605	0.1954
Residuals	1533	1259.66	0.8217	
Total	1536	1260.14		

the randomization test conducted using the code below, which supports that conclusion.

```
-----------------------------------------------------------------
library(AlgDesign)
D = gen.factorial(2,2) ## ME contrast columns in reverse order
D[, c(1:2)] = D[,c(2:1)] ## Reverse the order of the columns
colnames(D) = c("F1","F2")
L = model.matrix(~ F1 + F2 + F1*F2,D)
ITER=20000
Fobs=0.1954
Farray=NULL
set.seed(100)
for (count in 1:ITER) {
wnew = sample(c(rep(1,N_j[1]),rep(2,N_j[2]),
              rep(3,N_j[3]),rep(4,N_j[4])))

Fnew=anova(lm(y1 ~ factor(wnew)))$F[1]
Farray = c(Farray, Fnew)
}

hist(Farray, main=NULL,xlab="F")
abline(v=Fobs, lty=3, lwd=2)
pvalF = sum(Farray >= Fobs)/ITER
text(4,4000,paste("p-value= ",round(pvalF,2)))
-----------------------------------------------------------------
```

Table 6.11 shows the estimated factorial effects computed using the methodology described in Section 6.3, and asymptotic Neymanian inference under the assumption of strict additivity following the procedure outlined in Section 6.6. The estimated standard errors in the third column are the observed values of $\sqrt{\widehat{V}_{\text{Ney}}}$ defined in (6.37). The fourth column shows the observed values of the statistic $T_{\text{scaled}}(\widehat{\tau}(\cdot))$. The p-values are obtained as $Pr\left(|Z| \geq |T^{\text{obs}}_{\text{scaled}}|\right)$, where Z denotes the standard normal variable, which shows that none of these effects is significant.

The education experiment also recorded several covariates for each freshman who participated in the study, including the person's sex. The 1537 freshmen whose first year GPAs were available comprised 892 females and 645 males. Consider a hypothetical blocked experiment where the treatment assignment was done independently for female and male blocks, with the

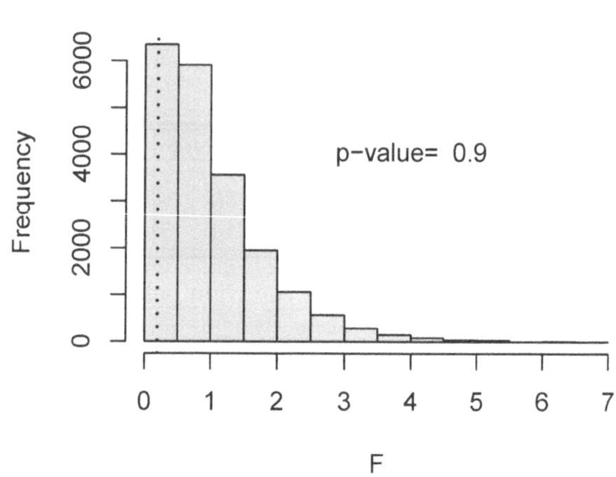

FIGURE 6.6
Fisher randomization test for outcome Y_1 (GPA after year 1) with statistic F_{Tre}; dashed line indicates the observed value of the test statistic

TABLE 6.11
Observed T_{scaled} statistic and approximate p-values for each estimated factorial effect

Factorial effect	Estimated effect	S.E.	$T^{\text{obs}}_{\text{scaled}}$	p-value
M.E. of SSP	.032	.059	.541	.59
M.E. of SFP	.026	.059	.434	.66
INT	.038	.059	.639	.52

pre-determined distribution shown in Table 6.12. Such an assignment allows us to perform inference separately for the two blocks. Using the same methodology as above, the estimated factorial effects, their standard errors, observed value of the $T^{\text{obs}}_{\text{scaled}}$ test statistic, and the p-values calculated using a standard normal distribution, are shown in Table 6.13, where the M.E. of factor SFP appears to be possibly interesting with a p-value of 0.11. Thus, taking covariate information into account may help detect effects that can otherwise

TABLE 6.12
Break-up of number of units assigned to treatments by sex

Block	Treatment combination $z_1 z_2$			
(Sex)	00	01	10	11
Female	540	145	132	75
Male	395	92	99	59
Total	935	237	231	134

go undetected due to heterogeneity in the study population, which can be interesting especially if these factors can be used in assigning treatments in the future.

TABLE 6.13
Observed T_{scaled} statistic and approximate p-values for each estimated factorial effect for female freshmen

Factorial effect	Estimated effect	S.E.	$T^{\text{obs}}_{\text{scaled}}$	p-value
M.E. of SSP	.025	.078	0.316	.75
M.E. of SFP	.121	.078	1.556	.11
INT	.043	.078	0 .552	.58

6.8 Problems

6.1 Complete the two missing columns in Table 6.1 with the calculated values of the contrast vectors $\boldsymbol{\lambda}$ for the CMEs $T_i(F_2|F_1 = 0)$ and $T_i(F_2|F_1 = 1)$ defined in (6.5) and (6.6) respectively

6.2 Define the unit-level interaction effect $\tau_i(F_2F_1)$ as a contrast of the unit-level potential outcomes, and show that it equals $\tau_i(F_1F_2)$ defined in (6.11), which means the interaction effect is symmetric with respect to the order of the factors.

6.3 For $i = 1, \ldots, N$ and $z_1, z_2 \in \{0, 1\}$, let $W_{i1}(z_1)$ and $W_{i2}(z_2)$ denote binary assignment variables such that

$$W_{i1}(z_1) = \begin{cases} 1 & \text{if the } i\text{th unit receives level } z_1 \text{ of factor } F_1 \\ 0 & \text{otherwise,} \end{cases} \tag{6.39}$$

and

$$W_{i2}(z_2) = \begin{cases} 1 & \text{if the } i\text{th unit receives level } z_2 \text{ of factor } F_2 \\ 0 & \text{otherwise,} \end{cases} \tag{6.40}$$

(i) Represent the assignment random variable $W_i \in \{1, \ldots, 2^K\}$ in terms of $W_{i1}(1)$ and $W_{i2}(2)$.

(ii) Consider a 2^2 factorial experiment in which the assignment of each of N units to the two levels of the two factors F_1 and F_2 is determined by two independent tosses of a fair coin, independent of assignments of the remaining units. Write the joint probability distribution of the $N \times 1$ random vectors $(W_{11}(1), \ldots, W_{N1}(1))^{\mathrm{T}}$, $(W_{12}(1), \ldots, W_{N2}(1))$ and $\mathbf{W} = (W_1, \ldots, W_N)^{\mathrm{T}}$, where $W_{i1}(z_1)$, $W_{i2}(z_2)$ are defined in (6.39), (6.40) respectively, for $i = 1, \ldots, N$ and $z_1, z_2 \in \{0, 1\}$. and W_i is as defined in part (i).

6.4 In the education experiment described at the beginning of this Chapter, suppose that each of the 1656 units received the four treatment combinations numbered 1, 2, 3 and 4 with probabilities 0.60, 0.15, 0.15 and 0.10 respectively, where the unit level assignments were independent of each other. Calculate:

(i) the probability that this assignment resulted in 1006, 250, 250 and 150 units being assigned to treatment combinations 1, 2, 3 and 4 respectively.

(ii) $E[W_{i1}(z_1)]$, $E[W_{i2}(z_2)]$ for $z_1, z_2 \in \{0,1\}$ where $W_{i1}(z_1)$, $W_{i2}(z_2)$ are defined in (6.39), (6.40) respectively for $i = 1, \dots, N$ and $z_1, z_2 \in \{0,1\}$.

(iii) $\text{Cov}[W_{i1}(0), W_{i2}(0)]$, $\text{Cov}[W_{i1}(0), W_{i2}(1)]$, $\text{Cov}[W_{i1}(1), W_{i2}(0)]$ and $\text{Cov}[W_{i1}(1), W_{i2}(1)]$ for $i = 1, \dots, N$.

6.5 Prove (6.27), for a balanced and replicated 2^2 experiment with r units receiving each treatment, using the decomposition (6.30).

6.6 Calculate the observed values of individual F-statistic for the three factorial effects - the MEs of F_1, ME of F_2 and the interaction between F_1 and F_2 - from the observed data of Table 6.5, and expand the ANOVA Table 6.6 into an ANOVA table like the one in Table 6.7 that shows the contribution of each factorial effect. Show that the square of T_{scaled} in Table 6.8 for a specific factorial effect is the F-ratio for that effect in the ANOVA table that you just obtained.

6.7 From the observed data of Table 6.5, calculate the two-sided p-values for testing the Neymanian hypotheses $\overline{\tau}(F_1) = 0$, $\overline{\tau}(F_2) = 0$ and $\overline{\tau}(F_1F_2) = 0$ individually. Also, construct 95% large-sample confidence intervals for each of the three factorial effects.

6.8 In the education experiment discussed in this Chapter and reported in [1], for the outcome Y_1, i.e., GPA of freshmen after year 1, assess the sharp null hypothesis of no treatment effect using first the test statistic

$$\widehat{\overline{\tau}}_{\max} = \max\left[|\widehat{\overline{\tau}}(F_1)|, |\widehat{\overline{\tau}}(F_2)|, |\widehat{\overline{\tau}}(F_1, F_2)|\right],$$

which has the largest absolute value among all the three estimated effects, and second its scaled version obtained by dividing it by its standard error.

6.9 In the education experiment, for the outcome Y_1, i.e., GPA of freshmen after year 1, obtain 95% fiducial intervals for the three factorial effects by inverting the p-value functions obtained by using the randomization test for several sharp null hypotheses.

6.10 In the education experiment discussed in this Chapter and reported in [1], one outcome of interest was the GPA of the freshmen after year 2 (Y_2), although the incentives were given in the first year only. In this problem, you will infer the causal effects of the two incentives on Y_2 using the data obtained from the 1347 units for which the outcome Y_2 was recorded.

(i) Construct the appropriate ANOVA table and compute the F statistic.

(ii) Perform a randomization test assuming a completely randomized treatment assignment mechanism. Report the p-value and state your conclusion.

(iii) Perform randomization tests for each of the three factorial effects using the T_{scaled} statistic and interpret your results.

(iv) Obtain 95% large-sample confidence intervals individually for each of the average factorial effects.

(v) Perform tasks (i)-(iv) separately for female and male students, assuming a randomized block design in which females and males were separately allocated to the treatment combinations through a completely randomized assignment mechanism.

7

2^K factorial designs

In Section 1.4.4 of Chapter 1 we introduced an experiment planned by the New York Department of Education (NYDE) to study five "incentive programs" to be introduced to high schools that needed performance improvement. These five programs were: (1) a quality review, (2) a periodic assessment, (3) inquiry teams, (4) a school-wide performance bonus program, and (5) an online resource program. The goal of the experiment was to examine the causal effects of these interventions and their possible interactions on the performances of the schools as measured by scores (numbers between 1 and 100) on the school's progress reports, issued annually near the start of each school year. These scores would be numbers between 1 and 100. Prior to designing the experiment, the NYDE collected data on several covariates suspected to be correlated with the school performance, such as total number of students, five different race variables (proportion of white, black, Asian, Native American and Latino students), proportion of female students, enrollment rate, and poverty rate.

Each school could receive either the "treatment" (active) level or the "control" (inactive) level of each of the above five factors depending on the experimental design. For example, the treatment and control levels of the "quality review" factor respectively were conducting the review, and not conducting the review. Thus the total number of possible treatment combinations is $2^5 = 32$. In Chapter 6 we discussed 2^2 factorial experiments, where the aim was to study the causal effects of two treatment factors, each at two levels, on the response of interest. In this Chapter we will extend the concepts, methods and tools for designing and analysing 2^2 experiments to 2^K factorial experiments where the goal is to study the causal effects of $K (> 2)$ two-level factors (and their interactions) on the response of interest. Such an experiment has 2^K combinations of factor levels. The NYDE experiment is an example of a 2^5 experiment.

7.1 Factorial effects in a 2^K experiment

Here, we introduce some key definitions and formal notation that are nearly identical to, different from those used by Dasgupta et al. (2015) [20]. Consider

DOI: 10.1201/9781003048794-7

a 2^K experiment with K factors denoted by $F_1, \ldots, F_K$ and N units, where the levels of each of the K factors are denoted by 0 and 1. Let the $J = 2^K$ treatment combinations $z_1 z_2 \ldots z_K$, where $z_k \in \{0, 1\}$ for $j = 1, \ldots, K$, be arranged lexicographically starting with $(0\ 0\ \ldots\ 0\ 0)$, $(0\ 0\ \ldots\ 0\ 1)$, and ending with $(1\ 1\ \ldots\ 1\ 1)$. Thus the labels of these treatment combinations arranged in lexicographic order are

$$j = 2^{K-1} z_1 + 2^{K-2} Z_2 + \cdots + z_K + 1,$$

i.e., one plus the binary representation of $z_1, \ldots, z_K$. For example, in a 2^3 experiment, the eight treatment combinations starting with $(0\ 0\ 0)$, followed by $(0\ 0\ 1)$ and ending with $(1\ 1\ 1)$ shown in Table 7.1 are numbered as $j = 1, 2, \ldots, 8$ respectively.

TABLE 7.1
Treatment combinations and potential outcomes for unit i in a 2^3 experiment

Treatment number (j)	1	2	3	4	5	6	7	8
Level combination	000	001	010	011	100	101	110	111
Potential outcomes	$Y_i(1)$	$Y_i(2)$	$Y_i(3)$	$Y_i(4)$	$Y_i(5)$	$Y_i(6)$	$Y_i(7)$	$Y_i(8)$

For $i = 1, \ldots, N$, the ith unit has $J = 2^K$ potential outcomes $Y_i(1), \ldots, Y_i(J)$ corresponding to the J treatment combinations. Let $\mathbf{Y}_i = (Y_i(1), \ldots, Y_i(J))^{\mathrm{T}}$ denote the $J \times 1$ vector of potential outcomes for unit i. For unit i, the unit-level main effect of factor F_k, $k = 1, \ldots, K$ is defined as the difference between the averages of potential outcomes for unit i for which the levels of factor k are at levels 0 versus levels 1. For example, in a 2^3 experiment, the ith unit has the 8 potential outcomes $Y_i(1), \ldots, Y_i(8)$ of Table 7.1. Using the same argument used to define unit-level main effects for a 2^2 design in (6.8) and (6.10), that is, comparisons of averages of potential outcomes corresponding to the two levels of each factor, the unit-level main effect of factors F_1, F_2 and F_3 are

$$\begin{aligned}
\tau_i(F_1) &= \frac{Y_i(5) + Y_i(6) + Y_i(7) + Y_i(8)}{4} - \frac{Y_i(1) + Y_i(2) + Y_i(3) + Y_i(4)}{4} \\
&= \frac{1}{4}(-1 - 1 - 1 - 1 + 1 + 1 + 1 + 1)\mathbf{Y}_i \qquad (7.1) \\
\tau_i(F_2) &= \frac{Y_i(3) + Y_i(4) + Y_i(7) + Y_i(8)}{4} - \frac{Y_i(1) + Y_i(2) + Y_i(5) + Y_i(6)}{4} \\
&= \frac{1}{4}(-1 - 1 + 1 + 1 - 1 - 1 + 1 + 1)\mathbf{Y}_i \qquad (7.2) \\
\tau_i(F_3) &= \frac{Y_i(2) + Y_i(4) + Y_i(6) + Y_i(8)}{4} - \frac{Y_i(1) + Y_i(3) + Y_i(5) + Y_i(7)}{4} \\
&= \frac{1}{4}(-1 + 1 - 1 + 1 - 1 + 1 - 1 + 1)\mathbf{Y}_i \qquad (7.3)
\end{aligned}$$

From (7.1)-(7.3), we see that in a 2^K factorial experiment, each unit-level main effect $\tau_i(F_k)$ for $k = 1, \ldots, K$, is a contrast of the unit-level potential outcomes $Y_i(1), \ldots, Y_i(J)$ of the form

$$\tau_i(F_k) = \frac{1}{2^{K-1}} \sum_{j=1}^{J} \lambda_j(F_k) Y_i(j), \tag{7.4}$$

where the contrast coefficient $\lambda_j(F_k) = -1$ for those treatment combinations j in which the level of factor F_k is zero and $\lambda_j = +1$ for those treatment combinatons j in which the level of factor F_k is one. Note that (7.4) can also be written in vector notation as:

$$\tau_i(F_k) = \frac{1}{2^{K-1}} \boldsymbol{\lambda}(F_k)^{\mathrm{T}} \mathbf{Y_i}, \tag{7.5}$$

where $\boldsymbol{\lambda}(F_k)$ is the $J \times 1$ contrast vector $(\lambda_1(F_k), \ldots, \lambda_J(F_k))^{\mathrm{T}}$.

Next, defining the two-factor interaction between two factors F_k and $F_{k'}$ as scaled differences between the conditional main effects, as in (6.3) - (6.6), and following the line of argument leading to (6.11), we can express the two factor interaction in the form

$$\tau_i(F_k F_{k'}) = \frac{1}{2^{K-1}} \sum_{j=1}^{J} \lambda_j(F_k F_{k'}) Y_i(j), \tag{7.6}$$

where the contrast coefficients $\lambda_j(F_k F_{k'})$ can be obtained by multiplying the contrast coefficients of the main effects of F_k and $F_{k'}$, i.e., $\lambda_j(F_k F_{k'}) = \lambda_j(F_k)\lambda_j(F_{k'})$ for $j = 1, \ldots, J$.

Equation (7.6) can also be written in vector notation as follows:

$$\tau_i(F_k F_{k'}) = \frac{1}{2^{K-1}} \boldsymbol{\lambda}(F_k F_{k'})^{\mathrm{T}} \mathbf{Y_i}, \tag{7.7}$$

where $\boldsymbol{\lambda}(F_k F_{k'})$ is the $J \times 1$ contrast vector for the interaction between factors F_k and $F_{k'}$, obtained by multiplying the contrast vectors $\boldsymbol{\lambda}(F_k)$ and $\boldsymbol{\lambda}(F_{k'})$ elementwise. Let $\circ$ denote the operation of multiplying two vectors of equal dimension element-wise. Then, we can write:

$$\boldsymbol{\lambda}(F_k F_{k'}) = \boldsymbol{\lambda}(F_k) \circ \boldsymbol{\lambda}(F_{k'}) \tag{7.8}$$

For example, in a 2^3 experiment, the contrast vectors for the main effect of F_1 and the main effect of F_2, obtained respectively from (7.1) and (7.2), are:

$$\boldsymbol{\lambda}(F_1) = \begin{pmatrix} -1 \\ -1 \\ -1 \\ -1 \\ +1 \\ +1 \\ +1 \\ +1 \end{pmatrix}, \quad \boldsymbol{\lambda}(F_2) = \begin{pmatrix} -1 \\ -1 \\ +1 \\ +1 \\ -1 \\ -1 \\ +1 \\ +1 \end{pmatrix}.$$

Hence the contrast vector for the two-factor interaction between F_1 and F_2 is:

$$\boldsymbol{\lambda}(F_1F_2) = \boldsymbol{\lambda}(F_1) \circ \boldsymbol{\lambda}(F_2) = \begin{pmatrix} +1 \\ +1 \\ -1 \\ -1 \\ -1 \\ -1 \\ +1 \\ +1 \end{pmatrix}$$

Higher order interactions of order three and above can be defined in an analogous manner in terms of conditional lower order interactions, and can be expressed as contrasts of the unit-level potential outcomes, where the contrast coefficients are obtained by multiplying the contrast coefficients for the main effects of the component factors involved in the interaction. Thus, in a 2^3 experiment, the contrast vector for the three-factor interaction $F_1F_2F_3$ is:

$$\boldsymbol{\lambda}(F_1F_2F_3) = \boldsymbol{\lambda}(F_1) \circ \boldsymbol{\lambda}(F_2) \circ \boldsymbol{\lambda}(F_3) = \begin{pmatrix} -1 \\ +1 \\ +1 \\ -1 \\ +1 \\ -1 \\ -1 \\ +1 \end{pmatrix}$$

To summarize, a 2^K experiment has 2^{K-1} factorial effects comprising K main effects, $\binom{K}{2}$ two-factor interactions, $\binom{K}{3}$ three-factor interactions, and so on, ending with one K-factor interaction. We can thus define 2^{K-1} unit-level factorial effects of the form:

$$\tau_i(\cdot) = \frac{1}{2^{K-1}} \sum_{j=1}^{J} \lambda_j(\cdot) Y_i(j) = \frac{1}{2^{K-1}} \boldsymbol{\lambda}(\cdot)^{\mathrm{T}} \mathbf{Y}_i, \tag{7.9}$$

as a contrast of the unit-level potential outcomes $Y_i(1), \ldots, Y_i(J)$, where the dot $(\cdot)$ in (7.9) represents any main effect (e.g., F_1) or interaction of any order (e.g., $F_1F_2F_4$ in a 2^4 experiment). The contrast vector $\boldsymbol{\lambda}(\cdot)$ for any factorial effect can be obtained by the procedure described earlier in this section. As in Chapter 6, for completeness, unit i's average potential outcome is

$$\tau_i(0) = \frac{1}{2^{K-1}} \sum_{j=1}^{J} Y_i(j) = \frac{1}{2^{K-1}} \boldsymbol{\lambda}(0)^{\mathrm{T}} \mathbf{Y}_i,$$

where $\boldsymbol{\lambda}(0)$ is a $J \times 1$ column vector of ones.

A 2^3 experiment has three main effects, three two-factor interactions and one three-factor interaction. The contrast coefficients for the seven factorial effects are shown in Table 7.2 without the common divisor 4 (in Exercise 7.2, the readers are asked to fill out the incomplete columns in Table 7.2). As in Table 6.2, the seven contrast vectors represented by the seven columns are orthogonal to each other. Upon completion of Exercise 7.2, it can be seen that the seven contrast columns in Table 7.2), preceeded by the column of ones for the unit average, forms an 8×8 matrix **L**, which in an extension of the 4×4 matrix defined in (6.26) for a 2^2 experiment to a 2^3 experiment.

TABLE 7.2
Contrast vectors $\lambda(F_j)$'s for the seven factorial effects in a 2^3 experiment (without divisor 4)

Tr.	Level		Contrast vectors for						
	comb.	Avg	Main effects			2-factor Int			3-factor Int
(j)	$(z_1 z_2 z_3)$	$\boldsymbol{\lambda}(0)$	$\boldsymbol{\lambda}(F_1)$	$\boldsymbol{\lambda}(F_2)$	$\boldsymbol{\lambda}(F_3)$	$\boldsymbol{\lambda}(F_1F_2)$	$\boldsymbol{\lambda}(F_1F_3)$	$\boldsymbol{\lambda}(F_2F_3)$	$\boldsymbol{\lambda}(F_1F_2F_3)$
1	000	1	−1	−1	−1	+1			−1
2	001	1	−1	−1	+1	+1			+1
3	010	1	−1	+1	−1	−1			+1
4	011	1	−1	+1	+1	−1			−1
5	100	1	+1	−1	−1	−1			+1
6	101	1	+1	−1	+1	−1			−1
7	110	1	+1	+1	−1	+1			−1
8	111	1	+1	+1	+1	+1			+1

As shown in Section 6.3, the matrix **L**, referred to as the *model matrix* in many experimental design textbooks, is easy to generate in R using the function gen.factorial() available in the package "AlgDesign" [92]. Generation of the **L** matrix for a 2^3 factorial experiment is shown below:

```
-----------------------------------------------------------
library(AlgDesign)
D = gen.factorial(2,3) ## ME contrast columns in reverse order

D[, c(1:3)] = D[,c(3:1)] ## Reverse the order of the columns
colnames(D) = c("F1","F2","F3")
L = model.matrix(~ . + .*. + .*.*.,D)

> L
  (Intercept) F1 F2 F3 F1:F2 F1:F3 F2:F3 F1:F2:F3
1           1 -1 -1 -1     1     1     1       -1
2           1 -1 -1  1     1    -1    -1        1
3           1 -1  1 -1    -1     1    -1        1
4           1 -1  1  1    -1    -1     1       -1
5           1  1 -1 -1    -1    -1     1        1
6           1  1 -1  1    -1     1    -1       -1
7           1  1  1 -1     1    -1    -1       -1
8           1  1  1  1     1     1     1        1
-----------------------------------------------------------
```

The factorial effects (causal estimands) at the *population level* are the averages of all unit level factorial effectz over the units in the population, and are defined as

$$\overline{\tau}(\cdot) = \frac{1}{N}\sum_{i=1}^{N} \tau_i(\cdot), \tag{7.10}$$

where as before, in (7.10) the dot $(\cdot)$ represents any main effect or interaction. Let

$$\overline{Y}(j) = \frac{1}{N}\sum_{i=1}^{N} Y_i(j) \tag{7.11}$$

denote the average of all N potential outcomes for each treatment combination $j = 1, \ldots, J$. Denoting the $J \times 1$ vector $\left(\overline{Y}(1), \ldots, \overline{Y}(J)\right)^{\mathrm{T}}$ by $\overline{\mathbf{Y}}$, the population level factorial effect can be alternatively represented as

$$\overline{\tau}(\cdot) = \frac{1}{2^{K-1}}\sum_{j=1}^{J} \lambda_j(\cdot)\overline{Y}(j) = \frac{1}{2^{K-1}}\boldsymbol{\lambda}(\cdot)^{\mathrm{T}}\overline{\mathbf{Y}}, \tag{7.12}$$

by averaging both sides of (7.9) over $i = 1, \ldots, N$. The expression on the extreme right of equation (7.12) will be the most useful when estimating the factorial effects from observed outcomes.

In a 2^3 experiment, the two factor interaction between F_1 and F_2 is

$$\overline{\tau}(F_1F_2) = \frac{1}{4}\boldsymbol{\lambda}(F_1F_2)^{\mathrm{T}}\overline{\mathbf{Y}},$$

where $\boldsymbol{\lambda}(F_1F_2)^{\mathrm{T}}$ is the row vector $(+1+1-1-1-1-1+1+1)$ as seen from the fifth column of Table 7.2. See Exercise 7.3 for more development of the notation.

At this point we suggest that the readers solve Exercise 7.4 to bolster understanding of the notation introduced above, and the causal estimands in 2^K factorial experiments. Note that we have not yet mentioned observed data. The calculations discussed are on potential outcomes, and out of J potential outcomes for each unit, only one will eventually be observed and the rest will be missing.

Finally, as in (2.25)-(2.26) and in Section 6.6, we also define the variances of all N potential outcomes for each treatment combination $j = 1, \ldots, J$.

$$S_j^2 = \frac{1}{N-1}\sum_{i=1}^{N}\left(Y_i(j) - \overline{Y}(j)\right)^2. \tag{7.13}$$

7.2 Design of 2^K experiments: the assignment mechanism

In the 2^5 factorial experiment described at the beginning of this chapter, the NYDE provided a dataset that includes 50 pretreatment covariates for 1376 schools. The initial design question was, how should the NYDE allocate the schools to the 32 treatment combinations such that the effects of the incentive programs and their combinations will be well estimated?

The most straightforward way is to simply randomize the schools to the 32 treatment combinations. However, as discussed elaborately in Chapters 3 and 4, randomized experiments can yield "unlucky allocations", where some covariates are not well balanced across treatment groups. In Chapter 3 we discussed the strategy of blocking to avoid undesirable allocations. This idea was generalized in Chapter 4 to using a pre-defined acceptance criterion for allocations. In Section 5.6 of Chapter 5, this idea was extended to the case of $J(\geq 2)$ treatments, using the Mahalanobis distance criterion to ensure that the J treatment groups did not reveal undesirable covariate imbalance.

Classic experimental design textbooks, like Box Hunter and Hunter (2005) [10] and Wu and Hamada (2009) [93], propose using blocking to improve covariate balance in factorial experiments. However, as noted earlier in Chapter 4, blocking is most applicable when the number of covariates is small all covariates are categorical. In contrast, the NYDE experiment has a large number of covariates, most of which are not categorical. Extending earlier work by Morgan and Rubin (2012) [61], Branson et al. (2017) [12] proposed a Mahalanobis-distance-based acceptance criterion, and suggested using a rerandomization algorithm until an acceptable allocation is found.

We illustrate the idea proposed in Branson et al. (2017) [12] with a small example. Consider designing a 2^3 experiment with $N = 22$ units. A 2^3 experiment has 8 treatment combinations shown in Table 7.1. Assume that the experimenter decides to allocate 3 units each to treatment combinations $1, \ldots, 6$ and 2 units each to treatment combinations 7 and 8. Further, assume that there are three pre-measured covariates x_1, x_2 and x_3 shown in Table 7.3. In fact, these covariate values are simulated from a multivariate normal distribution with mean vector $(10, 8, 3)$ and covariance matrix

$$\begin{pmatrix} 2 & 1 & 1 \\ 1 & 2 & 1 \\ 1 & 1 & 2 \end{pmatrix}.$$

As in previous chapters, we denote the $N \times 1$ assignment vector by $\mathbf{W} = (W_1, \ldots, W_N)^{\mathrm{T}}$ where W_i denotes the treatment combination index $j = 1, \ldots, J$ assigned to unit i. In the current example, a 22×1 assignment vector for a completely randomized design can be generated by randomly permuting a vector of the form

$$(1, 1, 1, 2, 2, 2, 3, 3, 3, 4, 4, 4, 5, 5, 5, 6, 6, 6, 7, 7, 8, 8)^{\mathrm{T}}.$$

The reader, in Exercise 7.5, is asked to enumerate the total number of possible allocations under this assignment mechanism.

Consider one such assignment vector

$$\mathbf{w}^* = (6, 4, 7, 6, 5, 1, 3, 4, 1, 8, 2, 5, 2, 8, 5, 7, 4, 1, 3, 3, 6, 2)^{\mathrm{T}}.$$

TABLE 7.3
Simulated covariates in a hypothetical 2^3 experiment

Unit	x_1	x_2	x_3
1	7.80	6.57	2.60
2	10.69	8.41	3.93
3	8.66	5.75	0.41
4	12.26	10.22	3.14
5	11.35	8.52	2.68
6	9.03	8.31	1.69
7	14.23	12.64	4.14
8	10.98	9.45	3.58
9	10.69	8.61	3.30
10	11.73	6.97	4.22
11	10.62	6.62	3.83
12	10.22	7.37	3.93
13	11.81	8.96	2.41
14	8.14	7.95	0.32
15	11.09	10.74	4.44
16	10.80	7.48	4.24
17	8.93	6.79	1.92
18	9.91	7.70	3.30
19	9.78	8.83	2.68
20	8.84	9.00	1.82
21	10.70	6.85	2.63
22	8.81	9.11	3.58

Let us first describe the impact of using this assignment vector for treatment allocation on the estimation of the main effect of factor F_1, i.e., $\overline{\tau}(F_1)$. This (and any) assignment vector partitions the 22 units into two groups that correspond to the treated for factor F_1 (for which the contrast coefficient $\lambda_j(F_1)$ is +1) and control (for which the contrast coefficient $\lambda_j(F_1)$ is −1) units for factor F_1. Thus, from vector $\mathbf{w}^*$, we can generate a vector

$$\mathbf{w}^*_{F_1} = (+1-1+1+1+1-1-1-1-1+1-1+1-1+1+1+1-1-1-1-1+1-1)^{\mathrm{T}}$$

for the main effect of factor F_1 using the mapping between the treatment combination j and the contrast coefficient $\lambda_j(F_1)$ in Table 7.2. An easy way to do this is to use the appropriate (in this case the second) column of the matrix $\mathbf{L}$.

```
-------------------------------------------------------
w = c(6,4,7,6,5,1,3,4,1,8,2,5,2,8,5,7,4,1,3,3,6,2)
w1 = numeric(22)
for (i in 1:22) {
  w1[i] = L[w[i],2]
}
```

```
>w1
[1]  1 -1  1  1  1 -1 -1 -1 -1  1 -1  1 -1  1  1  1 -1 -1 -1 -1
[21]  1 -1
-------------------------------------------------------
```

The proportion of units receiving the treated and control levels of factor F_1 are $p(+1) = 10/22 = 0.45$ and $p(-1) = 12/22 = 0.55$ respectively. The vectors $\overline{\mathbf{x}}(+1)$ and $\overline{\mathbf{x}}(-1)$ of covariate means in the treated and control groups are respectively $(10.275, 7.842, 2.861)^{\mathrm{T}}$ and $(10.3600, 8.7025, 3.0150)^{\mathrm{T}}$, as shown below. Note that in the code, $\mathbf{X}$ represents the 22×3 matrix of covariates, which is partitioned into $\mathbf{X}(+)$ and $\mathbf{X}(-)$ based on $\mathbf{w}_1^*$.

```
-------------------------------------------------------
Xplus = X[w1==+1,] # covariate matrix for treated units
Xminus = X[w1==-1,] # covariate matrix for control units
Xplusbar = as.vector(apply(Xplus,2,mean)) # covariate means in
treated group
Xminusbar = as.vector(apply(Xminus,2,mean)) # covariate means in
control group
> Xplusbar
[1] 10.275  7.842  2.861
> Xminusbar
[1] 10.3600  8.7025  3.0150
-------------------------------------------------------
```

Finally, using the definition (4.4) of Mahalanobis distance between covariate means of the treated and control groups of units, we can compute the Mahalanobis distance between the covariate means of the treated and control groups for factor F_1 as

$$
\begin{aligned}
D_M(F_1) &= Np(+1)p(-1)\left(\overline{\mathbf{x}}(+1) - \overline{\mathbf{x}}(-1)\right)^{\mathrm{T}} \left[\mathrm{Cov}(\mathbf{X})\right]^{-1} \left(\overline{\mathbf{x}}(+1) - \overline{\mathbf{x}}(-1)\right) \\
&= 2.39,
\end{aligned}
$$

substituting $p(+1) = 0.45$, $p(-1) = 0.55$, $\overline{\mathbf{x}}(+1)^{\mathrm{T}} = (10.275, 7.842, 2.861)$ and $\overline{\mathbf{x}}(-1)^{\mathrm{T}} = (10.3600, 8.7025, 3.0150)$.

The Mahalanobis distances between covariate means of the treated and control groups of units for main effects of factors F_2 and F_3 can be computed similarly, and these distances can be denoted as $D_M(F_2)$ and $D_M(F_3)$ respectively.

Now turning to the interactions, we have seen throughout Section 7.1, that each interaction $\overline{\tau}(\cdot)$ can be represented in the same form as a main effect using the contrast vector $\lambda(\cdot)$ shown in the appropriate column of Table 7.2. Thus, we can represent the +1's and 1's in $\lambda(\cdot)$ as the treatment and control labels for an interaction. Consequently, the assignment vector $\mathbf{w}^*$ can be converted to a vector $\mathbf{w}^*(\cdot)$ using the mapping between j and the appropriate column of Table 7.2, and the Mahalanobis distance can be computed using the same procedure as described above. Below, we show two R functions that compute a vector of Mahalanobis distances corresponding to each factorial effect. The first function MDl(X,Wl) computes the Mahalanobis distance for covariate matrix X and assignment vector Wl. The function is the same as the function MD(X,W) used in Chapter 4, except for the fact that Wl is a vector of $+1$'s and -1's. The second function MDall(X,W) uses a vector W with eight levels as in a 2^3 experiment, converts W into seven vectors Wl of $+1$'s and -1's, and computes the MD for each factorial effect using MDl(X,Wl) for each Wl.

```
-----------------------------------------------------------
MD1 = function(X,W1) {
  N = length(W1)
  p1plus = sum(W1==+1)/N
  p1minus = sum(W1==-1)/N
  sigmainv = solve(cov(X))
  Xplus = X[W1==+1,]
  Xminus = X[W1==-1,]
  Xplusbar = as.vector(apply(Xplus,2,mean))
  Xminusbar = as.vector(apply(Xminus,2,mean))

  MD = N*p1plus * p1minus * t(Xplusbar - Xminusbar) %*%
  sigmainv %*% (Xplusbar - Xminusbar)
  return(MD)
}

# Given W (vector of integer treatment combination index j)
# and X,
# Calculate MD for each factorial effect

MDall = function(X,W) {
MDvec = NULL
  for (j in 1:7){
  W1 = numeric(22)
  for (i in 1:22) {
    W1[i] = L[W[i],j+1]
  } # end i
MD1 <- MD1(X,W1)
 MDvec = c(MDvec, MD1)
  } # end j
return(MDvec)
}

> round(MDall(Xmatrix, W),2)
[1] 2.39 0.89 1.04 2.50 1.88 0.40 2.05
-----------------------------------------------------------
```

Table 7.4 shows the Mahalanobis distances between covariate means for the seven factorial effects $\overline{\tau}(F_1), \ldots \overline{\tau}(F_1F_2F_3)$ for the assignment vector $\mathbf{w}^*$. The values suggest that among the 7 factorial effects, this particular allocation will result in maximum imbalance ($D_M(F_1F_2) = 2.50$) between the covariate averages for the two levels of the interaction between F_1 and F_2, and minimum imbalance ($D_M(F_2F_3) = 0.40$) between the covariate averages for the two levels of the interaction between F_2 and F_3.

TABLE 7.4
Mahalanobis distance between covariate means for factorial effects for assignment vector $\mathbf{w}^*$

Factorial effect ℓ	F_1	F_2	F_3	F_1F_2	F_1F_3	F_2F_3	$F_1F_2F_3$
$D_M(\cdot)$	2.39	0.89	1.04	2.50	1.88	0.40	2.05

We need to define acceptable allocations in terms of Mahalanobis distances to determine whether an assignment vector, like $\mathbf{w}^*$, is acceptable. The criteria proposed by Branson et al. (2017) [12] was

$$D_M(\cdot) \leq a(\cdot), \tag{7.14}$$

where as before, the dot $(\cdot)$ in (7.14) represents any factorial effect and $a(\cdot)$ is a predetermined threshold for that factorial effect. The threshold criterion can be the same for all factorial effects, or they can be different. For example, to place more importance on main effects than interactions, thresholds corresponding to main effects should be tighter (smaller) than those corresponding to interactions. Also, some factorial effects (e.g., higher order interactions) could be unimportant, making $a(\cdot) = \infty$ for such effects $\overline{\tau}(\cdot)$.

Recall from Section 4.3 of Chapter 4, that the Mahalanobis distance $D_M(\cdot)$ for factorial effect $\overline{\tau}(\cdot)$ is approximately distributed as a χ^2 random variable with k degrees of freedom, where k denotes the number of covariates. If the seven χ^2 variables would have been mutually independent, then the proportion of allocations that will be acceptable according to criterion (7.14) would be the product of the $J-1$ probabilities $\Pr\left(\chi^2_k \leq a(\cdot)\right)$. However, the actual acceptance probability will be smaller due to the mutual dependence of the χ^2 variables.

Suppose in the current example, we choose the Mahalanobis distance threshold for each effect as 2.50, making assignment vector $\mathbf{w}^*$ acceptable. If the χ^2 variables were independent, the total proportion of allocations that would be acceptable according to this criterion is

$$\left[\Pr\left(\chi^2_3 \leq 2.50\right)\right]^7 = [\text{pchisq}(2.5, 3)]^7 = 0.01,$$

where pchisq(q,k) is the R function that computes the probability that a χ^2 random variable with k degrees of freedom is smaller than q. However, as seen below, 10000 unrestricted draws from the distribution of $\mathbf{W}$ (generated by randomly permuting the observed assignment vector 10000 times) yields only 39 draws that are acceptable based on the threshold of 2.5 for each factorial effect, making the proportion of acceptable draws approximately 0.004.

```
---------------------------------------------------------------
## Proportion acceptable

iter=10000
set.seed(20)
acceptable = 0
for (l in 1:iter) {
  Wrep = sample(c(1,1,1,2,2,2,3,3,3,4,4,4,5,5,5,6,6,6,7,7,8,8))
  if (max(MDall(Xmatrix,Wrep))<=2.5) {
    acceptable = acceptable + 1
  } # end if
} # end ITER

> acceptable
[1] 39
> acceptable/iter
[1] 0.0039
---------------------------------------------------------------
```

Instead of choosing the same threshold for each factorial effect, we can choose a threshold of 2 for the main effects, 3 for the two-factor interactions and 4 for the three-factor interaction. This choice will make the assignment vector $\mathbf{w}^*$ unacceptable because the Mahalanobis distance $D_M(F_1)$ for the main effect of factor F_1 exceeds the threshold $a(F_1) = 2$. The proportion of allocations that will be acceptable according to this criterion is left as an exercise (problem 7.7).

7.3 Observed outcomes in 2^K experiments

Consider a 2^K factorial experiment with N_j units assigned to the jth treatment combination, $j = 1, \ldots, J$. Let $\mathbf{W} = (W_1, \ldots, W_N)^{\mathrm{T}}$ denote the assignment vector where W_i is an integer between 1 and J representing the treatment combination assigned to unit $i = 1, \ldots, N$. As in Chapters 5 and 6, denote the observed outcome corresponding to the ith experimental unit by $y_i, i = 1, \ldots, N$, so that as in (6.19),

$$y_i = Y_i(W_i) = \sum_{i=1}^{N} \mathbb{I}_{\{W_i=j\}} Y_i(j),$$

and $\mathbf{y}$ is the $N \times 1$ column vector of observed outcomes.

Suppose the observed treatment assignment vector $\mathbf{w}$ in the example of Table 7.14 in Exercise 7.5 based on a completely randomized assignment mechanism that assigns 3 units each to treatments $1-6$ and 2 units each to treatments 7 – 8 is

$$\mathbf{w}^* = (6, 4, 7, 6, 5, 1, 3, 4, 1, 8, 2, 5, 2, 8, 5, 7, 4, 1, 3, 3, 6, 2)^{\mathrm{T}}.$$

The corresponding *observed* outcomes are displayed in Table 7.5; the missing potential outcomes are represented by question marks.

TABLE 7.5
Observed outcomes in a hypothetical 2^3 experiment

i	W_i	Treatment combinations (in lexicographic order)							
		1	2	3	4	5	6	7	8
1	6	?	?	?	?	?	12.72	?	?
2	4	?	?	?	11.52	?	?	?	?
3	7	?	?	?	?	?	?	15.54	?
4	6	?	?	?	?	?	13.68	?	?
5	5	?	?	?	?	14.07	?	?	?
6	1	10.39	?	?	?	?	?	?	?
7	3	?	?	11.60	?	?	?	?	?
8	4	?	?	?	11.78	?	?	?	?
9	1	8.37	?	?	?	?	?	?	?
10	8	?	?	?	?	?	?	?	15.30
11	2	?	10.97	?	?	?	?	?	?
12	5	?	?	?	?	15.23	?	?	?
13	2	?	8.62	?	?	?	?	?	?
14	8	?	?	?	?	?	?	?	14.53
15	5	?	?	?	?	12.96	?	?	?
16	7	?	?	?	?	?	?	16.10	?
17	4	?	?	?	12.71	?	?	?	?
18	1	9.80	?	?	?	?	?	?	?
19	3	?	?	11.17	?	?	?	?	?
20	3	?	?	12.34	?	?	?	?	?
21	6	?	?	?	?	?	15.43	?	?
22	2	?	10.83	?	?	?	?	?	?
Average	$\overline{y}.(j)$	9.52	10.14	11.70	12.00	14.09	13.94	15.82	14.91
Variance	s_j^2	1.08	1.74	0.35	0.39	1.29	1.89	0.16	0.30

7.3.1 Estimation of factorial effects

As earlier, for $j = 1, \ldots, J$, let

$$\overline{y}.(j) = \frac{1}{N_j} \sum_{i:W_i=j} y_i, \tag{7.15}$$

denote the average of all observed outcomes for treatment combination j, where N_j is the number of observations for treatment j, and let

$$s_j^2 = \frac{1}{N_j - 1} \sum_{i:W_i=j} (y_i - \overline{y}.(j))^2, \tag{7.16}$$

denote the sample variance of all observed outcomes for treatment combination j. The last two rows of Table 7.5 shows the computed values of $\overline{y}.(j)$ and s_j^2 for $j = 1, \ldots, 8$ using the code shown below.

```
------------------------------------------------------------
# Calculation of the vectors of averages and variances

w = c(6,4,7,6,5,1,3,4,1,8,2,
      5,2,8,5,7,4,1,3,3,6,2)
y = c(12.72,11.52,15.54,13.68,14.07,
      10.39,11.60,11.78,8.37,15.30,
      10.97,15.23,8.62,14.53,12.96,
      16.10,12.71,9.80,11.17,12.34,
      15.43,10.83)
ybarvector = varyvector = NULL
for (l in 1:8) {
  ybarvector = c(ybarvector,mean(y[w==l]))
  varyvector = c(varyvector,var(y[w==l]))
}
> round(ybarvector,2)
[1]  9.52 10.14 11.70 12.00 14.09 13.94 15.82 14.91
> round(varyvector,2)
[1] 1.08 1.74 0.35 0.39 1.29 1.89 0.16 0.30
------------------------------------------------------------
```

Under a completely randomized assignment mechanism, $\overline{y}.(j)$ is an unbiased estimator (with respect to its randomization distribution) of $\overline{Y}(j)$, the average potential outcome for treatment combination j, defined in (7.11). Denoting the vector of averages $(\overline{y}.(1), \ldots, \overline{y}.(1))^{\mathrm{T}}$ by $\overline{\mathbf{y}}$, we can obtain an unbiased estimator of a factorial effect $\overline{\tau}(\cdot)$ by substituting $\overline{\mathbf{y}}$ in place of $\overline{\mathbf{Y}}$ in (7.12), that is,

$$\widehat{\overline{\tau}}(\cdot) = \frac{1}{2^{K-1}} \boldsymbol{\lambda}(\cdot)^{\mathrm{T}} \overline{\mathbf{y}}. \tag{7.17}$$

Computation of the main effect of F_1 using (7.19) in the current example is shown below:

```
------------------------------------------------------------
# Estimate of ME(F1)

> (t(L[,2]) %*% ybarvector)/4
         [,1]
[1,] 3.849583
------------------------------------------------------------
```

An efficient way of computing estimates of all factorial effects simultaneously using R would be to use a procedure similar to the one described in Section 6.3. As in Section 6.3, define a $J \times 1$ vector $\overline{\boldsymbol{\tau}}_{\mathrm{adj}}$ in which the first element is $\overline{\tau}(0)$, the average of all JN potential outcomes, and the remaining $J-1$ elements are equal to half of the $J-1$ factorial effects $\overline{\tau}(\cdot)$. Then, from (7.12), it can be shown that (Exercise 7.8)

$$\overline{\boldsymbol{\tau}}_{\text{adj}} = \frac{1}{2^K}\mathbf{L}^{\mathrm{T}}\overline{\mathbf{Y}}, \tag{7.18}$$

An unbiased estimator of the vector $\overline{\boldsymbol{\tau}}_{\text{adj}}$ can be obtained by substituting $\overline{\mathbf{y}}$ in place of $\overline{\mathbf{Y}}$ in (7.18), i.e.,

$$\widehat{\overline{\boldsymbol{\tau}}}_{\text{adj}} = \frac{1}{2^K}\mathbf{L}^{\mathrm{T}}\overline{\mathbf{y}}. \tag{7.19}$$

Recall that by definition, elements $2,\ldots,J$ of the vector $\overline{\boldsymbol{\tau}}_{\text{adj}}$ are half of the factorial effects $\overline{\tau}(\cdot)$. Thus, estimates of the factorial effects are the last $J-1$ elements of $\widehat{\overline{\boldsymbol{\tau}}}_{\text{adj}}$, multiplied by 2. In the current example, the factorial effects are estimated through the following calculations, which estimate the main effect of factor F_1 as the largest factorial effect, followed by the main effect of factor F_2. The other factorial effects appear to be small. The next task will be to perform inference to assess which of the effects are regarded as important.

```
--------------------------------------------------------
# Estimation of factorial effects

taubarhatadj = (t(L) %*% ybarvector)/8 # L is the model matrix
taubarhatobs = 2*taubarhatadj[2:8] # adjust factorial effects by 2

> round(taubarhatobs,2)
[1]  3.85  1.69 -0.03 -0.34 -0.49 -0.27 -0.11
--------------------------------------------------------
```

7.3.2 ANOVA decomposition in 2^K factorial experiments

The ANOVA decomposition in 2^K factorial experiments follows along similar lines as described in Section 6.4. Because a 2^K experiment can be visualized as one with $J = 2^K$ treatment arms, the observed data can be projected onto three orthogonal spaces - the overall average of dimension 1, the treatment space of dimension $J-1$ and residual space of dimension $N-J$ for balanced designs, as discussed and explained in Section 5.4.1.1.

As before, we can write each observed outcome as

$$y_i = \overline{y}.(\cdot) + \{\overline{y}.(W_i) - \overline{y}.(\cdot)\} + \{y_i - \overline{y}.(W_i)\},$$

exactly as in (5.15), leading to the decomposition (5.17), i.e.,

$$\text{SSTot} = \text{SSTre} + \text{SSRes},$$

where SSTot, SSTre and SSRes are defined in (5.18), (5.19) and (5.20) respectively. Following the argument in Section 5.4.1.1, SSTre and SSRes have degrees of freedom $J-1$ and $N-J$ respectively, and we can construct an ANOVA table like the one in Table 5.5. Thus, the statistic

$$F_{\text{Tre}} = \text{MSTre}/\text{MSRes} = \frac{\text{SSTre}/(J-1)}{\text{SSRes}/(N-J)}$$

measures the contributions of the treatment factors in the factorial experiment relative to the unexplained variation to the total variation in the observed data. It can easily be seen (problem 7.9) that the mean square residual MSRes can be expressed as

$$\text{MSRes} = \sum_{j=1}^{J} (N_j - 1)s_j^2,$$

where s_j^2 , defined in (7.16), denotes the variance of all observed outcomes for treatment combination $j = 1, \ldots, J$.

The ANOVA Table obtained from the observed data given in Table 7.5 is shown in Table 7.6.

TABLE 7.6
ANOVA Table obtained from the data in Table 7.5

Source	DF	SS	MS	F-ratio
Treatment	7	93.49	13.36	13.43
Residuals	14	13.92	0.99	
Total	21	104.41		

The calculations in R are shown below (note that we do not need the p-value in the output, which, in our case will follow from a randomization test).

```
---------------------------------------------------------
> W
 [1] 6 4 7 6 5 1 3 4 1 8 2 5 2 8 5 7 4 1 3 3 6 2
> anova(lm(y ~ factor(W)))
Analysis of Variance Table

Response: y
          Df Sum Sq Mean Sq F value    Pr(>F)
factor(W)  7 93.486 13.3551  13.429 3.081e-05 ***
Residuals 14 13.923  0.9945
---------------------------------------------------------
```

7.4 Fisherian inference for 2^K experiments

For a 2^K experiment, Fisher's sharp null hypothesis of no treatment effect on any unit can be written as:

$$H_0 : Y_i(1) = Y_i(2) = \ldots = Y_i(J), \quad i = 1, \ldots, N,$$

or equivalently as

$$\tau_i(F_1) = \tau_i(F_2) = \ldots = \tau_i(F_1 \ldots F_K), \quad i = 1, \ldots, N.$$

that is, all the J unit-level factorial effects are equal for each unit.

If the interest lies in assessing if there is a deviation from the sharp null hypothesis without a particular direction in mind, as in Chapters 5 and 6, a commonly used test statistic is the F_{Tre} = MSTre/MSRes ratio, associated with the decomposition of the total sum of squares of the observed outcomes and defined in Section 7.3.2. Using the square root of this F-statistic, denoted by $T_{\text{Tre}} = \sqrt{F_{\text{Tre}}}$, is another option. Under the sharp null hypothesis of no treatment effect on any unit, the missing potential outcomes for each unit can be imputed by the single observed outcome in that row. The randomization distribution of F_{Tre} can be obtained by taking hypothetical draws from the distribution of $\boldsymbol{W}$, and calculating the statistic from the ANOVA decomposition for each draw.

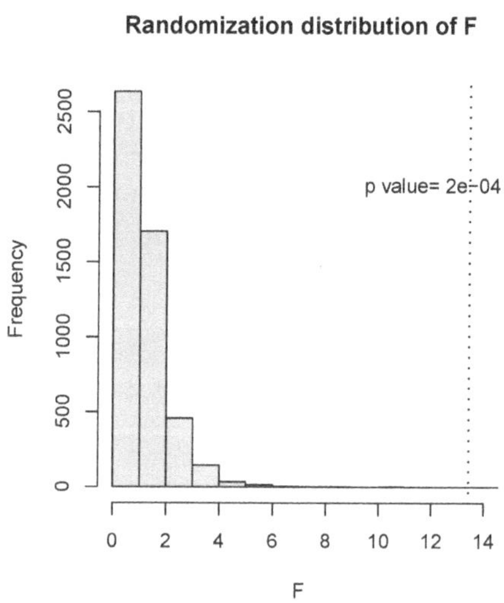

FIGURE 7.1
Randomization test for the data in Table 7.5 using the F_{Tre} statistic assuming a completely randomized assignment mechanism, where the dashed vertical line represents the observed value, and the number next to them the corresponding p-value

Figure 7.1 shows results of a randomization test for the data in Table 7.5 using the F_{Tre} statistic assuming a completely randomized assignment mechanism to approximate the randomization distribution. The small p-value suggests a significant effect of one or more of the factorial effects. As the R code below shows, 5000 draws were generated from the distribution of $\mathbf{W}$, which means for a completely randomized assignment mechanism, randomly permuting the observed assignment vector $\mathbf{w}^*$ 5000 times.

```
------------------------------------------------------------
# Randomization test using F-statistic

Fobs = anova(lm(y~factor(W)))$F[1]
iter = 5000
F_array = NULL
set.seed(200) # to reproduce exactly the same set of draws
for (i in 1:iter) {
  Wrep = sample(c(1,1,1,2,2,2,3,3,3,4,4,4,5,5,5,6,6,6,7,7,8,8))
   Frep = anova(lm(y ~ factor(Wrep)))$F[1]
   F_array = c(F_array, Frep)
} # end ITER
pvalF = sum(F_array >= Fobs)/iter
hist(F_array, nclass=20, main="Randomization distribution of F",
xlim=c(0,14),xlab="F")
abline(v=Fobs, lty=3, lwd=2)
text(12,2000,paste("p value=",pvalF))
------------------------------------------------------------
```

As in the case of 2^2 experiments described in Chapter 6, each of the $J-1$ estimated absolute factorial effects $|\widehat{\overline{\tau}}(\cdot)|$, can be used as a test statistic because each defines a specific type of deviation from the null hypothesis. Therefore, a Fisher randomization test can be obtained by comparing each estimated factorial effect against its randomization distribution under the sharp null hypothesis of no treatment effects. Due to the symmetry of the definition of all factorial effects, the reference distribution under the sharp null hypothesis of no effects is the same for all $J-1$ factorial effects. In this approach, one can adjust for multiple comparisons, for example, by using the ensemble-adjusted p-value described in Section 5.4.1.6 of Chapter 5, which simply modifies the p-value for each test to $\max\{Jp, 1\}$. Scaled versions of the estimated factorial effects $\widehat{\overline{\tau}}(\cdot)$, obtained by dividing them by their respective standard errors, i.e.,

$$T_{\text{scaled}}(\overline{\tau}(\cdot)) = \frac{\widehat{\overline{\tau}}(\cdot)}{\sqrt{\frac{\text{MSRes}}{J}\sum_{j=1}^{J}\frac{1}{N_j}}}. \tag{7.20}$$

can also be used.

Figure 7.2 shows results of a randomization test for the data in Table 7.5 using the absolute estimated factorial effects $|\widehat{\overline{\tau}}(\cdot)|$ (unscaled) as test statistics, assuming a completely randomized assignment mechanism. The p-values are shown in the R output below, from which it is clear that the main effect of F_1 is the strongest and significant by traditional standards, followed by the

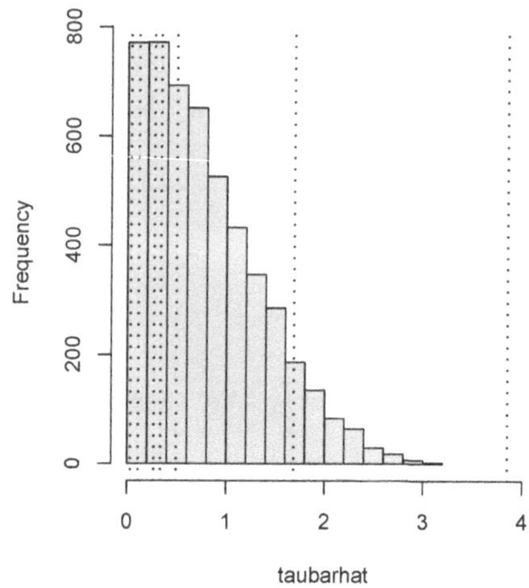

FIGURE 7.2
Randomization test for the data in Table 7.5 using the absolute values of estimated factorial effects $\widehat{\overline{\tau}}(\cdot)$ (unscaled) as test statistics, assuming a completely randomized assignment mechanism, where the dashed vertical lines represents the observed values

main effect of F_2. As shown in the R code, 5000 draws were generated from the distribution of $\mathbf{W}$ to approximate its randomization distribution under the sharp null hypothesis of no treatment effect.

```
-----------------------------------------------------------------------
# Randomization test using the unscaled estimated factorial effects
# as test statistic

iter=5000

taubarhat1_array = NULL
set.seed(10) # to reproduce exactly the same set of draws
for (i in 1:iter) {
Wrep = sample(c(1,1,1,2,2,2,3,3,3,4,4,4,5,5,5,6,6,6,7,7,8,8))
ybarvector_rep = NULL
for (l in 1:8) {
  ybarvector_rep = c(ybarvector_rep,mean(y[Wrep==l]))
} # end l
  taubarhatrep_adj = (t(L) %*% ybarvector_rep)/8
  taubarhatrep = 2*taubarhatrep_adj[2:8]
  taubarhat1_array = c(taubarhat1_array, taubarhatrep[1])
} # end ITER
```

```
hist(abs(taubarhat1_array),xlim=c(0,4),main=NULL, xlab="taubarhat")
abline(v=abs(taubarhatobs),lwd=2, lty=3)

pval = NULL
for (i in 1:7){
  pval = c(pval, sum(abs(taubarhat1_array) >=
  abs(taubarhatobs[i]))/iter)
}
> pval
[1] 0.0000 0.0850 0.9742 0.7402 0.6272 0.7924 0.9156
----------------------------------------------------------------------
```

Randomization test using the scaled versions $T_{\text{scaled}}(\overline{\tau}(\cdot))$ of the estimated factorial effects for the data in Table 7.5 under a completely randomized assignment mechanism is left as an exercise (Problem 7.9).

Another reasonable option is to use the absolute value of the estimated factorial effect $\widehat{\overline{\tau}}_{\max}(\cdot)$ that has the largest absolute value among all the J estimated effects. A scaled version of $\widehat{\overline{\tau}}_{\max}(\cdot)$ obtained by dividing it by its standard error, can also be used. See exercise 7.9.

If the assignment mechanism was not completely randomized, and instead based on a predefined acceptability criterion, like the one based on Mahalanobis distance, then the randomization distribution of any test statistic will be obtained from repeated draws from the distribution of $\mathbf{W}$ that satisfies the acceptability criterion. All other steps in the randomization test will remain the same. We demonstrate such a test, assuming that the observed outcomes in Table 7.5 were generated using an assignment mechanism based on the acceptance criterion (7.14) with $a(\cdot) = 4$ for all factorial effects, i.e., a randomized allocation will be acceptable only when $D_M(\cdot) \leq 4$ for all factorial effects. From Table 7.4, it is immediately obvious that the assignment vector $\mathbf{w}^*$ is acceptable according to this criterion. As shown in the R code below, 5000 *acceptable* draws were generated from the rerandomized distribution of $\mathbf{W}$ to approximate its randomization distribution under the sharp null hypothesis of no treatment effect. In order to obtain 5000 acceptable draws, a total of 61359 draws from the unrestricted randomization distribution of $\mathbf{W}$ had to be generated, making the proportion of acceptable draws 8.1%. Exercise 7.10 asks the readers to conduct a re-randomization test based on different acceptance thresholds for different factorial effects.

```
----------------------------------------------------------------------
# Re-randomization test using the unscaled estimated factorial effects
# as test statistic

iter=5000

taubarhat1_array = NULL
set.seed(10)
draws = acceptable = 0
```

```
while (acceptable < iter) {
  draws = draws + 1
  Wrep = sample(c(1,1,1,2,2,2,3,3,3,4,4,4,5,5,5,6,6,6,7,7,8,8))
  if (max(MDall(Xmatrix,Wrep))<=4) {
    ybarvector_rep = NULL
    for (l in 1:8) {
      ybarvector_rep = c(ybarvector_rep,mean(y[Wrep==l]))
    } # end l
    taubarhatrep_adj = (t(L) %*% ybarvector_rep)/8
    taubarhatrep = 2*taubarhatrep_adj[2:8]
    taubarhat1_array = c(taubarhat1_array, taubarhatrep[1])
    acceptable = acceptable + 1
  } # end if
} # end ITER

hist(abs(taubarhat1_array),xlim=c(0,4),main=NULL, xlab="taubarhat")
abline(v=abs(taubarhatobs),lwd=2, lty=3)

pval = NULL
for (i in 1:7){
  pval = c(pval, sum(abs(taubarhat1_array) >=
abs(taubarhatobs[i]))/acceptable)
}
> draws
[1] 61359
> acceptable
[1] 5000
> pval
[1] 0.0000 0.0786 0.9738 0.7292 0.6132 0.7796 0.9074
------------------------------------------------------------------
```

7.5 Neymanian inference from completely randomized 2^K experiments

Neymanian inference in a 2^K factorial design is a straightforward extension of Section 6.6. Under a completely randomized assignment mechanism, it is straightforward to see that the vector of estimated factorial effects $\widehat{\overline{\boldsymbol{\tau}}}_{\text{adj}}$ is an unbiased estimator of the vector $\overline{\boldsymbol{\tau}}_{\text{adj}}$ (see exercise 7.11). Hence each estimated factorial effect $\widehat{\overline{\tau}}(\cdot)$ is an unbiased estimator of $\overline{\tau}(\cdot)$. As shown in Dasgupta et al. (2015) [20] and Lu (2016) [54], the variance of $\widehat{\overline{\tau}}(\cdot)$ is

$$\text{Var}(\widehat{\overline{\tau}}(\cdot)) = \frac{1}{J}\sum_{j=1}^{J}\frac{S_j^2}{N_j} - \frac{S^2(\overline{\tau}(\cdot))}{N}, \tag{7.21}$$

where S_j^2, defined in (7.13), denotes the variance (with divisor $N-1$) of all potential outcomes corresponding to the treatment combination $j = 1, \ldots, J$, and $S^2(\overline{\tau}(\cdot))$ denotes the variance of unit level factorial effects $\tau_i(\cdot)$, $i = 1, \ldots, N$, i.e.,

$$S^2(\overline{\tau}(\cdot)) = \frac{1}{N-1} \sum_{i=1}^{N} \left(\tau_i(\cdot) - \overline{\tau}(\cdot)\right)^2, \qquad (7.22)$$

that are not estimable from observed data (see part (iv) of problem 7.4).

As in Section 6.6, $\text{Var}(\widehat{\overline{\tau}}(\cdot))$ can be conservatively estimated by assuming strict additivity, as in Definition 5.1 of Chapter 5, which means that the unit-level factorial effects $\tau_1(\cdot), \ldots, \tau_N(\cdot)$ are all equal for each factorial effect, making S_j^2 equal for all $j = 1, \ldots, J$ and $S^2(\overline{\tau}(\cdot)) = 0$. Consequently, from (7.21), for each factorial effect,

$$\text{Var}(\widehat{\overline{\tau}}(\cdot)) = \frac{S^2}{J} \sum_{j=1}^{J} \frac{1}{N_j},$$

which can be unbiasedly estimated as

$$\widehat{V}_{\text{Ney}} = \frac{\text{MSRes}}{J} \sum_{j=1}^{J} \frac{1}{N_j}, \qquad (7.23)$$

thereby justifiying the denominator of the statistic $T_{\text{scaled}}(\overline{\tau}(\cdot))$ defined in (7.20).

To test the Neymanian null hypothesis that a particular population-level factorial effect is zero, that is,

$$H_{0\text{N}} : \overline{\tau}(\cdot) = 0,$$

we can use the statistic $T_{\text{scaled}}(\overline{\tau}(\cdot))$ defined in (7.20). The large sample distribution of this statistic can be approximated by a standard normal distribution, whence a two-sided p-value can be obtained as $Pr\left(|Z| \geq |T_{\text{scaled}}^{\text{obs}}(\cdot)|\right)$, where $T_{\text{scaled}}^{\text{obs}}(\cdot)$ is the observed value of the test statistic. The smaller the p-value, the stronger is the evidence against $H_{0\text{N}}$.

A large sample $100(1-\alpha)\%$ confidence interval for $\overline{\tau}(\cdot)$ is:

$$\widehat{\overline{\tau}}(\cdot) \pm z_{\alpha/2}\sqrt{\widehat{V}_{\text{Ney}}}. \qquad (7.24)$$

The two-sided p-values for the Neymanian hypothesis of zero average effect for the seven factorial effects in the hypothetical 2^3 experiment with observed values shown in Table 7.5 are summarized in Table 7.7. The R code used to obtain these p-values is shown below. Computation of 95% large-sample confidence intervals for these effects is left as an exercise.

TABLE 7.7
Two-sided p-values for the Neymanian hypothesis of zero average effect

Index	Factorial effect	p-value (two-sided)
1	Main effect of F_1	0.0000
2	Main effect of F_2	0.0024
3	Main effect of F_3	0.9570
4	Interaction effect of F_1, F_2	0.5408
5	Interaction effect of F_1, F_3	0.3781
6	Interaction effect of F_2, F_3	0.6272
7	Interaction effect of F_1, F_2, F_3	0.8431

```
-----------------------------------------------------
# Neymanian p-values (two-sided)

MSRes = anova(lm(y ~ factor(W)))$'Mean Sq'[2] # MSRes
Vhat_Ney = MSRes/8*(6/3+2/2)
SE = sqrt(Vhat_Ney) # compute standard error
T_scaled = taubarhatobs/SE
pval = numeric(7)
for (i in 1:7){
pval[i] = 2* min(pnorm(Tobs[i]),1-pnorm(Tobs[i]))
}
>round(data.frame(taubarhatobs,SE=rep(SE,7),Tobs,pval),4)

  taubarhatobs      SE    Tobs   pval
1         3.85 0.5559  6.9260 0.0000
2         1.69 0.5559  3.0402 0.0024
3        -0.03 0.5559 -0.0540 0.9570
4        -0.34 0.5559 -0.6116 0.5408
5        -0.49 0.5559 -0.8815 0.3781
6        -0.27 0.5559 -0.4857 0.6272
7        -0.11 0.5559 -0.1979 0.8431
-----------------------------------------------------
```

We conclude this section with two important remarks. First, the Neymanian inference described here is valid only under a completely randomized treatment assignment. For general assignment mechanisms for other predefined acceptance rules for allocations, Neymanian analysis is not straightforward. For randomized block designs, inference can be performed along the lines of Section 3.5.1. However, for assignment mechanisms in which acceptance rules are defined in terms of Mahalanobis distance, Neymanian inference is complex and beyond the scope of this book. Interested readers may refer to Li et al. (2020) [52].

Second, the estimated factorial effects $\widehat{\overline{\tau}}(F_1), \widehat{\overline{\tau}}(F_2), \dots, \widehat{\overline{\tau}}(F_1 F_2 \dots F_K)$ are correlated with each other. An expression for the $J \times J$ covariance matrix of

$\widehat{\overline{\boldsymbol{\tau}}}_{\text{adj}}$ was derived by Lu (2016) [54], which can be used to perform joint large-sample inference on multiple factorial effects using their asymptotic multivariate normal distribution. We omit these details here.

7.6 Factorial experiments with binary outcomes

As discussed and emphasized througout this book, the methodology described so far in this chapter is applicable to experiments with any type of response variable Y - continuous, or binary, which is a great feature of randomization based inference. However, as discussed in Section 2.8, there are some interesting aspects about experiments with binary outcome variables that are worthwhile noting. For such outcome variables, we can denote the potential outcome $Y_i(j) = 1$ as a "success" (e.g., survival) and $Y_i(j) = 0$ as a "failure" (e.g., death). Then the average potential outcome $\overline{Y}(j)$ represent the proportion P_j of successful potential outcomes for treatment combination $j = 1, \dots, J$, and are estimands (not causal, though) of interest.

Similarly, the variance S_j^2 of the potential outcomes corresponding to treatment combination $j = 1, \dots, J$, with divisor $N-1$, can be expressed as

$$S_j^2 = \frac{N}{N-1} P_j(1-P_j). \tag{7.25}$$

Consequently, the vector $\overline{\mathbf{Y}}$ of average potential outcomes equals the $J \times 1$ vector of proportions $\mathbf{P} = (P_1, \dots, P_J)^{\mathrm{T}}$, so that the vector of adjusted factorial effects defined in (7.18) can be expressed as

$$\overline{\boldsymbol{\tau}}_{\text{adj}} = \frac{1}{J}\mathbf{L}^{\mathrm{T}}\overline{\mathbf{P}}, \tag{7.26}$$

Using similar notation as in Table 2.5, the observed binary outcomes $y_1, \dots, y_N$ can be summarized in Table 7.8, where

$$n_{j1} = \sum_{i:W_i=j} y_i$$

denotes the number of observed successes for treatment combination j, so that the estimated proportion of successes for treatment j is $p_j = n_{j1}/N_j$.

Replacing the vector $\mathbf{P}$ in (7.26) by its observed counterpart $\mathbf{p} = (p_1, \dots, p_J)^{\mathrm{T}}$, an unbiased estimator of $\overline{\boldsymbol{\tau}}_{\text{adj}}$ is readily obtained as:

$$\widehat{\overline{\boldsymbol{\tau}}}_{\text{adj}} = \frac{1}{J}\mathbf{L}^{\mathrm{T}}\overline{\mathbf{p}}. \tag{7.27}$$

To test the sharp null hypothesis of no treatment effect on any unit, we can use the F-statistic, estimators of individual factorial effects $\widehat{\overline{\tau}}(\cdot)$ or their scaled

TABLE 7.8
2×2 table of observed binary outcomes

Treatment	$y = 1$	$y = 0$	Total
1	n_{11}	n_{10}	N_1
2	n_{21}	n_{20}	N_2
$\vdots$	$\vdots$	$\vdots$	$\vdots$
J	n_{J1}	n_{J0}	N_J
Total	n_{+1}	n_{+0}	N

versions $T_{\text{scaled}}(\overline{\tau}(\cdot))$, defined in (7.20). With a little algebra, the following can be shown (exercise 7.13):

$$\text{SSTre} = \sum_{j=1}^{J} n_{j1} p_j - \frac{n_{+1}^2}{N}, \tag{7.28}$$

$$\text{SSRes} = \sum_{j=1}^{J} N_j p_j (1 - p_j), \tag{7.29}$$

from which the F statistic can be easily derived.

Substituting (7.29) into (7.23), the Neymanian estimator of the variance of the estimated factorial effect $\widehat{\tau}(\cdot)$ defined in (7.23) can be obtained as

$$\widehat{V}_{\text{Ney}} = \frac{\sum_{j=1}^{J} N_j p_j (1 - p_j)}{J(N - J)} \sum_{j=1}^{J} \frac{1}{N_j}, \tag{7.30}$$

the square root of which yields the denominator of $T_{\text{scaled}}(\overline{\tau}(\cdot))$.

7.6.1 Example: audit experiment to identify effect of racial discrimination

Social scientists have increasingly turned to"audit" experiments (or "correspondence" experiments) to investigate discrimination in markets for jobs, loans, housing, and other opportunities (Baert 2018 [4]). The interpretation, analysis and implications of these experiments is discussed in Gaddis (2018) [37]. In a typical audit experiment, researchers assign "signals" (the treatment) to applications (experimental units) at random and compare success rates across treatment conditions. The signal provides information about the latent traits of the applicant, such as their race, gender, or other demographic information. The applications are kept absolutely identical except for this "signal", which reveals the experimental treatment condition. For example, in a study by Bertrand and Mullainathan (2004) [6] on discrimination in the labor market, the name in the resume was "Emily" versus "Lakisha", or some names presumably suggestive of race, thereby manipulating the *presentation* of race.

A key issue that often raises questions about inference drawn from audit experiments as above is confounded signaling of latent traits involved. Such confounding arises because names that are predictive of race are also predictive of other factors, importantly socio-economic status. An example is given in Fryer (2004) [35]. Names may also convey information about the likely age, nationality, religion, and other factors.

As pointed out in Pashley et al. (2025) [67], a design-based approach to address the challenge arising from confounded signaling is to use factorial experiments, applying multiple interventions or level combinations of factors simultaneously on a population of experimental units using a randomized assignment mechanism. As discussed in this Chapter, in addition to letting the experimenter identify the "main" or marginal effect of each factor, factorial designs also permit assessment of interactions among factors. Such interactions can be meaningful in the context of discrimination experiments. For example, how different is the effect of race signaling on access to lawyers for people belonging to low income versus high income groups?

Libgober (2020) [53] reported a study to determine whether black individuals and white individuals face similar barriers in acquiring legal services from lawyers. The experimental units were lawyers in private practice, selected from the California bar directory. Requests for representation from potential clients who claimed to be wrongfully accused of driving under the influence (DUI) were sent to lawyers by email. In these emails, race was signaled through the use of a name highly correlated with a specific race - black (denoted by 0) or white (denoted by 1). Two other treatment factors - income (low, denoted by 0 and high, denoted by 1) and gender (female, denoted by 0 and male, denoted by 1 - were also signaled in the emails. Thus, each lawyer received an email that incorporated one of the eight possible level combinations of the three treatments 000, 001, 010, 011, 100, 101, 110 and 111. In our notation, these eight treatment combinations are indexed $1, \ldots, 8$ respectively. Ninety six lawyers were selected at random from the California bar directory for the study, and each of the eight treatment combinations were assigned to 12 lawyers using a completely randomized assignment mechanism. The primary outcome of interest was binary, taking value 1 if the lawyer replied and 0 otherwise. The 96 lawyers were thus randomly divided into 8 groups of 12 each, and all lawyers in each group j received treatment combination $j = 1, \ldots, 8$. The experiment generated a 96×1 vector of observed binary outcomes $\mathbf{y}$, part of which is shown in Table 7.9.

A randomization-based analysis of the experimental data was done by Pashley et al. (2025) and will be described next. The observed outcomes from the are summarized in Table 7.10 following the notation introduced earlier in this section and in Table 7.8. A quick visualization of the observed factorial effects can be provided by plots of the main-effects and two-factor interactions plot shown in Pashley et al. (2025) [67].

TABLE 7.9
Partial data from the Libgober (2020) study

Unit (i)	Tr (j)	Race	Gender	Income	y
1	1	0	0	0	0
2	1	0	0	0	1
3	1	0	0	0	0
4	1	0	0	0	0
5	1	0	0	0	0
6	1	0	0	0	0
7	1	0	0	0	0
8	1	0	0	0	0
9	1	0	0	0	0
10	1	0	0	0	0
11	1	0	0	0	1
12	1	0	0	0	0
13	2	0	0	1	0
14	2	0	0	1	0
⋮	⋮	⋮	⋮	⋮	
24	2	0	0	1	1
25	3	0	1	0	0
⋮	⋮	⋮	⋮	⋮	
96	8	1	1	1	1

TABLE 7.10
Summary of observed outcomes from the Libgober (2020) experiment

	Treatment combination (j)							
Measure	1	2	3	4	5	6	7	8
N_j	12	12	12	12	12	12	12	12
n_{j1}	2	2	2	3	5	2	5	6
n_{j0}	10	10	10	9	7	10	7	6
p_j	0.167	0.167	0.167	0.250	0.417	0.167	0.417	0.500
s_j^2	0.1515	0.1515	0.1515	0.2045	0.2652	0.1515	0.2652	0.2727

Substituting the observed vector of proportions $\mathbf{p}$ in (7.27), we obtain $\widehat{\overline{\boldsymbol{\tau}}}_{\text{adj}}$ and consequently the estimated factorial effects, which are shown in the second column of Table 7.11. The scaled version of each effect, i.e., statistic $T_{\text{scaled}}(\overline{\tau}(\cdot))$ computed by dividing the estimates by the standard error (S.E.) $\sqrt{\widehat{V}_{\text{Ney}}}$ where $\widehat{V}_{\text{Ney}}$ is given by (7.30) as shown in the third column.

The steps involved in the calculation of the entries in Table 7.11 are as follows: first we create the model matrix $\mathbf{L}$ and the vector $\mathbf{p}$ from the second-last row of Table 7.10.

TABLE 7.11
Estimates of factorial effects (unscaled and scaled)

Factorial effect	Estimate	S.E.	$T_{\text{scaled}}(\overline{\tau}(\cdot))$
Race	0.19	0.092	2.05
Gender	0.10	0.092	1.13
Income	-0.02	0.092	-0.23
Race × Gender	0.06	0.092	0.68
Race × Income	-0.06	0.092	-0.68
Gender × Income	0.10	0.092	1.13
Race × Gender × Income	0.06	0.092	0.68

```
---------------------------------------------------------------
# Model matrix
library(AlgDesign)
D = gen.factorial(2,3) ## ME contrast columns in reverse order

D[, c(1:3)] = D[,c(3:1)] ## Reverse the order of the columns
colnames(D) = c("F1","F2","F3")
L = model.matrix(~ . + .*. + .*.*.,D)
> L
  (Intercept) F1 F2 F3 F1:F2 F1:F3 F2:F3 F1:F2:F3
1           1 -1 -1 -1     1     1     1       -1
2           1 -1 -1  1     1    -1    -1        1
3           1 -1  1 -1    -1     1    -1        1
4           1 -1  1  1    -1    -1     1       -1
5           1  1 -1 -1    -1    -1     1        1
6           1  1 -1  1    -1     1    -1       -1
7           1  1  1 -1     1    -1    -1       -1
8           1  1  1  1     1     1     1        1

# vector p of estimated proportions
p <- c(0.167,0.167,0.167,0.250,0.417,0.167,0.417,0.500)
---------------------------------------------------------------
```

Next the estimated factorial effects are computed using (7.27), the MSRes from (7.29), and finally the S.E. as the square root of $\widehat{V}_{\text{Ney}}$ given by (7.30).

```
---------------------------------------------------------------
# Estimator of taubarhat_adj
 taubarhatadj = (t(L) %*% p)/8
 taubarhatobs = 2*taubarhatadj[2:8] # adjust factorial effects by 2
> taubarhatobs
[1]  0.1875  0.1040 -0.0210  0.0625 -0.0625  0.1040  0.0625
 ### MSRes, SE and Tscaled #######
 MSRes = (12/88)*sum(p*(1-p))
 Vhat_Ney = MSRes/24
 SE = sqrt(Vhat_Ney)
```

```
> round(SE,4)
[1] 0.0917
 T_scaled_obs = taubarhatobs/SE
> round(T_scaled_obs,3)
 2.045  1.134 -0.229  0.682 -0.682  1.134  0.682
-----------------------------------------------------------------
```

Results of a randomization test to assess the sharp null hypothesis of no treatment effect on any unit using $T_{\text{scaled}}(\overline{\tau}(F_1))$, i.e., scaled version of the estimated effect of race, is shown in Figure 7.3. Note that testing the null hypothesis against a one-sided alternative for race makes more sense than a two-sided one, because because if an average discriminatory effect exists, it is expected to be uni-directional (e.g., emails signaling sender as black being less likely to receive a response than emails signaling sender as white). The p-value of 0.03 suggests a probable effect of such discrimination.

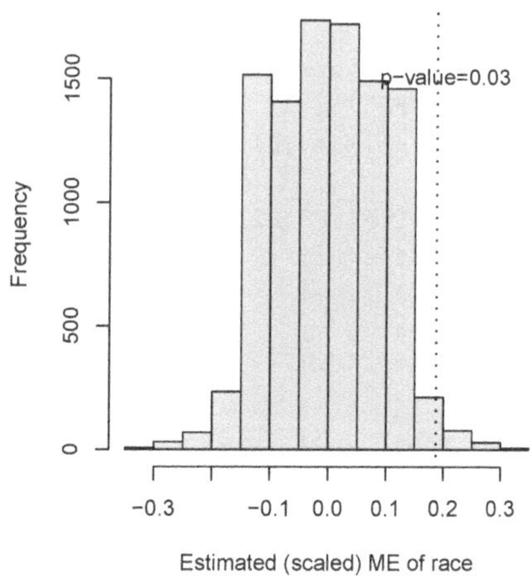

FIGURE 7.3
Randomization test for the Libgober(2020) experiment using the estimated main effect (scaled) of race, where the dashed vertical line represents the observed value, and the number next to it the corresponding p-value

Fisherian inference with other test statistics and Neymanian inference, for the effects of race, gender and income on discrimination from the Libgober (2020) experiment, are left as exercises (problem 7.14).

7.7 Optimal allocation of units to treatments

So far we have assumed that whatever assignment mechanism we adopt (completely randomized or a pre-defined acceptance rule), the number of units N_j

to be allocated to treatment $j = 1, \ldots, J$ is fixed, such that $\sum_{j=1}^{j} N_j = N$. A common practice is to use balanced designs if N is a multiple of J. However, as discussed in Section 2.7, balanced designs need not always be optimal. Ravichandran et al. (2024) [72] proposed exact theoretical solutions for optimal allocation in 2^K factorial experiments under complete and block randomization under different optimality criteria. The idea generalizes the approach outlined in Section 2.7 in which the optimal allocation was obtained by minimizing the estimable part of the variance of the estimated average treatment effect.

As mentioned in the concluding remark of Section 7.5, an expression for the $J \times J$ covariance matrix of $\widehat{\boldsymbol{\tau}}_{\text{adj}}$ was derived by Lu (2016) [54], which consisted of an estimable part, a matrix $\mathbf{V}$, and a non-estimable part. In the spirit of classical optimal designs Atkinson (2007) [3], Ravichandran et al. (2024) [72] defined a design optimality criterion as a scalar function of the matrix $\mathbf{V}$. Specifically, the D-optimality criterion, which aims to minimize the determinant of the covariance matrix, or the A-optimality criterion, which aims to minimize the trace (sum of the diagonal elements) of the covariance matrix, or the E-optimality criterion which aims to minimize the maximum eigenvalue of the covariance matrix, were considered. An A-optimal allocation minimizes the average variance of estimators of the factorial effects $\widehat{\overline{\tau}}(\cdot)$. Readers interested in the interpretations of the D-optimality and E-optimality criteria are referred to Ravichandran et al. (2024) [72] for details.

Denoting by π_j as the proportion of units to be assigned to treatment $j = 1, \ldots, J$, so that $\sum_{j=1}^{J} \pi_j = 1$, the A-optimal, D-optimal and E-optimal designs for a completely randomized 2^K factorial experiment are shown in Table 7.12, where S_j^2, defined in (7.13), denotes the variance of all N potential outcomes for treatment combination j.

TABLE 7.12
Optimal allocations of units in completely randomized 2^K experiments

Criterion	Optimal proportion p_j
A-optimal	$p_j = S_j / (\sum_j S_j)$
D-optimal	$p_j = 1/J$
E-optimal	$p_j = S_j^2 / (\sum_j S_j^2)$

As noted in Section 7.5, under the assumption of strict additivity (Definition 5.1 of Chapter 5), the unit-level factorial effects $\tau_1(\cdot), \ldots, \tau_N(\cdot)$ are all equal, making S_j^2 equal for all $j = 1, \ldots, J$. Consequently the A-optimal, D-optimal and E-optimal allocations in Table 7.12 are identical, each being a balanced allocation with $p_j = 1/J$ for $j = 1, \ldots, J$.

In order to implement the results of Table 7.12 for A-optimal and E-optimal allocations, researchers need to "guess" the variances S_j^2, $j = 1, \ldots, J$ and substitute them into the expressions for p_j. A common practice conducts pilot studies to obtain preliminary estimates of the S_j^2's.

Assuming that the observed data of Table 7.5 were from a pilot experiment, the A-optimal, D-optimal and E-optimal allocations for a similar experiment conducted on a similar population on units are shown in Table 7.13.

TABLE 7.13
Optimal allocations of units in a future 2^3 experiment using Table 7.5 as pilot data

Criterion	Optimal proportion $\mathbf{p} = (p_1, \ldots, p_8)$
A-optimal	(0.15, 0.19, 0.08, 0.09, 0.16, 0.20, 0.06, 0.08)
D-optimal	(0.125, 0.125, 0.125, 0.125, 0.125, 0.125, 0.125, 0.125)
E-optimal	(0.15, 0.24, 0.05, 0.05, 0.18, 0.26, 0.02, 0.04)

7.8 Problems

7.1 Write the sixteen treatment combinations in a 2^4 factorial experiment in lexicographic order (i.e., starting with 0000, 0001 and ending with 1110, 1111).

7.2 Complete Table 7.2 by entering the missing contrast coefficients for the two-factor interactions between F_1 and F_3, and between F_2 and F_3.

7.3 In a 2^4 experiment,

(i) How many two-factor, three-factor and four-factor interactions can be defined?

(ii) Create a Table similar to Table 7.2 for the contrast coefficients.

(iii) Express (mathematically) the unit-level three-factor interaction $\tau_i(F_1F_2F_4)$ as a contrast of the unit-level potential outcomes $Y_i(1), \ldots, Y_i(16)$ in forms (7.9) and (7.12).

(iv) Express (mathematically) the population-level three-factor interaction $\overline{\tau}(F_1F_2F_4)$ as (a) a contrast of the average potential outcomes $\overline{Y}(1), \ldots, \overline{Y}(16)$ and (b) average over unit-level three-level interactions $\tau_i(F_1F_2F_4)$.

7.4 The potential outcomes for $N = 22$ units in a hypothetical 2^3 factorial experiment are shown in Table 7.14. Note that these are not observed data. The purpose of this exercise is to solidify the understanding of the unit-level and population-level causal estimands, i.e., the factorial effects.

TABLE 7.14
Potential outcomes in a hypothetical 2^3 experiment

Unit	Treatment combinations (in lexicographic order)							
	1	2	3	4	5	6	7	8
1	10.02	9.33	11.03	12.74	12.82	12.72	14.00	15.13
2	9.82	7.88	11.97	11.52	13.93	15.13	15.32	13.68
3	8.63	8.73	12.23	12.56	13.58	13.54	15.54	16.61
4	9.40	9.63	11.70	10.75	13.81	13.68	15.02	17.15
5	10.29	9.31	11.32	12.38	14.07	14.92	16.50	14.80
6	10.39	9.13	12.66	10.57	15.16	14.08	16.73	14.42
7	8.79	9.90	11.60	10.95	14.59	15.04	16.67	16.65
8	9.64	9.75	11.67	11.78	12.58	14.74	16.95	15.45
9	8.37	8.15	13.37	10.51	12.39	15.26	14.32	16.52
10	9.74	9.92	14.14	13.17	14.89	14.95	14.79	15.30
11	11.10	10.97	12.51	10.52	14.15	13.52	14.04	15.56
12	10.76	10.18	12.79	11.57	15.23	14.20	17.47	15.32
13	9.76	8.62	11.10	10.95	13.24	13.97	16.37	16.96
14	10.99	8.56	12.53	13.52	14.42	12.80	17.07	14.53
15	10.74	10.36	11.35	12.59	12.96	14.62	16.53	16.18
16	10.09	8.24	12.29	11.78	14.71	13.09	16.10	14.56
17	9.05	9.68	10.76	12.71	13.37	14.25	17.34	14.86
18	9.80	9.35	11.54	12.72	14.56	12.94	16.09	15.59
19	10.93	11.09	11.17	12.44	14.66	13.64	15.61	16.14
20	10.48	9.24	12.34	12.16	12.34	12.79	15.75	17.06
21	9.40	9.17	13.07	12.66	15.03	15.43	17.16	15.43
22	7.81	10.83	13.22	14.22	15.13	14.63	15.14	17.28
Average								
Variance								

(i) Calculate the average and variance of the potential outcomes for each treatment combination and enter the calculated values in the last two rows of Table 7.14.

(ii) Calculate the unit-level factorial effects $\tau_i(F_1)$ (main effect of factor F_1), $\tau_i(F_2F_3)$, (2 factor interaction between factors F_2 and F_3) and $\tau_i(F_1F_2F_3)$ (3 factor interaction between the three factors) for $i = 1, \ldots, 22$.

(iii) Calculate the population-level counterparts of the unit-level factorial effects in (ii), i.e., $\overline{\tau}(F_1)$, $\overline{\tau}(F_2F_3)$, and $\overline{\tau}(F_1F_2F_3)$.

(iv) Find the variances $S^2(\overline{\tau}(F_2))$ and $S^2(\overline{\tau}(F_2F_3))$ of unit-level factorial effects $\tau_i(2)$ and $\tau_i(6)$ respectively, using formula (7.22).

(v) Generate the model matrix $\mathbf{L}$ and use it to obtain the vector of factorial effects $\overline{\boldsymbol{\tau}}_{\text{adj}}$ and $\overline{\boldsymbol{\tau}}$.

7.5 In the hypothetical 2^3 factorial experiment described in Section 7.2 with $N = 22$ units, how many possible allocations exist if a completely randomized allocation is employed when the experimenter decides to allocate

3 units each to treatment combinations $1, \ldots, 6$ and 2 units each to treatment combinations 7 and 8? Obtain an expression for $\Pr(\mathbf{W} = \mathbf{w})$ for a specific realization $\mathbf{w} = (w_1, \ldots, w_N)^{\mathrm{T}}$.

7.6 In the hypothetical 2^3 factorial experiment described in Section 7.2 with $N = 22$ units and the 22×3 matrix $\mathbf{X}$ of covariates shown in Table 7.3, compute the Mahalanobis distances between the vectors of covariate averages for each of the seven factorial effects and report your results in a table similar to Table 7.4 for the following two vectors:

$$(5, 1, 8, 5, 4, 4, 6, 3, 2, 4, 7, 2, 7, 2, 6, 3, 5, 3, 6, 8, 1, 1)$$
$$(8, 8, 7, 7, 6, 6, 6, 5, 5, 5, 4, 4, 4, 3, 3, 3, 2, 2, 2, 1, 1, 1)$$

7.7 In a 2^3 factorial experiment, if the following acceptance criterion is adopted for randomized allocations:

$$\begin{aligned} \max\{D_M(F_1), D_M(F_2), D_M(F_3)\} &\le 2, \\ \max\{D_M(F_1F_2), D_M(F_1F_3), D_M(F_2F_3)\} &\le 3, \\ D_M(F_1F_2F_3) &\le 4 \end{aligned}$$

determine the approximate proportion of allocations that will be acceptable among all possible randomized allocations using simulations. Will the two allocations in problem 7.6 be accepted under this criterion?

7.8 Prove (7.18) for a 2^3 factorial experiment.

7.9 Show that the mean square residual, MSRes, in a 2^K factorial experiment can be expressed as

$$\text{MSRes} = \sum_{j=1}^{J} (N_j - 1) s_j^2,$$

where

$$s_j^2 = \frac{1}{N_j - 1} \sum_{i:W_i = j} (y_i - \overline{y}.(j))^2$$

denotes the sample variance of all observed outcomes for treatment combination $j = 1, \ldots, J$.

7.10 Conduct randomization tests for the data in Table 7.5 to assess the sharp null hypothesis of no treatment effect on any unit, *assuming a completely randomized assignment mechanism*, using the following test statistics:

(i) The $T_{\text{Tre}} = \sqrt{F_{\text{Tre}}}$ statistic.

(ii) Absolute values of the scaled estimated factorial effects $T_{\text{scaled}}(\overline{\tau}(\cdot))$ for all seven factorial effects.

(iii) Absolute value of the estimated factorial effect $\widehat{\overline{\tau}}_{\max}$ that has the largest absolute value among all seven estimated effects

7.11 Conduct randomization tests for the data in Table 7.5 to assess the sharp null hypothesis of no treatment effect on any unit, *assuming that the assignment mechanism was based on the acceptance criterion defined in Problem 7.7*, using the following test statistics:

(i) The F_{Tre} statistic.

(ii) Absolute values of the scaled estimated factorial effects $T_{\text{scaled}}(\overline{\tau}(\cdot))$ for all seven factorial effects.

(iii) Absolute value of the estimated factorial effect $\widehat{\overline{\tau}}_{\max}$ that has the largest absolute value among all seven estimated effects

7.12 Consider a completely randomized 2^K factorial experiment with N_j units assigned to treatment $j = 1, \ldots, J$. Let $W_i \in \{1, \ldots, J\}$ denote the treatment assignment variable for unit i.

(i) Define the binary random variable $W_i(j)$ taking value 1 if unit i receives treatment j, and zero otherwise. Express W_i in terms of the $W_i(j)$'s.

(ii) Calculate $\mathbb{E}[W_i]$.

(iii) Show that $\mathbb{E}[\overline{y}.(j)] = \overline{Y}(j)$.

(iv) Show that the vector of estimated factorial effects, $\widehat{\overline{\boldsymbol{\tau}}}_{\text{adj}}$, is an unbiased estimator of the vector $\overline{\boldsymbol{\tau}}_{\text{adj}}$.

7.13 From the observed data of Table 7.5, construct 95% large sample confidence intervals for the seven factorial effects $\overline{\tau}(\ell)$ for $\ell = 1, \ldots, 7$.

7.14 For a 2^K experiment with binary outcomes, show that the sample variance $s_j^2 = \frac{1}{N_j - 1} \sum_{i:W=j} (y_i - \overline{y}.(j))^2$ for treatment $j = 1, \ldots, J$ can be expressed as

$$s_j^2 = \frac{N_j}{N_j - 1} p_j (1 - p_j).$$

Hence prove (7.28) and (7.29).

7.15 Data from the experiment reported by Libgober (2020) [53] and described in Section 7.6.1 is available in the file Libgober_Pilot_data.csv. Perform the following analysis with the data:

(i) Conduct a randomization test to assess the effect of no treatment effect on any unit using the F_{Tre} statistic.

(ii) Obtain 95% large sample confidence intervals for each factorial effect using the Neymanian approach outlined in Section 7.5 and 7.6.

7.16 Considering the experiment in Libgober(2020) [53] and discussed in Section 7.6.1 to be a pilot experiment, determine the A-optimal, D-optimal and E-optimal allocations for a future 2^3 experiment done with the same objective with a similar population of lawyers as units.

8

Design and analysis of factorial experiments with constraints

8.1 Introduction

So far we have assumed that it is possible to conduct 2^K factorial experiments presuming two conditions: (i) *all* the 2^K treatment combinations can be administered without any practical or resource constraints, and (ii) the levels of each treatment factor can be assigned to the experimental units using the same assignment mechanism for each unit. In reality, both of these assumptions can be violated. In the previous chapter, we described an experiment planned by the New York Department of Education (NYDE) to study five incentive programs to be introduced to high schools, which needed performance improvement. Such a 2^5 factorial experiment has 32 treatment combinations. In 1.4.4, we discussed how the above two assumptions can be violated.

Assumption (i) can be violated due to resource constraints that do not permit conducting an experiment with all 32 treatment combinations. Running an experiment with a fraction, say half or one-quarter, of these 32 treatment combinations could be considerably easier from administrative and resource perspectives, e.g., requiring fewer experimental units. However, eliminating some treatment combinations from the experiment would result in a reduction of the number of potential outcomes for each unit, and consequently a reduced potential outcomes matrix with fewer columns than the matrix associated with a full factorial design. Such a reduction will not permit the experimenter to *define* all the $2^5 - 1 = 31$ factorial effects the way they were defined in a full factorial experiment. Such experiments, called *fractional factorial designs*, are discussed in Section 8.2. A formal exposition of 2-level fractional factorial designs in the context of causal inference can be found in Pashley and Bind (2023) [68].

To see how assumption (ii) can be violated, consider a simplified version of this school experiment with two factors: a mid-year quality review by a team of experts, and a bonus scheme for teachers. Each factor has two levels denoted by 1 (application) and 0 (non-application). A completely randomized assignment of the 40 schools to the four treatment combinations $00, 01, 10, 11$ is likely to disperse the schools assigned to level 1 of the first factor (i.e., schools to undergo review) all over the state. However, such a design may be awkward

DOI: 10.1201/9781003048794-8

to implement because of prohibitive travel cost. A more practical alternative could be to partition these 40 schools by geographic proximity into four groups of ten *whole-plots*, where the terminology comes from the agricultural origins of the design. Two of these whole-plots would then be randomly assigned to level 0 (no review) and the other two to level 1 (reviewed) of the first factor. The teacher bonus scheme can then be applied to half of the schools chosen randomly within each whole plot. A design with such a double (or, in general, multiple) randomization scheme is called a *split-plot design* and is discussed in Section 8.3.

8.2 Two-level fractional factorial designs

As discussed in Chapter 7, the causal estimands in a 2^5 factorial experiment constitute 31 factorial effects, each with one degree of freedom. The decomposition of these degrees of freedom is shown in Table 8.1.

TABLE 8.1
Break-up of degrees of freedom

Effect	Main effects	2-FI	3-FI	4-FI	5-FI
DF	5	10	10	5	1

A 2^5 full factorial experiment permits an experimenter to define all of these thirty-one factorial effects that comprise five main effects (e.g., $\overline{\tau}(F_1)$), ten two-factor interactions (2-FI) (e.g., $\overline{\tau}(F_1F_2)$), five three-factor interactions (3-FI) (e.g., $\overline{\tau}(F_1F_2F_3)$), five four-factor interactions (4-FI) (e.g., $\overline{\tau}(F_1F_2F_3F_4)$), and one five-factor interactions (5-FI) (e.g., $\overline{\tau}(F_1F_2F_3F_4F_5)$) as contrasts of the averages of the potential outcomes $\overline{Y}(1), \ldots, \overline{Y}(32)$ for the 32 treatment combinations. In many real-life experiments, higher order interactions can be assumed to be small, are hard to interpret, and thus less important than main effects and lower order interactions. Note that sixteen out of the thiry one degrees of freedom shown in the second row of Table 8.1 are associated with three, four and five-factor interactions. If the experimenters can assume away these interactions (by virtue of prior experimental results or available domain knowledge), then they are left with fifteen factorial effects (five main effects and ten 2-FIs) of interest that can be uniquely defined from only sixteen potential outcomes for which it suffices to choose half of the thirty two combinations of a 2^5 full factorial experiment. Such an experiment is called a *fractional factorial experiment*. In this chapter we consider $1/2^P$ fractions of 2^K experiments. Because such experiments involve 2^{K-P} treatment combinations, they are called 2^{K-P} fractional factorial designs.

There is a price that the experimenter has to pay by eliminating some treatment combinations from the experiment. This price is the *aliasing* of factorial effects, which means certain factorial effects, as defined in a full factorial design, are indistinguishable from certain others as a consequence of fractionating the experiment. In order to unbiasedly estimate an effect unambiguously, all effects that are aliased with that particular effect need to be zero.

The methodology for the construction of 2^{K-P} fractional factorial designs that will be discussed in this chapter is based on the following two principles:

1. *Effect sparsity principle* (Box and Meyer 1986 [11]), which assumes that only a small subset of all the $2^K - 1$ factorial effects, which can be defined from a 2^K full factorial experiment, are non negligible.

2. *Effect hierarchy principle* (Wu and Hamada 2009 [93]), which assumes that lower order effects are more important than higher order effects. That is, main effects are most important, followed by 2-FIs, and so on.

Whereas it is not necessary that the above two principles will hold in every experimental setting, they do hold in many realistic situations and therefore are often useful to consider when designing fractionated experiments.

8.2.1 Understanding fractional factorial with a 2^{3-1} experiment

Suppose an experiment involves three factors F_1, F_2 and F_3 each with two levels 0 and 1, but the experimenter has resources to conduct only four out of the eight *runs* (treatment combinations) listed in Table 8.2.

TABLE 8.2
A 2^3 full factorial design

Lexicographic index (j)	F_1	F_2	F_3	$Y_i(j)$
1	0	0	0	$Y_i(1)$
2	0	0	1	$Y_i(2)$
3	0	1	0	$Y_i(3)$
4	0	1	1	$Y_i(4)$
5	1	0	0	$Y_i(5)$
6	1	0	1	$Y_i(6)$
7	1	1	0	$Y_i(7)$
8	1	1	1	$Y_i(8)$

The observable potential outcomes for each unit from such an experiment will correspond only to the four treatment combinations in the chosen fraction. Suppose we choose the first four combinations. In that case the four observable potential outcomes $Y_i(1)$, $Y_i(2)$, $Y_i(3)$ and $Y_i(4)$ associated with the for any unit will not permit us to define the main effect of factor F_1 as any contrast of these potential outcomes, because all the treatment combinations will have F_1

TABLE 8.3
A 2^{3-1} fractional factorial design

Lexicographic index(j)	F_1	F_2	F_3	$Y_i(j)$
2	0	0	1	$Y_i(2)$
3	0	1	0	$Y_i(3)$
5	1	0	0	$Y_i(5)$
8	1	1	1	$Y_i(8)$

at level zero. Similarly choosing combinations 1,2,5,6 will not be a good idea because such choices will not permit definition of the main effect of factor F_2. However, consider the fraction in Table 8.3, obtained by choosing combinations 2,3,5,8 from Table 8.2.

Following (7.5), the unit-level main effects (for unit i) of F_1, F_2 and F_3 are defined as:

$$\tau_i(F_1) = \frac{1}{2}\boldsymbol{\lambda}(F_1)^{\mathrm{T}}\boldsymbol{Y}_i, \quad \tau_i(F_2) = \frac{1}{2}\boldsymbol{\lambda}(F_2)^{\mathrm{T}}\boldsymbol{Y}_i, \quad \tau_i(F_3) = \frac{1}{2}\boldsymbol{\lambda}(F_3)^{\mathrm{T}}\boldsymbol{Y}_i, \tag{8.1}$$

where $\boldsymbol{Y}_i = (Y_i(2), Y_i(3), Y_i(5), Y_i(8))^{\mathrm{T}}$ denotes the four-component column-vector consisting of the potential outcomes in the last column of Table 8.3, and the contrast vectors $\boldsymbol{\lambda}(F_1)$, $\boldsymbol{\lambda}(F_2)$ and $\boldsymbol{\lambda}(F_3)$ are shown in Table 8.4, obtained by removing four rows of Table 7.2.

TABLE 8.4
Contrast vectors $\boldsymbol{\lambda}(F_j)$'s for the seven factorial effects in a 2^{3-1} experiment (without divisor 2)

Tr.	Level	Contrast vectors for							
	comb.	Avg	Main effects			2-FI			3-FI
(j)	($z_1z_2z_3$)	$\boldsymbol{\lambda}(0)$	$\boldsymbol{\lambda}(F_1)$	$\boldsymbol{\lambda}(F_2)$	$\boldsymbol{\lambda}(F_3)$	$\boldsymbol{\lambda}(F_1F_2)$	$\boldsymbol{\lambda}(F_1F_3)$	$\boldsymbol{\lambda}(F_2F_3)$	$\boldsymbol{\lambda}(F_1F_2F_3)$
2	001	1	−1	−1	+1	+1			+1
3	010	1	−1	+1	−1	−1			+1
5	100	1	+1	−1	−1	−1			+1
8	111	1	+1	+1	+1	+1			+1

To define the three 2-FIs, we need to define the appropriate $\boldsymbol{\lambda}(\cdot)$ vector by multiplying the relevant $\boldsymbol{\lambda}(\cdot)$ vectors in Table 8.4 element-wise. Recalling from Section 7.1 the notation $\circ$, representing the operation of multiplying two vectors of equal dimension element-wise, we have:

$$\boldsymbol{\lambda}(F_1F_2) = \boldsymbol{\lambda}(F_1) \circ \boldsymbol{\lambda}(F_2) = \begin{pmatrix} +1 \\ -1 \\ -1 \\ +1 \end{pmatrix} = \boldsymbol{\lambda}(F_3). \tag{8.2}$$

This means the unit-level factorial effect for the 2-FI F_1F_2, defined as

$\tau_i(F_1F_2) = \boldsymbol{\lambda}(F_1F_2)^{\mathrm{T}}\mathbf{Y}_i$, equals

$$\tau_i(F_3) = \boldsymbol{\lambda}(F_3)^{\mathrm{T}}\mathbf{Y}_i = \frac{Y_i(2) + Y_i(8) - Y_i(3) - Y_i(5)}{2},$$

the main effect of factor F_3. Consequently the 2^{3-1} design chosen in Table 8.3 is *unable to distinguish* between the 2-FI between factors F_1 and F_2, and the main effect of factor F_3. In such a case, the 2-FI F_1F_2 is said to be *aliased* with the main effect of factor F_3.

According to this definition, after completion of Exercise 8.1, it can be seen that: (i) The 2-FI F_1F_3 is aliased with the main effect of factor F_2, and (ii) the 2-FI F_2F_3 is aliased with the main effect of factor F_1. Representing each factorial effect by the index ℓ representing the subscripts of the factors involved, (e.g., main effect of factor F_1 by 1, main effect of factor F_2 by 2, 2-FI F_1F_2 by 12, and so on) the aliasing relationships for the 2^{3-1} design in Table 8.3 can be represented as:

$$3 \triangleq 12, \quad 2 \triangleq 13, \quad 1 \triangleq 23, \tag{8.3}$$

where the symbol $\triangleq$ represents "delta-equivalence", meaning equivalence after a certain (delta) operation.

What happens to the 3-FI 123? As seen in the last column of Table 8.4, the element-wise multiplication of vectors $\boldsymbol{\lambda}(F_1)$, $\boldsymbol{\lambda}(F_2)$ and $\boldsymbol{\lambda}(F_3)$ yields

$$\boldsymbol{\lambda}(F_1F_2F_3) = \boldsymbol{\lambda}(F_1) \circ \boldsymbol{\lambda}(F_2) \circ \boldsymbol{\lambda}(F_3) = \begin{pmatrix} +1 \\ +1 \\ +1 \\ +1 \end{pmatrix} = \boldsymbol{\lambda}(0).$$

Therefore the 3-FI 123 is indistinguishable from the mean, and we can add one more aliasing relationship to the three already listed in (8.3) to obtain a complete set of aliasing relationships:

$$I \triangleq 123, \quad 3 \triangleq 12, \quad 2 \triangleq 13, \quad 1 \triangleq 23, \tag{8.4}$$

where I represents the mean of potential outcomes of unit i.

Why did we use the notation I for the mean effect? Note that each element of the set of symbols $\mathcal{S} = \{I, 1, 2, 3, 12, 13, 23, 123\}$ actually represents a column vector of positive ones and negative ones. This set is an *algebraic group* under the operation $\circ$ with identity element I. This means, the elements of this set satisfy the following properties:

1. *Closure*: For any two of the elements a and b belonging to the group, $a \circ b$ also belongs to the group. For example, by operating 12 with 13 (i.e., $\boldsymbol{\lambda}(F_1F_2) \circ \boldsymbol{\lambda}(F_1F_3)$), we get 23, which also belongs to the group.

2. *Associativity*: $a \circ (b \circ c) = (a \circ b) \circ c$ for all a, b belonging to the group.

3. *Identity*: There exists an element I in the group such that for any element a in the group, $a \circ I = a$. The mean effect, represented by I, is the identity element.

4. *Invertibility*: For any element a in the group, there exists an element b also in the group satisfying the inverse property $a \circ b = b \circ a = I$. Note that every element is an inverse of itself, because for any element a in the group, $a \circ a = I$.

The design has four aliasing relationships, each of which involves subsets of two elements of $\mathcal{S}$. The subset $\{I, 123\}$ is also a *subgroup* (a subgroup is a subset of a group that is itself a group). Thus, the aliasing relationship that involves I, the identity element of $\mathcal{S}$, is called the *defining contrast subgroup* in factorial design literature. All the other aliasing groups can be obtained from the defining contrast subgroup by using a simple multiplicative operation with modulo 2 powers. For example, multiplying both sides of $I = 123$ with 1 yields: $1 = 1^2 23 = 23$.

Construction of the 2^{3-1} design in Table 8.3 is based on the general theory of fractional factorial designs by Mukerjee and Wu (2006) [64]. An easier construction method from Wu and Hamada, (2009) [93], that is applicable only to 2^{K-P} designs, is given below:

Procedure 8.1 ***Steps in constructing a*** 2^{3-1} ***design:***

Step 1: Create the columns for factors F_1, F_2 *as the four treatment combinations* $(z_1 z_2)$ *of a* 2^2 *full factorial design in lexicographic order, i.e.* $(00, 01, 10, 11)$. *Note that the first two columns of Table 8.3 are created this way.*

F_1	F_2
0	*0*
0	*1*
1	*0*
1	*1*

Step 2: Change the label 0 in each column to -1, *thereby yielding the contrasts for the main effects* F_1 *and* F_2.

F_1	F_2
–1	*–1*
–1	*+1*
+1	*–1*
+1	*+1*

Step 3: Create the column for F_3 by multiplying the two columns for F_1 and F_2 (with levels labeled –1 and 1) element-wise, thereby yielding the contrast for the main effect of F_3. This immediately explains why this choice of fraction leads to $\boldsymbol{\lambda}(F_{12}) = \boldsymbol{\lambda}(F_3)$ – it follows from the construction of the design.

F_1	F_2	F_3
–1	–1	+1
–1	+1	–1
+1	–1	–1
+1	+1	+1

Step 4: Re-convert the label –1 to 0 in each column, yielding Table 8.3.

The relation $3 \triangleq 12$ is thus called the *defining relation* of this 2^{3-1} fraction. If we would have started with a 2^2 full factorial design with factors F_2 and F_3 and generated the column of factor F_1 by multiplying the contrast columns for these factors element-wise, the defining relation of the design would have been $1 \triangleq 23$. However, such a design would have led to exactly the same aliasing structure as (8.4). We now formally define two important terms that will be used throughout this chapter:

Definition 8.1 Defining relation for a 2^{K-P} design: *A 2^{K-P} design is constructed by first constructing a 2^{K-P} full factorial design consisting of $K-P$ columns, and then by generating the remaining Kth columns from subsets of the previous columns using the $\circ$ operation. This generating relationship that connects the new columns to the original $K-P$ columns are called the defining relations or generators of the design.*

Definition 8.2 Defining contrast subgroup for a 2^{K-P} design: *The group of aliased effects that contains the identity element (i.e., the mean effect) is called the defining contrast subgroup.*

We are now in a position to compare a 2^3 full factorial design and the 2^{3-1} fraction chosen in Table 8.3. A single replicate of the 2^3 full factorial design has eight degrees of freedom, each defining an effect. These eight effects include the mean effect I and seven factorial effects $1, 2, 3, 12, 13, 23, 123$. A half fraction of a 2^3 design with defining relation $3 \triangleq 12$ has four degrees of freedom, and splits the eight effects of a 2^3 full factorial design into four aliasing groups as shown in (8.4).

Noting that steps 1 and 2 of Procedure 8.1 essentially involve swapping of factor labels between 0 and –1, for practical purposes, for the remainder of this section, we will denote the two levels of each factor by –1 and +1. This labeling makes the actual design matrix (as in Table 8.3) *exactly* identical to the matrix of contrasts for the main effects (as in Table 8.4).

8.2.2 Criteria for choosing "good" fractions: The maximum resolution criterion

Let us return to the problem of designing the school experiment that has five factors, each at two levels. Designing a balanced full factorial experiment will involve assigning 224 schools to the thirty-two treatment combinations, with seven schools assigned to each treatment combination. However, thirty two combinations may be deemed too large, and if the experimenters care less about 3-FIs, 4-FIs and 5-FIs, then the required number of estimable degrees of freedom is only fifteen (five main effects and ten 2 FIs). Because sixteen treatment combinations are adequate to define fifteen factorial effects, a 2^{5-1} experiment should suffice. Constructing a 2^{5-1} design means constructing a 16×5 matrix, in which each column has eight zeros and eight ones. To construct such a design, we start with a 2^4 full factorial design for factors $F_1, \ldots, F_4$, and construct the fifth column F_5 from these columns.

Consider two different 2^{5-1} designs D_1 and D_2 that are generated by the two defining relations:

$$D_1 : 5 \triangleq 123, \quad D_2 : 5 \triangleq 1234.$$

This means in the first design (D_1), the contrast column $\boldsymbol{\lambda}(F_5)$ for the main effect of F_5 is obtained as $\boldsymbol{\lambda}(F_5) = \boldsymbol{\lambda}(F_1) \circ \lambda(F_2) \circ \boldsymbol{\lambda}(F_3)$, whereas in the second design (D_2), the column $\boldsymbol{\lambda}(F_5)$ is obtained as $\boldsymbol{\lambda}(F_5) = \boldsymbol{\lambda}(F_1) \circ \boldsymbol{\lambda}(F_2) \circ \boldsymbol{\lambda}(F_3) \circ \boldsymbol{\lambda}(F_4)$. The two designs are shown in Table 8.5

TABLE 8.5
Two 2^{5-1} designs with different defining relations

Index	Design D_1: 5=123					Design D_2: 5=1234				
(l)	F_1	F_2	F_3	F_4	F_5	F_1	F_2	F_3	F_4	F_5
1	−1	−1	−1	−1	−1	−1	−1	−1	−1	+1
2	−1	−1	−1	+1	−1	−1	−1	−1	+1	−1
3	−1	−1	+1	−1	+1	−1	−1	+1	−1	−1
4	−1	−1	+1	+1	+1	−1	−1	+1	+1	+1
5	−1	+1	−1	−1	+1	−1	+1	−1	−1	−1
6	−1	+1	−1	+1	+1	−1	+1	−1	+1	+1
7	−1	+1	+1	−1	−1	−1	+1	+1	−1	+1
8	−1	+1	+1	+1	−1	−1	+1	+1	+1	−1
9	+1	−1	−1	−1	+1	+1	−1	−1	−1	−1
10	+1	−1	−1	+1	+1	+1	−1	−1	+1	+1
11	+1	−1	+1	−1	−1	+1	−1	+1	−1	+1
12	+1	−1	+1	+1	−1	+1	−1	+1	+1	−1
13	+1	+1	−1	−1	−1	+1	+1	−1	−1	+1
14	+1	+1	−1	+1	−1	+1	+1	−1	+1	−1
15	+1	+1	+1	−1	+1	+1	+1	+1	−1	−1
16	+1	+1	+1	+1	+1	+1	+1	+1	+1	+1

Which design is better – D_1 or D_2? To answer this question, let us list all the aliasing relationships of the two designs. Note that, in each design, one

TABLE 8.6
Aliasing relations for designs D_1 and D_2

Design D_1	Design D_2
$I \triangleq 1235$	$I \triangleq 12345$
$1 \triangleq 235$	$1 \triangleq 2345$
$2 \triangleq 135$	$2 \triangleq 1345$
$3 \triangleq 125$	$3 \triangleq 1245$
$4 \triangleq 12345$	$4 \triangleq 1235$
$5 \triangleq 123$	$5 \triangleq 1234$
$12 \triangleq 35$	$12 \triangleq 345$
$13 \triangleq 25$	$13 \triangleq 245$
$14 \triangleq 2345$	$14 \triangleq 235$
$15 \triangleq 23$	$15 \triangleq 245$
$24 \triangleq 1345$	$23 \triangleq 145$
$34 \triangleq 1245$	$24 \triangleq 135$
$45 \triangleq 1234$	$25 \triangleq 134$
$124 \triangleq 345$	$34 \triangleq 125$
$134 \triangleq 245$	$35 \triangleq 124$
$145 \triangleq 234$	$45 \triangleq 123.$

factorial effect will be aliased with the mean. Because there are thirty more effects to account for, and we have fifteen degrees of freedom available, there should be fifteen aliasing groups, each containing two factorial effects. The aliasing relations for designs D_1 and D_2 are shown in Table 8.6.

If all the five factors are assumed to be equally important, then design D_2 is clearly better because all the 2-FIs are aliased with 3-FIs that are presumed unimportant. On the contrary, in design D_1, each 2-FI that does not involve factor F_4 is aliased with another 2-FI. Such an aliasing structure is probably not desirable, because if one such 2-FI is found important from the analysis, it will not be clear whether the importance can be attributed to that 2-FI or to the one it is aliased with, leading to an ambiguity. It is therefore intuitively clear that if a first design has its main effects and 2-FIs aliased with interactions of higher order than a second design, then the former design is better. To formalize such comparisons, we need some formal criteria. One criterion is called the *Resolution* of a design and was first proposed by Box (1961) [9].

Definition 8.3 *The resolution of a design is the length of the shortest word in its defining contrast subgroup.*

In this definition, "word" refers to an element of the group of treatment effects, e.g., $1, 13, 345$. The "length of a word" means the number of digits in the word, e.g., 345 had length three, 12345 had length five, and so on. The phrase "shortest element" may not be quite clear, because so far we have only considered 2^{K-1} designs, which have only one defining relation and thus the only element besides the identity element in the defining contrast subgroup. However, the definition will make more sense once we introduce 2^{K-P} experiments with $P \geq 1$.

Because design D_1 has defining contrast subgroup $I \triangleq 1235$ and design D_2 has defining contrast subgroup $I \triangleq 12345$, their resolutions are IV and V respectively, and by virtue of having a higher resolution, design D_2 is considered better than design D_1. Are there situations where design D_1 may be more desirable than design D_2? The answer is left as an exercise (Problem 8.2).

Let us now discuss the construction of a 2^{K-P} design with $P \geq 2$ using an example. Consider a 2^{5-2} design, which is a one-fourth fraction of a 2^5 full factorial design and consists of eight rows and five columns. This design can be constructed by first creating a 2^3 full factorial design, and then creating two additional columns for factors F_4 and F_5 from the original three columns. Construction of the design will thus require two defining relations. Let $4 \triangleq 12$ and $5 \triangleq 23$ be these defining relations. Then the defining contrast subgroup of the design is $I \triangleq 124 \triangleq 235$. However, note that, because this relationship means $\boldsymbol{\lambda}(F_1F_2F_4) = \boldsymbol{\lambda}(F_2F_3F_5) = \boldsymbol{\lambda}(0)$, a vector of ones, it follows that $\boldsymbol{\lambda}(F_1F_2F_4) \circ \boldsymbol{\lambda}(F_2F_3F_5)$ should also equal $\boldsymbol{\lambda}(0)$. But by definition of factorial effects, we have that $\boldsymbol{\lambda}(F_1F_2F_4) \circ \boldsymbol{\lambda}(F_2F_3F_5) = \boldsymbol{\lambda}(F_1F_3F_4F_5)$, the contrast vector representing the 4-FI among factors F_1, F_3, F_4 and F_5. Consequently, the defining contrast subgroup is

$$I \triangleq 124 \triangleq 235 \triangleq 1345. \tag{8.5}$$

The defining contrast subgroup has three "words" – 124, 235 and 1345 – and the length of the shortest word (either 124 or 235) is three. Therefore by Definition 8.3, the resolution of the design is III. The aliasing relations of this design can be derived from (8.5) by multiplying throughout by $1, 2, \ldots$, and using the modulo 2 system for the power:

$$\begin{aligned}
I &\triangleq 24 \triangleq 1235 \triangleq 345, \\
2 &\triangleq 14 \triangleq 35 \triangleq 12345, \\
3 &\triangleq 1234 \triangleq 25 \triangleq 145, \\
4 &\triangleq 12 \triangleq 2345 \triangleq 135, \\
5 &\triangleq 1245 \triangleq 23 \triangleq 134, \\
13 &\triangleq 234 \triangleq 125 \triangleq 45, \\
15 &\triangleq 245 \triangleq 123 \triangleq 34.
\end{aligned}$$

Clearly, because a 2^{5-2} design has eight degrees of freedom that need to account for 32 effects including the mean, we should have 8 aliasing relationships (sets) with four effects in each set. One of these sets is the defining contrast subgroup.

We can now summarize some important properties of 2^{K-P} designs. The readers are referred to Wu and Hamada (2009) [93] for more details:

- A 2^{K-P} design has 2^{K-P} treatment combinations and K factors, and is therefore a $2^{K-P} \times K$ matrix of negative ones and ones. Each row represents a treatment combination and each column a factor.

- The fraction is constructed from a 2^{K-P} full factorial design with factors $F_1, \ldots, F_{K-P}$ by generating the remaining P columns using P independent defining relations.

- The defining contrast subgroup has $2^P - 1$ words in addition to the identity element I.

- There are 2^{K-P} aliasing relationships (including the defining contrast subgroup). Each relationship involves 2^P effects aliased with each other.

8.2.3 Clear effects, word-length pattern and the minimum aberration criterion

In the previous subsection we have seen how the resolution of a fractional factorial design can be used to help discriminate between good and bad designs. However, some additional criteria help to further discriminate between designs with the same resolution. All the new terminology and definitions that will be introduced in this subsection are obtained from Wu and Hamada (2009) [93].

Higher resolution is a desirable criterion under the effect hierarchy principle because a design with a higher resolution has fewer main effects and 2-FIs aliased with other 2-FIs than a design with lower resolution. Therefore the set of main effects and 2-FIs that are not aliased with other 2-FIs comprise an important characteristic of a 2^{K-P} design. Such effects are called *clear effects*. The greater the number of clear main effects and clear 2-FIs in a fractional factorial design, the more desirable is the design. It is easy to see that in a resolution IV design, all main effects are clear. Similarly, in a resolution V design, all main effects and 2-FIs are clear. A clear effect can be defined (and estimated) unambiguously if all interactions of order three and above can be assumed away.

To understand how we can discriminate between two designs with the same resolution, consider two 2^{7-2} designs D_3 and D_4 with the following defining relations:

$$D_3 : 6 \triangleq 1234, 7 \triangleq 1235, \quad D_4 : 6 \triangleq 123, 7 \triangleq 145.$$

Their defining contrast subgroups are:

$$D_3 : I \triangleq 12346 \triangleq 12357 \triangleq 4567, \quad D_4 : I \triangleq 1236 \triangleq 1457 \triangleq 234567.$$

Both the designs have the same resolution (IV), but different aliasing structures. Whereas design D_3 has one word of length four in the defining contrast subgroup, design D_4 has two words of length four in its defining contrast subgroup. Therefore it appears that although all the main effects in both designs are clear, design D_4 has fewer clear 2-FIs than design D_3, and therefore is less desirable under the effect hierarchy principle. To define a criterion that formalizes this comparison, we first define the word-length pattern of a 2^{K-P} design:

Definition 8.4 *Let A_r be the number of words of length r in the defining contrast subgroup of a design. The vector $\boldsymbol{W} = (A_3, A_4, \ldots,)$ is called the word length pattern of the design.*

Note that for any reasonable 2^{K-P} fractional factorial design, A_2 must be equal to zero, otherwise at least one pair of main effects will be aliased with each other. Also, note that the resolution of a design can be defined as the smallest integer r for which $A_r > 0$. We now define the *aberration criterion* that discriminates between designs on the basis of their word-length patterns.

Definition 8.5 *For any two 2^{K-P} designs, D_1 and D_2, let r be the smallest integer such that $A_r(D_1) \neq A_r(D_2)$. Then D_1 is said to have less aberration than D_2 if $A_r(D_1) < A_r(D_2)$. If there is no design with less aberration than D_1, then the design D_1 has minimum aberration.*

The minimum aberration criterion is justified by the effect hierarchy principle. Further discussion of the minimum aberration criterion can be found in Wu and Hamada (2009) [93], and readers interested in the underlying mathematical theory are referred to Mukerjee and Wu (2006) [64]. To see how this criterion can discriminate between two designs, note that the word length patterns of the two designs D_3 and D_4 defined earlier are $\{0, 1, 2, 0, 0, \ldots\}$ and $\{0, 2, 0, 1, 0, \ldots\}$ respectively. Because $A_4(D_3)(= 1) < A_4(D_4)(= 2)$, by the minimum aberration criterion, design D_3 is better than design D_4. Tables of minimum aberration 2^{K-P} designs for different values of K and P appear in Wu and Hamada (2009) [93], along with the defining relations and a complete listing of the clear main effects and 2-FIs.

8.2.4 Factorial effects, assignment mechanism and inference

The population-level estimands in a 2^{K-P} fractional factorial designs are analogous to the estimands defined in Section 7.1 and comprise the population-level averages $\overline{\tau}(\cdot)$ of the unit-level factorial effects $\overline{\tau}_i(\cdot)$. For example, in a 2^{3-1} fractional factorial, the causal estimands would comprise seven population factorial effects denoted by $\overline{\tau}(F_1)$, $\overline{\tau}(F_2)$, $\overline{\tau}(F_3)$, $\overline{\tau}(F_1F_2)$, $\overline{\tau}(F_1F_3)$, $\overline{\tau}(F_2F_3)$, $\overline{\tau}(F_1F_2F_3)$. However, only three of them can be estimated apart from the mean $\overline{\tau}(0)$ depending on the defining relation and aliasing structure of the design. Specifically, we can estimate one effect (typically the lowest-order effect) from each aliasing group under the assumption that the other effects in that group are negligible. For example, for a 2^{3-1} fractional factorial design with defining relation $3 \triangleq 12$ and aliasing relation given by (8.4), we can estimate the three main effects $\overline{\tau}(F_1)$, $\overline{\tau}(F_2)$ and $\overline{\tau}(F_3)$ under the assumption that the 2-FIs and

3-FIs are negligible. Averaging (8.1) over $i = 1, \ldots, N$, these three factorial effects are defined as:

$$\overline{\tau}(F_1) = \frac{1}{2}\boldsymbol{\lambda}(F_1)^{\mathrm{T}}\overline{\boldsymbol{Y}}, \quad \overline{\tau}(F_2) = \frac{1}{2}\boldsymbol{\lambda}(F_2)^{\mathrm{T}}\overline{\boldsymbol{Y}}, \quad \overline{\tau}(F_3) = \frac{1}{2}\boldsymbol{\lambda}(F_3)^{\mathrm{T}}\overline{\boldsymbol{Y}}, \qquad (8.6)$$

where $\overline{\boldsymbol{Y}} = (\overline{Y}(2), \overline{Y}(3), \overline{Y}(5), \overline{Y}(8))^{\mathrm{T}}$ denotes the four-component column-vector comprising of the average potential outcomes for treatment combinations 2,3,5 and 8, and the contrast vectors $\boldsymbol{\lambda}(F_1)$, $\boldsymbol{\lambda}(F_2)$ and $\boldsymbol{\lambda}(F_3)$ are shown in Table 8.4.

As mentioned earlier, in a 2^{K-P} experiment, there are 2^{K-P} aliasing relationships or groups (including the defining contrast subgroup). Each relationship involves 2^P effects aliased with each other. One factorial effect from each group can be estimated. For example, for the 2^{5-1} design D_2 with defining relation $5 = 1234$ defined in Table 8.5, the aliasing relationships in Table 8.6 show that in addition to the mean, this design will permit estimation of 15 factorial effects comprising five main effects and ten 2-FIs, under the assumption that the higher order interactions are negligible.

Once the fraction and the factorial effects to be estimated are defined, the assignment of the treatment combinations in the chosen fractions are assigned to the experimental units using any randomized treatment assignment mechanism as described in this book, and specifically, in Section 7.2. Fisherian and Neymanian inference can then be performed using the methods described in Sections 7.4 and 7.5 respectively.

Consider the hypothetical 2^3 experiment in the example of Table 7.14 (Exercise 7.5). Suppose, due to resource constraints, the experimenter cannot assign all the eight treatment combinations to the experimental units, and decides to choose a half fraction with defining relation $3 \triangleq 12$. As seen from Table 8.3, this means only treatment combinations 2, 3, 5 and 8 will be assigned to the treatments, and no unit will receive any of the treatment combinations 1, 4, 6 and 7. In a completely randomized experiment where it is pre-decided that 6,6,5 and 5 units will be assigned to treatment combinations 2, 3, 5 and 8 respectively, the assignment vector $\mathbf{w} = (w_1, \ldots w_{22})$ is obtained as:

```
------------------------------------------------------------
# Assignment
set.seed(50)
w = sample(c(rep(2,6),rep(3,6),rep(5,5),rep(8,5)))
> w
 [1] 5 3 8 8 5 2 3 3 2 8 2 5 2 8 5 8 3 2 3 3 5 2
------------------------------------------------------------
```

The observed outcomes, as obtained from the appropriate columns of Table 7.14 are shown in Table 8.7. From these observed outcomes, we compute the values of the observed averages $\overline{y}.(j)$ and variances s_j^2 for treatment combination $j \in \{2, 3, 5, 8\}$, where $\overline{y}.(j)$ and s_j^2 are defined in (7.15) and (7.16), as follows:

```
------------------------------------------------------------
y=c(12.82,11.97,16.61,17.15,14.07,9.13,11.60,11.67,8.15,15.30,10.97,15.23,
8.62,14.53,12.96,14.56,10.76,9.35,11.17,12.34,15.03,10.83)

T = c(2,3,5,8)
ybarvector = varyvector = NULL
for (l in 1:4) {
  ybarvector = c(ybarvector,mean(y[w==T[l]]))
  varyvector = c(varyvector,var(y[w==T[l]]))
}
> round(ybarvector,2)
[1]  9.51 11.59 14.02 15.63
> round(varyvector,2)
[1] 1.34 0.32 1.26 1.43
------------------------------------------------------------
```

TABLE 8.7
Observed outcomes in a hypothetical 2^{3-1} experiment

i	w_i	y_i
1	5	12.82
2	3	11.97
3	8	16.61
4	8	17.15
5	5	14.07
6	2	9.13
7	3	11.60
8	3	11.67
9	2	8.15
10	8	15.30
11	2	10.97
12	5	15.23
13	2	8.62
14	8	14.53
15	5	12.96
16	8	14.56
17	3	10.76
18	2	9.35
19	3	11.17
20	3	12.34
21	5	15.03
22	2	10.83

We now find unbiased estimates of the factorial effects in (8.6) by substituting the vector of observed averages $\overline{\mathbf{y}}$ in place of $\overline{\boldsymbol{Y}}$, i.e.,

$$\widehat{\tau}(F_1) = \frac{1}{2}\boldsymbol{\lambda}(F_1)^{\mathrm{T}}\overline{\boldsymbol{y}}, \ \ \widehat{\tau}(F_2) = \frac{1}{2}\boldsymbol{\lambda}(F_2)^{\mathrm{T}}\overline{\boldsymbol{y}}, \ \ \widehat{\tau}(F_3) = \frac{1}{2}\boldsymbol{\lambda}(F_3)^{\mathrm{T}}\overline{\boldsymbol{y}}, \tag{8.7}$$

where the contrast vectors $\boldsymbol{\lambda}(F_1)$, $\boldsymbol{\lambda}(F_2)$ and $\boldsymbol{\lambda}(F_1F_2)$ are shown in Table 8.4. These computations are shown below:

```
-----------------------------------------------------------
tau1barhat = (t(c(-1,-1,1,1)) %*% ybarvector)/2
tau2barhat = (t(c(-1,1,-1,1)) %*% ybarvector)/2
tau3barhat = (t(c(1,-1,-1,1)) %*% ybarvector)/2
> c(tau1barhat, tau2barhat, tau3barhat)
[1]  4.2793333  1.8423333 -0.2343333
-----------------------------------------------------------
```

Analogous to (7.19), an efficient way to compute estimates of all factorial effects in a 2^{K-P} design simultaneously using R would be to define a $2^{K-P} \times 1$ vector $\overline{\boldsymbol{\tau}}_{\text{adj}}$ in which the first element is $\overline{\tau}(0)$, the average of all $2^{K-P}N$ potential outcomes, and the remaining $2^{K-P}-1$ elements are equal to half of the $2^{K-P}-1$ factorial effects $\overline{\tau}(\cdot)$. Then, from (7.19), an unbiased estimator of $\overline{\boldsymbol{\tau}}_{\text{adj}}$ is:

$$\widehat{\overline{\boldsymbol{\tau}}}_{\text{adj}} = \frac{1}{2^{K-P}} \mathbf{L}^{\mathrm{T}} \overline{\mathbf{y}}, \tag{8.8}$$

where $\mathbf{L}$ is the $2^{K-P} \times 1$ model matrix whose columns constitute the contrast vectors for the mean and the 2^{K-P} factorial effects to be estimated from the experiment. In the current example, $\mathbf{L} = (\boldsymbol{\lambda}(0), \boldsymbol{\lambda}(F_1), \boldsymbol{\lambda}(F_2), \boldsymbol{\lambda}(F_3))$ where $\boldsymbol{\lambda}(0), \boldsymbol{\lambda}(F_1), \boldsymbol{\lambda}(F_2)$ and $\boldsymbol{\lambda}(F_3)$ are shown as columns in Table 8.4. The model matrix $\mathbf{L}$ in a 2^{K-P} fractional factorial design can be constructed using the defining relations and implemented in R for the current example as follows:

```
-----------------------------------------------------------
library(AlgDesign)
D = gen.factorial(2,2) ## ME contrast columns in reverse order

D[, c(1:2)] = D[,c(2:1)] ## Reverse the order of the columns
colnames(D) = c("F1","F2")
F3 = D$F1 * D$F2
Int = rep(1,4)
L = cbind(Int, D, F3)
> L
  Int F1 F2 F3
1   1 -1 -1  1
2   1 -1  1 -1
3   1  1 -1 -1
4   1  1  1  1
-----------------------------------------------------------
```

Consequently, the factorial effects can be estimated using (8.8):

```
-----------------------------------------------------------
# Estimation of all factorial effects
taubarhatadj = (t(L) %*% ybarvector)/4 # L is the model matrix
taubarhatobs = 2*taubarhatadj[2:4] # adjust factorial effects by 2
> round(taubarhatobs,2)
[1]  4.28  1.84 -0.23
-----------------------------------------------------------
```

The ANOVA decomposition of the observed outcomes in 2^{K-P} experiments is similar to that described in Section for 2^K full factorial experiments, the only difference being that the degrees of freedom due to treatment is $2^{K-P}-1$ instead of 2^K-1, and the degrees of freedom for residuals are $N-2^{K-P}$. The calculations for the current example are shown below. The p-value in the output can be interpreted as an approximation of the exact p-value obtained from the randomization test of the sharp null hypothesis of no treatment effect on any unit. The randomization test is left as an exercise.

```
--------------------------------------------------------
# ANOVA decomposition in 2^(3-1) fractional factorial
> anova(lm(y ~ factor(w)))
Analysis of Variance Table

Response: y
          Df  Sum Sq Mean Sq F value    Pr(>F)
factor(w)  3 119.289  39.763  37.573 5.871e-08 ***
Residuals 18  19.049   1.058
--------------------------------------------------------
```

Neymanian inference for the estimable factorial effects will be analogous to the description in Section 7.5 and is also left as an exercise.

8.3 Split-plot Designs

Split-plot designs originated in agricultural experiment. Consider an experiment in which two methods of irrigation (current, labeled 0 and new, labeled 1) and two fertilizers (existing, labeled 0 and new, labeled 1) are being compared with respect to their ability to improve yield of crops. The experimental units are eight small plots of land arranged in a single row, so that each of the four treatment combinations are assigned to two units. A completely randomized assignment may yield an allocation shown in Table 8.8, in which plots 1,3,5 and 8 receive the old irrigation and the remaining ones the new irrigation. However, plots are irrigated by large sprinklers that irrigate a large area of land, making it difficult to apply different irrigation methods to small adjacent plots of lands. In this example, irrigation is a *difficult-to-change factor.* A practical solution to this problem is to consider plots $1-4$ and $5-8$ as two large plots (referred to as *whole plots*), and randomly assign the two irrigation methods to these two whole plots. Within each whole plot, the two levels of fertilizer are assigned at random to the four individual plots, referred to as the *sub plots*. In this example, irrigiation is called the whole-plot treatment factor and fertilizer the sub-plot treatment factor.

TABLE 8.8
Impractical assignment of four treatment combinations to eight plots of land

Plot	1	2	3	4	5	6	7	8
Treatment combination	00	10	01	11	10	00	11	01

8.3.1 Estimands, treatment assignment and estimators

We explain the key ideas of a split-plot experiment with a simplified hypothetical version of the school experiment with 20 schools and two factors: a mid-year quality review by a team of experts (F_1), and a bonus scheme for teachers (F_2). Suppose the schools are divided into $H = 4$ groups or whole-plots (WPs) based on geographical proximity. WPs 1, 2, 3 and 4 comprise 4, 4, 6 and 6 schools respectively, which are called sub-plots (SPs). Hypothetical potential outcomes for the four treatment combinations are shown in Table 8.9.

TABLE 8.9
Potential outcomes and factorial effects in a hypothetical 2^2 school experiment

Subplot	WP	Treatment combinations				Factorial effects		
(i)	No.	00	01	10	11	$\tau_i(F_1)$	$\tau_i(F_2)$	$\tau_i(F_1F_2)$
1	1	60	50	70	80	20	0	10
2	1	60	50	70	80	20	0	10
3	1	60	50	70	80	20	0	10
4	1	60	50	70	80	20	0	10
WP Avg		60	50	70	80	20	0	10
5	2	70	60	80	90	20	0	10
6	2	70	60	80	90	20	0	10
7	2	70	60	80	90	20	0	10
8	2	70	60	80	90	20	0	10
WP Avg		70	60	80	90	20	0	10
9	3	70	60	80	90	20	0	10
10	3	70	60	80	90	20	0	10
11	3	70	60	80	90	20	0	10
12	3	70	60	80	90	20	0	10
13	3	70	60	80	90	20	0	10
14	3	70	60	80	90	20	0	10
WP Avg		70	60	80	90	20	0	10
15	4	80	80	90	90	10	0	0
16	4	80	80	90	90	10	0	0
17	4	80	80	90	90	10	0	0
18	4	80	80	90	90	10	0	0
19	4	80	80	90	90	10	0	0
20	4	80	80	90	90	10	0	0
WP Avg		80	60	90	90	10	0	0
Pop Avg		71.0	64.0	81.0	88.0	17.0	0.0	7.0

TABLE 8.10
Treatment assignment and observed outcomes in a hypothetical 2^2 school split-plot experiment

Subplot (i)	WP No.	Treatment combinations 00	01	10	11	Assignment W_{1i}	W_{2i}	W_i	Obs outcome y
1	1	?	?	70	?	1	0	3	70
2	1	?	?	70	?	1	0	3	70
3	1	?	?	?	80	1	1	4	80
4	1	?	?	?	80	1	1	4	80
WP Avg				70	80				75
5	2	70	?	?	?	0	0	1	70
6	2	?	60	?	?	0	1	2	60
7	2	?	60	?	?	0	1	2	60
8	2	70	?	?	?	0	0	1	70
WP Avg		70	60						65
9	3	?	60	?	?	0	1	2	60
10	3	?	60	?	?	0	1	2	60
11	3	?	60	?	?	0	1	2	60
12	3	70	?	?	?	0	0	1	70
13	3	70	?	?	?	0	0	1	70
14	3	70	?	?	?	0	0	1	70
WP Avg		70	60						65
15	4	?	?	90	?	1	0	3	90
16	4	?	?	?	90	1	1	4	90
17	4	?	?	90	?	1	0	3	90
18	4	?	?	?	90	1	1	4	90
19	4	?	?	90	?	1	0	3	90
20	4	?	?	?	90	1	1	4	90
WP Avg				90	90				90
Sample Avg		70	60	82	86				74.5

Table 8.9 also shows the unit level factorial effects $\tau_i(F_1)$, $\tau_i(F_2)$ and $\tau_i(F_1F_2)$ for $i = 1, \ldots, 20$. The unit level main effects of factor F_1 and the two factor interactions are not the same for all 20 units, whereas the unit level main effects of factor F_2 are all zero. Thus the potential outcomes are not strictly additive. However, within each WP, the unit-level factorial effects are identical for all units. Thus, in this case, strict additivity holds within each WP.

WP-level factorial effects for WP $h = 1, \ldots, H$ can be defined as:

$$\overline{\tau}_h(\cdot) = \frac{1}{M_h} \sum_{i \in h} \tau_i(\cdot), \tag{8.9}$$

where M_h is the number of units in WP h. Here $M_1 = M_2 = 8$ and $M_3 = M_4 = 6$. For example, the main effect of F_1 for WP 2 is $\overline{\tau}_2(F_1) = 20$, the main effect of F_2, for WP 3 is $\overline{\tau}_3(F_2) = 0$ and the interaction effect between F_1 and F_2 in WP 4 is $\overline{\tau}_4(F_1F_2) = 0$.

The population-level factorial effects, as before, are defined as

$$\overline{\tau}(\cdot) = \frac{1}{N} \sum_{i=1}^{N} \tau_i(\cdot), \tag{8.10}$$

which can also be expressed as weighted averages of the WP-level factorial effects, i.e.,

$$\overline{\tau}(\cdot) = \frac{1}{N} \sum_{h=1}^{H} \overline{\tau}_h(\cdot). \tag{8.11}$$

The assignment mechanism for the split-plot design involves a two-step process - first, the four WPs are randomly assigned to the two levels of factor F_1. Thus, factor F_1 is the WP-factor. Second, for $h = 1, \ldots, H$, the units in WP h are randomly assigned to the two levels of factor F_2, such that predetermined numbers N_{0h} and N_{1h} of SPs or units receive treatment levels 0 and 1 respectively. Factor F_2 is thus the SP-factor. Let $\mathbf{W}_1$ and $\mathbf{W}_2$ denote the $N \times 1$ binary assignment vectors for factors F_1 and F_2 respectively, and $\mathbf{W}$ denote the $N \times 1$ assignment vector whose elements are the treatment numbers 1, 2, 3, 4 in the lexicographic order (i.e., corresponding to 00, 01, 10, and 11 respectively). Then $W_i = 2W_{1i} + W_{2i}$ for $i = 1, \ldots, N$.

One such assignment is shown below, where an equal number of units is assigned to the two levels of factor F_2 within each WP. Thus, $N_{0h} = N_{1h} = 2$ for $h = 1, 2$ and $N_{0h} = N_{1h} = 3$ for $h = 3, 4$. The observed outcomes generated by this assignment generates the shown in Table 8.10

```
-------------------------------------------------------------
# Assignment of the 20 units to the four treatment combinations
M = c(4,4,6,6)
NO = c(2,2,3,3)
N1 = c(2,2,3,3)
set.seed(99)
r1 = sample(c(0,0,1,1))
W1 = numeric(20)
W2 = NULL
 for (h in 1:4) {
 W1[POT$WP==h] = r1[h]
 W2 = c(W2, sample(c(rep(0,NO[h]),rep(1,N1[h]))))
 }
W = 2*W1 + W2 + 1
> W1
 [1] 1 1 1 1 0 0 0 0 0 0 0 0 0 0 1 1 1 1 1 1
> W2
 [1] 0 0 1 1 0 1 1 0 1 1 1 0 0 0 0 1 0 1 0 1
> W
 [1] 3 3 4 4 1 2 2 1 2 2 2 1 1 1 3 4 3 4 3 4
-------------------------------------------------------------
```

The point estimates of the factorial effects are obtained using the methodology described in Section 6.3. First the vector of observed averages $\overline{\mathbf{y}}$ is computed as $(70, 60, 82, 86)^{\mathrm{T}}$, and premultiplied by $\mathbf{L}^{\mathrm{T}}/4$, where $\mathbf{L}$ is the model matrix. The calculation shown below yields the following estimates of factorial effects as $\widehat{\overline{\tau}}(F_1) = 19$, $\widehat{\overline{\tau}}(F_2) = -3$, $\widehat{\overline{\tau}}(F_1F_2) = 7$. Mukerjee and Dasgupta (2022) [63] proved that these estimates are unbiased estimates of the true factorial effects.

```
--------------------------------------
# Calculate ybarvector
y = c(70,70,80,80,70,60,60,70,60,60,60,70,70,70,90,90,90,90,90,90)
w = c(3,3,4,4,1,2,2,1,2,2,2,1,1,1,3,4,3,4,3,4)
ybarvector = NULL
for (l in 1:4) {
  ybarvector = c(ybarvector,mean(y[w==l]))
}

> ybarvector
[1] 70 60 82 86

# Generate L matrix
library(AlgDesign)
D = gen.factorial(2,2) ## ME contrast columns in reverse order
D[, c(1:2)] = D[,c(2:1)] ## Reverse the order of the columns
colnames(D) = c("F1","F2")
L = model.matrix(~ F1 + F2 + F1*F2,D)

# Obtain taubarhatadj
taubarhatadj = (t(L) %*% ybarvector)/4  # adjusted tau bar hat
taubarhat = 2*taubarhatadj[2:4] # adjust factorial effects by 2

> taubarhat
[1] 19 -3  7
--------------------------------------
```

8.3.2 Fisherian inference for split-plot designs

To test Fisher's sharp null hypothesis of no treatment effect on any unit:

$$H_0 : Y_i(1) = Y_i(2) = Y_i(3) = Y_i(4), \quad i = 1, \ldots, N,$$

or equivalently,

$$H_0 : \tau_i(F_1) = \tau_i(F_2) = \tau_i(F_1, F_2) = 0, , \quad i = 1, \ldots, N.$$

a methodology analogous to the one described in Section 6.5 can be used. Specifically, a Fisher randomization test can be obtained by comparing each estimated factorial effect (or its absolute value) against its randomization distribution under the sharp null hypothesis of no treatment effects. As earlier, under this sharp null hypothesis, the missing potential outcomes in each row of Table 8.10 can be imputed by the single observed outcome in that row. The randomization distribution of an estimated factorial effect can be obtained by taking draws from the distribution of $\mathbf{W}$ (now defined by the split-plot assignment mechanism), and calculating the estimated factorial effect for each draw.

The major difference from the procedure described in Section 6.5 is that, due to different assignment mechanisms associated with the WP and the SP factors, the reference distribution under the sharp null hypothesis of no effects is *not* the same for all factorial effects. In the current example, the total number of possible allocations of units to levels of the WP factor F_1 is $\binom{4}{2} = 6$, whereas for the SP-factor F_2, this number is $\binom{4}{2}^2 \times \binom{6}{3}^2 = 36 \times 400$. Consequently, in this example, the randomization distribution of the main effect of F_1 will have only six unique support points (i.e., values with non-zero probability), whereas the randomization distributions of the main effect of F_2 and the interaction between F_1 and F_2 will have many more. The smallest p-value for the WP-factor in this example can therefore only be $1/6$ for a one-sided test and $1/3$ for a two-sided test, whereas the smallest p-value for an SP-factor (and its interaction with the WP-factor) can be $1/(36 \times 400)$ for a one-sided test and $2/(36 \times 400)$ for a two-sided test.

The R code used to implement the randomization test to assess Fisher's sharp null hypothesis using the absolute values of the estimates of the factorial effects $\widehat{\overline{\tau}}(F_1)$, $\widehat{\overline{\tau}}(F_2)$ and $\widehat{\overline{\tau}}(F_1F_2)$, is shown below. In each iteration, a new assignment vector $\mathbf{w}_{\text{rep}}$ is drawn, the vector $\overline{\mathbf{y}}_{\text{rep}}$ of observed treatment averages is computed, and eventually $\widehat{\overline{\tau}}_{\text{rep}}(F_1), \widehat{\overline{\tau}}_{\text{rep}}(F_2)$ and $\widehat{\overline{\tau}}_{\text{rep}}(F_1F_2)$ are calculated obtained using (6.25). The p-values for the for the main effects of F_1, F_2 and interaction F_1F_2 are obtained as the proportion of values of the randomization distributions of their absolute values that are greater than or equal to the observed absolute estimated values of the three factorial effects (19, 3 and 7 respectively). The p value for the main effect of the WP-factor F_1 is 0.33, which is the smallest one can obtain in this example. The fact the observed value of the test statistic is extreme can also be visualized from the left panel of Figure 8.1. The p-values for the main effect of F_2 and the interaction between F_1 and F_2 are 0.31 and 0.003 respectively, and suggest possible existence of an interaction between factors F_1 and F_2, but essentially no evidence for the main effect of the SP-factor F_2.

```
---------------------------------------
y = c(70,70,80,80,70,60,60,70,60,60,60,70,70,70,90,90,90,90,90,90)
w = c(3,3,4,4,1,2,2,1,2,2,2,1,1,1,3,4,3,4,3,4)
N = length(y)
N0 = N1 = c(2,2,3,3)
library(AlgDesign)
# L matrix
D = gen.factorial(2,2) ## ME contrast columns in reverse order
```

```
D[, c(1:2)] = D[,c(2:1)] ## Reverse the order of the columns
colnames(D) = c("F1","F2")
L = model.matrix(~ F1 + F2 + F1*F2,D)

ITER=5000 # Number of iterations

tau1hatobs= 19
tau2hatobs=-3
tau12hatobs=7

tau1hatarray=tau2hatarray = tau12hatarray = NULL

for (count in 1:ITER) {
  r1 = sample(c(0,0,1,1))
  w1 = numeric(20)
  w2 = NULL
  for (h in 1:4) {
    w1[POT$WP==h] = r1[h]
    w2 = c(w2, sample(c(rep(0,N0[h]),rep(1,N1[h]))))
  }
  wrep = 2*w1 + w2 + 1

  ybarvectorrep = NULL
  for (l in 1:4) {
    ybarvectorrep = c(ybarvectorrep,mean(y[wrep==l]))
  }
  taubarhatadjrep = (t(L) %*% ybarvectorrep)/4
  tau1barhatrep = 2*taubarhatadjrep[2]
  tau2barhatrep = 2*taubarhatadjrep[3]
  tau12barhatrep = 2*taubarhatadjrep[4]
  tau1hatarray = c(tau1hatarray,abs(tau1barhatrep))
  tau2hatarray = c(tau2hatarray,abs(tau2barhatrep))
  tau12hatarray = c(tau12hatarray,abs(tau12barhatrep))
}

# p-values
(pval1 = sum(tau1hatarray >= abs(tau1hatobs))/ITER)
(pval2 = sum(tau2hatarray >= abs(tau2hatobs))/ITER)
(pval3 = sum(tau12hatarray >= abs(tau12hatobs))/ITER)

# histograms
par(mfrow=c(1,3))
hist(tau1hatarray, main=NULL, xlab=expression(hat(bar(tau))[1]))
abline(v=abs(tau1hatobs), lwd=1.5, lty=2)
text(17,1500,round(pval1,2),cex=1.1)
hist(tau2hatarray, main=NULL, xlab=expression(hat(bar(tau))[2]), nclass=
abline(v=abs(tau2hatobs), lwd=1.5, lty=2)
text(4.1,2500,round(pval2,2),cex=1.1)
hist(tau12hatarray, main=NULL, xlab=expression(hat(bar(tau))[12]), nclas
```

```
abline(v=abs(tau12hatobs), lwd=1.5, lty=2)
text(5.5,2500,round(pval3,3),cex=1.1)
-----------------------------------------
```

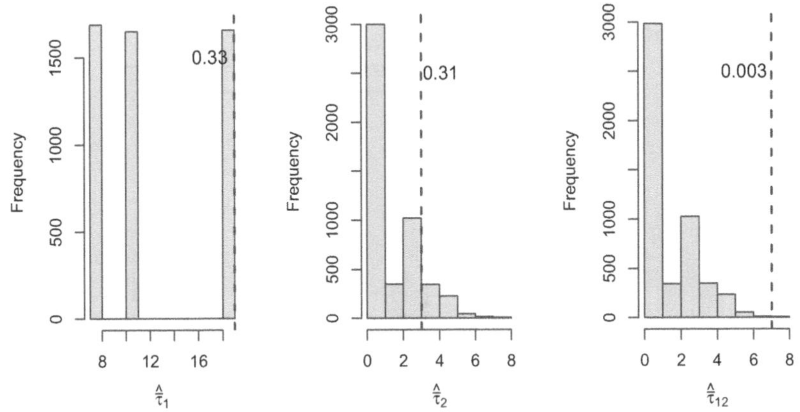

FIGURE 8.1
Randomization test for the data in Table 8.10 using $|\widehat{\overline{\tau}}(F_1)|$, $|\widehat{\overline{\tau}}(F_2)|$ and $|\widehat{\overline{\tau}}(F_1F_2)|$ as the test statistics, where the dashed vertical lines represent the observed values of the test statistics, and the numbers next to them the corresponding p-values

In this approach, one can adjust the p-value for multiple comparisons, for example, by using the ensemble-adjusted p-value described in Section 5.4.1.6 of Chapter 5, which simply modifies the raw p-value for each test to $\max\{3p, 1\}$.

8.4 Problems

8.1 Complete Table 8.4 by entering the missing contrast vectors for the two-factor interactions between F_1 and F_3, and between F_2 and F_3.

8.2 Consider two different 2^{5-1} designs D_1 and D_2 that are generated by the two defining relations:

$$D_1 : 5 \triangleq 123, \quad D_2 : 5 \triangleq 1234.$$

The aliasing relations for designs D_1 and D_2 are shown in Table 8.6. In the absence of any prior information on the relative importance of the factors and their effects, which of these two fractional factorial designs would you prefer and why? Is there any situation where design D_1 may be more desirable than design D_2?

8.3 Table 8.11 shows a design matrix for a 2^{K-P} experiment.

TABLE 8.11
Design matrix for a 2^{K-P} experiment

Combination	1	2	3	4	5	6
1	−1	−1	−1	−1	−1	−1
2	−1	−1	−1	+1	−1	+1
3	−1	−1	+1	−1	+1	−1
4	−1	−1	+1	+1	+1	+1
5	−1	+1	−1	−1	+1	+1
6	−1	+1	−1	+1	+1	−1
7	−1	+1	+1	−1	−1	+1
8	−1	+1	+1	+1	−1	−1
9	+1	−1	−1	−1	+1	+1
10	+1	−1	−1	+1	+1	−1
11	+1	−1	+1	−1	−1	+1
12	+1	−1	+1	+1	−1	−1
13	+1	+1	−1	−1	−1	−1
14	+1	+1	−1	+1	−1	+1
15	+1	+1	+1	−1	+1	−1
16	+1	+1	+1	+1	+1	+1

(i) Identify K and P.

(ii) Write down the defining contrast subgroup of the design.

(iii) What is the resolution of the design?

(iv) Write down the word-length pattern of the design.

(v) Which main effects and two factor interactions are clear?

(vi) Which effects are aliased with the four factor interaction 3456?

(vii) How does this design compare with another 2^{6-2} design having defining relation $5 \triangleq 123$, $6 \triangleq 234$?

8.4 An experimenter considers running a 2^{7-3} design, and is contemplating two possibilities: design A with generators $5 \triangleq 123, 6 \triangleq 124, 7 \triangleq 1234$, and design B with generators $5 \triangleq 123, 6 \triangleq 124, 7 \triangleq 134$.

(a) Which design, A or B, is better? Why?

(b) Write down the full design matrix for the design you would recommend (you may use R to generate the design matrix).

(c) Which factorial effects of your recommended design are aliased with the mean?

(d) Write down the contrast column for estimating the two factor interaction 12. With which other factorial effects is this contrast aliased?

(e) Write down the contrast column for estimating the six factor interaction 123456. With which other factorial effects is this contrast aliased?

8.5 Using the observed outcomes in Table 8.7 from the hypothetical 2^{3-1} experiment described in Section 8.2.4, assess the sharp null hypothesis of no treatment effect on any experimental unit using (i) the F-statistic, (ii) unscaled estimates of the factorial effects $\widehat{\overline{\tau}}(\cdot)$ and (iii) scaled (divided by the standard errors) estimates of the factorial effects, i.e. $\widehat{\overline{\tau}}(\cdot)/S.E.(\widehat{\overline{\tau}}(\cdot))$.

8.6 Using the observed outcomes in Table 8.7 from the hypothetical 2^{3-1} experiment described in Section 8.2.4, obtain approximate 95% Neymanian confidence intervals for the three factorial effects $\widehat{\overline{\tau}}(F_1)$, $\widehat{\overline{\tau}}(F_2)$ and $\widehat{\overline{\tau}}(F_3)$.

9

Model-based analysis of designed experiments and superpopulation inference

9.1 Introduction

In all the previous chapters of this book, the inference for treatment effects was restricted to a finite population of experimental units, where the potential outcomes were considered at least partially unknown, but not random variables. The randomness in the observed outcomes was generated only through the fully known assignment mechanism.

In this chapter, we will consider a different approach that views the potential outcomes themselves as random variables. This approach has different advantages than the Fisherian and Neymanian approaches used to draw inference on causal inferences, an issue discussed in Chapter 8 of Imbens and Rubin (2015) [44]. First, this approach allows us to draw inference on a wide variety of estimands other than the average treatment effect considered thus far. Second, and most important, the model-based approach can not only draw inference for any finite population estimands, but it can also easily accommodate unit-level and super population estimands (introduced in Section 9.3) that are more awkward to consider from the Fisherian and Neymanian perspectives.

9.2 Potential outcomes: transitioning from fixed quantities to random variables

Recall the tomato experiment discussed in Chapter 2 where the experimental units comprise eleven adjacent plots of land on which two fertilizers A (standard) and B (new) are compared with respect to their ability to increase the yield of tomato plants. Under the assumption of fixed potential outcomes $Y_i(0)$ and $Y_i(1)$ for the ith unit under SUTVA, using two forms of inference – Fisherian and Neymanian – both driven by the completely randomized assignment mechanism, no treatment effect was found. The estimand of interest was the

DOI: 10.1201/9781003048794-9

finite population average treatment effect $\overline{\tau} = \overline{Y}(1) - \overline{Y}(0)$, and its standard unbiased estimator $\widehat{\overline{\tau}} = \overline{y}(1) - \overline{y}(0)$ was considered as a natural statistic to assess both the Fisherian "strong" and Neymanian "weak" null hypotheses.

Using both forms of inference, we were able to obtain $100 \times (1 - \alpha)\%$ intervals for the true value of $\overline{\tau}$. The Neymanian approach described in Section 2.6 was based on the normal approximation of the randomization distribution of the estimator $\widehat{\overline{\tau}}$ and led to a 95% confidence interval for $\overline{\tau}$. On the other hand, in the more intuitive Fisherian approach described in Section 2.5.6, the missing potential outcomes were imputed under multiple sharp null hypotheses, and the 95% "fiducial interval" was obtained as $[-7, 11]$ by inverting the resultant p-value functions as shown in Figure 2.2.

Recall that to test the Fisherian sharp null hypothesis of a constant treatment effect of τ^* on any experimental unit, an important step (step 3 of Procedure 2.1) was to impute the missing potential outcomes under the sharp null hypothesis $H_{0F}^{\tau^*} : Y_i(1) - Y_i(0) = \tau^*, \quad i = 1, \ldots, N$ defined in (2.10), using the following rule:

$$Y_i^{\text{mis}} = \begin{cases} Y_i^{\text{obs}} + \tau^* & \text{if} \quad W_i = 0 \\ Y_i^{\text{obs}} - \tau^* & \text{if} \quad W_i = 1 \end{cases}$$

where $Y_i^{\text{obs}} = W_i Y_i(1) + (1 - W_i) Y_i(0)$ and $Y_i^{\text{mis}} = (1 - W_i) Y_i(1) + W_i Y_i(0)$ denote the observed and missing potential outcomes for unit i. In the new setting where the potential outcomes are random variables, the imputation model also needs to be a probablistic one, so that the missing potential outcomes, conditional on the observed outcomes and the observed treatment assignment, are generated from a probability law that follows from the assumed model on the potential outcomes and the assignment mechanism.

From the observed data in the tomato experiment shown in Table 2.2, we see that the observed averages of the outcomes in the control and treated groups were $\overline{y}(0) = 20.70$ and $\overline{y}(1) = 22.53$ respectively. The sample variances in these groups were $s_0^2 = 49.43$ and $s_1^2 = 29.51$, respectively. Based on these summaries, we may choose the following imputation model for the observed outcomes

$$Y_i^{\text{mis}} | \mathbf{y}, \mathbf{W} \sim \begin{cases} \mathcal{N}(20.70, 49.43) & \text{if} \quad W_i = 0 \\ \mathcal{N}(22.53, 29.51) & \text{if} \quad W_i = 1. \end{cases} \tag{9.1}$$

Every imputation using this model would generate a new set of missing outcomes and thus provide a complete set of potential outcomes. From each such complete set of potential outcomes, the value of $\overline{\tau}$ can be directly computed. A histogram obtained from 500 such values of $\overline{\tau}$ is shown in Figure 9.1. The lower and upper 0.025 quantiles computed from these values are –1.96 and 5.31 are shown as dotted lines. Thus we immediately obtain the 95% interval $[-1.69, 5.31]$ for $\overline{\tau}$ based on the imputation model 9.1.

The imputation model 9.1 is clearly not a good one because it does not incorporate any uncertainty about our knowledge about the distribution of the potential outcomes. This aspect will be further elaborated in Section 9.4

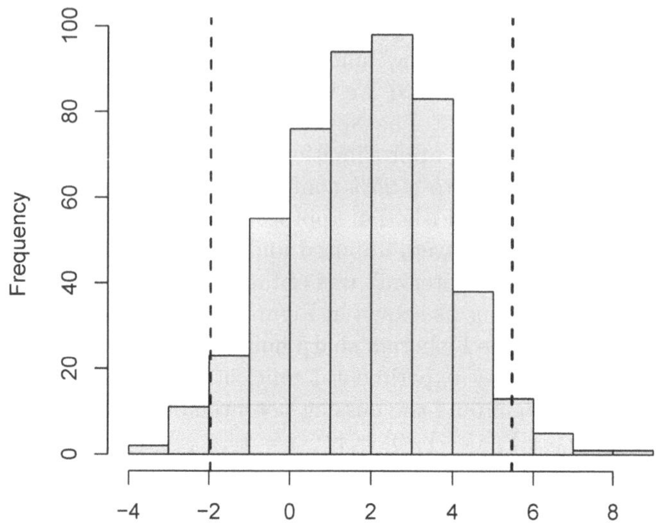

FIGURE 9.1
Histogram of $\overline{\tau}$ values obtained from 500 imputations of missing data using the model 9.1. The lower and upper 0.025-th quantiles are shown as vertical dashed lines

with another example, and a principled approach will be presented in Sections 9.5–9.7.

9.3 Potential outcomes, finite population and super population estimands

Consider a finite population of N units indexed by $i = 1, \ldots, N$ that can typically be considered to be a random sample drawn from a target or super population, and as in Chapter 2, let $Y_i(0)$ and $Y_i(1)$ denote the potential outcomes for unit i when exposed to control and treatment respectively. Let $\mathbf{X}_i$ denote a p-dimensional vector of pre-measured covariates for unit i.

Example 9.1 *To illustrate, consider a subset of 20 subjects (infants) from the LEAP trial described in Chapter 2, and assume that these were the experimental units randomly sampled from a super population of infants to which we would like to extend our inferences. The age in months (x_1) and sex (1 if female and 2 if male) of these 20 infants are the pre-treatment covariates shown in Table 9.1. Assume that a completely randomized assignment* *mechanism was used to assign 10 infants to treatment and the remaining 10 to control, as shown in the column labeled w in the table. The observed and missing (indicated by ?) potential outcomes are shown in columns 5 and 6, and the observed binary outcome y for each unit is also shown in the last column.*

TABLE 9.1
20 subjects in the LEAP randomized trial

Unit (infant)	x_1 (age)	x_2 (sex)	w	$Y(0)$	$Y(1)$	y
1	5.39	1	1	?	1	1
2	5.98	2	1	?	1	1
3	8.64	2	0	1	?	1
4	10.58	2	1	?	1	1
5	8.18	2	0	1	?	1
6	6.28	1	1	?	1	1
7	7.85	2	0	0	?	0
8	8.87	2	1	?	1	1
9	8.97	2	1	?	1	1
10	6.47	1	0	1	?	1
11	10.71	1	0	1	?	1
12	10.12	1	0	1	?	1
13	5.32	2	1	?	1	1
14	6.11	2	1	?	1	1
15	8.08	2	0	0	?	0
16	5.32	1	0	0	?	0
17	8.21	1	1	?	1	1
18	6.37	2	0	1	?	1
19	5.29	1	1	?	1	1
20	6.05	2	0	1	?	1

Recall the unit-level treatment effects $\tau_i = Y_i(1) - Y_i(0)$ defined in (2.1) and the finite population average treatment effect $\overline{\tau} = \overline{Y}(1) - \overline{Y}(0)$ defined in (2.3). We also define the super population average treatment effect as

$$\tau^{\text{SP}} = \mathbb{E}[Y_i(1)] - \mathbb{E}[Y_i(0)], \tag{9.2}$$

where the expectation $\mathbb{E}[\cdot]$ is defined with respect to the probability distribution of the potential outcomes in the target population of interest from which the 20 units were randomly sampled. This probability distribution will be defined more precisely in Section 9.4. We also define the conditional average treatment effect as the expectation of unit-level treatment effects of units with covariate value $\mathbf{x}$, i.e.,

$$\tau^{\text{SP}}(\mathbf{x}) = \mathbb{E}[Y_i(1)|\mathbf{X}_i = \mathbf{x}] - \mathbb{E}[Y_i(0)|\mathbf{X}_i = \mathbf{x}]. \tag{9.3}$$

To illustrate, consider the conditional average treatment effect for all female infants aged 8 months or older, i,e.,

$$\mathbb{E}[Y_i(1)|X_1 \geq 8, X_2 = 1] - \mathbb{E}[Y_i(0)|X_1 \geq 8, X_2 = 1].$$

Note that in the previous chapters, we were able to draw inference on $\overline{\tau}$, either by inverting Fisher's randomization test or using the Neymanian approximate repeated sampling inference. The model-based approach allows us to draw inference for causal effects defined with respect to individual units, the finite population of N units, and various super populations. In the following section, we describe this inferential approach without considering any covariates, but this approach will then be extended to the case where covariates are considered in the inference in Chapter 9.

We emphasize that the model-based approach allows us to draw inference (obtain point and interval estimates) for any causal estimand, including parameters of distributions of potential outcomes such as quantiles or measures of dispersion.

9.4 A naive approach and the desirability for models

Prior to the formal introduction of the model-based approach, we provide an intuitive idea of the inference for the unit-level treatment effects τ_i and the finite population average treatment effect $\overline{\tau}$ using this approach. Let us consider Example 9.1 and discuss a naive approach to impute each missing potential outcomes using the observed outcomes. A summary of the observed outcomes in Table 9.1 shows that all the 10 units who received the treatment did not have a peanut allergy, i.e., the observed proportion of successful cases in the treated group is $p_1 = 1$. Similarly, the observed proportion of success (not having peanut allergy) in the control group is $p_0 = 0.7$. Consequently, we could impute all the missing outcomes in the treated group by 1 (success with certainty), and those in the control group by a Bernoulli random variable by probability of success $p_0 = 0.7$. The function Imp_POT shown in the code below takes a data frame like the one shown in Table 9.1, p_0 and p_1 as the three arguments, and replaces the missing potential outcomes using the aforementioned strategy.

```
---------------------------------------------------------
# Function to impute missing outcomes from Bern(p0) and Bern(p1)
# Input: Dataframe with columnnames X1, .., Xp, W, Y0, Y1, y
# Y0 and Y1 has missing values depending on W

Imp_POT <- function(POT,p0,p1) {
N <- nrow(POT)
for (i in 1:N) {
if (POT$W[i] == 1) {
POT$Y0[i] = rbinom(1,1,p0)
```

```
} # end if
else {
POT$Y1[i] = rbinom(1,1,p1)
} # end else

} # end for
return(POT)
}

> set.seed(101)
> Imp_POT(Ypot,0.7,1)
      x1 x2 W Y0 Y1 y
1   5.39  1 1  1  1 1
2   5.98  2 1  1  1 1
3   8.64  2 0  1  1 1
4  10.58  2 1  0  1 1
5   8.18  2 0  1  1 1
6   6.28  1 1  1  1 1
7   7.85  2 0  0  1 0
8   8.87  2 1  1  1 1
9   8.97  2 1  1  1 1
10  6.47  1 0  1  1 1
11 10.71  1 0  1  1 1
12 10.12  1 0  1  1 1
13  5.32  2 1  1  1 1
14  6.11  2 1  1  1 1
15  8.08  2 0  0  1 0
16  5.32  1 0  0  1 0
17  8.21  1 1  1  1 1
18  6.37  2 0  1  1 1
19  5.29  1 1  1  1 1
20  6.05  2 0  1  1 1
---------------------------------------------------------
```

Note that all the imputed values in the column of $Y(1)$ are ones, because the hypothetical success probability p_1 equals 1. On the other hand, among units 1, 2, 4, 6, 8, 9, 13, 14, 17, 19 for which the values of $Y(0)$ were missing and thus imputed, the imputed outcome is zero for only one unit (unit 4) and is one for the remaining nine units. Thus, this imputation implies that units $i = 4, 7, 15, 16$ have a positive treatment effect $\tau_i = 1$, whereas the remaining 16 units have no treatment effect, i.e., $\tau_i = 0$. The proportion of ones in the columns of $Y(1)$ and $Y(0)$ in this imputed table are thus 1 and $16/20 = 0.8$, making the imputed value of $\overline{\tau} = 1 - 0.8 = 0.2$.

Thus, once we fill in the missing observed outcomes in Table 9.1 using this strategy, we can *simply* calculate the value of $\overline{\tau}$. We can repeat this process to generate multiple imputations of missing outcomes, and then assess the uncertainty in $\overline{\tau}$ from its realizations over these multiple imputations. Note that this is different from *estimating* $\overline{\tau}$ that we have done so far in the previous chapters - rather, this is computation of an actual $\overline{\tau}$ from the imputed matrix of potential outcomes. Under this naive strategy, the imputed values of potential outcomes for the treated group will always remain the same, but those for the control group will vary across imputations. Figure 9.2 shows a histogram of $\overline{\tau}$ values obtained from 100 imputations using the strategy described above, implemented through the following lines of code.

```
-----------------------------------------------------------
# Function to compute taubar from complete table of potential outcomes
taubar = function(POT){
mean(POT$Y1) - mean(POT$Y0)
}

# Multiple imputations
ITER=100
taubarvec=NULL
for (i in 1:ITER) {
  Ypotnew = Imp_POT(Ypot,0.7,1)
  taubarvec = c(taubarvec, taubar(Ypotnew))
}
hist(taubarvec, main=expression(paste("Histogram of ", bar(tau))),
 xlab=expression(bar(tau)))
-----------------------------------------------------------
```

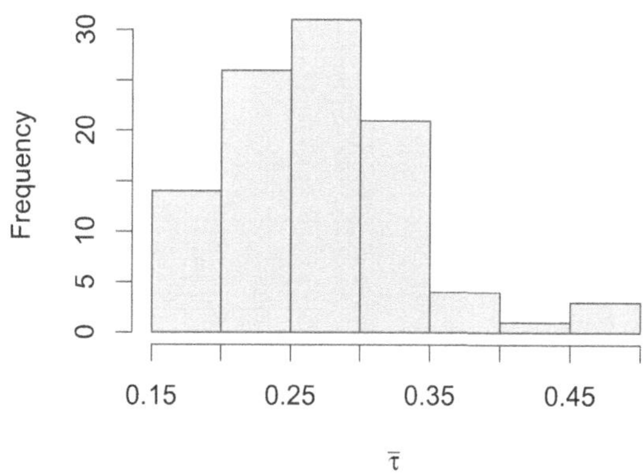

FIGURE 9.2
Histogram of $\overline{\tau}$ values obtained from 100 imputations of missing data using the naive strategy of Section 9.4

The naive approach for imputation of missing potential outcomes does not really address the uncertainty associated with estimation of $\overline{\tau}$, because the missing outcomes are imputed in a manner that suggests knowledge of the exact distribution of each potential outcome. In this example involving 20 units, the information for the distributions of treatment and control outcomes is entirely based on the 10 observations for each. In the treated group of 10 units, all outcomes are one, suggesting that all treated units pass the allergy test. However, when the set of all 530 observations in the LEAP data set is studied, we will see that some treated units fail the allergy test, yielding observed values equal to zero. Further, if we imagine a larger population of infants from which these 530 infants were randomly sampled and want to extend inference to such a population, we need to take into account additional sources of uncertainty. Such uncertainty can be incorporated through a model for the potential outcomes, which we discuss shortly. But before that, we note that in the context of finite population inference, the primary goal of this model-based approach is to build a probability model $\text{pr}(\mathbf{Y}^{\text{mis}}|\mathbf{Y}^{\text{obs}}, \mathbf{w})$ where $\mathbf{Y}^{\text{obs}} = \mathbf{y}$ and $\mathbf{Y}^{\text{mis}}$ denote, respectively, the vectors of observed and missing potential outcomes, and $\mathbf{w}$ is the observed assignment vector. The naive approach used earlier also used such a model for the missing potential outcome.

$$\text{pr}\left(Y_i^{\text{mis}} = y|\mathbf{Y}^{\text{obs}}, \mathbf{w}\right) = \begin{cases} 1 & \text{if} \quad w_i = 1, \\ 0.7^y(1-0.7)^{1-y} & \text{if} \quad w_i = 0. \end{cases}$$

However, the model did not allow for uncertainty in the estimation of the distribution of the missing potential outcomes, that is, the probability that Y_i^{mis} is 1 is fixed at the observed proportion, 0.7.

9.5 Model for potential outcomes, exchangeability of units and prior distribution

We first specify a model for the joint probability distribution of the two potential outcomes $(Y(0), Y(1))$ for all units in a target super population with no covariates. Let this probability distribution be denoted by $\text{pr}\,(Y(0), Y(1)|\theta)$ where $\text{pr}(\cdot|\theta)$ denotes a probability function indexed by unknown parameter θ. Because at this point, we have no way to distinguish between the units, indices $1, \ldots, N$ are simply a random permutation of labels $1, \ldots, N$; therefore, the units, i.e., the rows of $(\mathbf{Y}(0), \mathbf{Y}(1))$ have a row-exchangeable distribution, i.e., their joint distribution is unchanged under permutation of the row labels. This fact implies, without essentially no loss of generality, that the rows of $(\mathbf{Y}(0), \mathbf{Y}(1))$ can be modeled as independent and identically distributed, given some parent vector parameter θ;

$$\text{pr}\,(\mathbf{Y}(0), \mathbf{Y}(1)|\theta) = \prod_{i=1}^{N} \text{pr}\,(Y_i(0), Y_i(1)|\theta)\,.$$

This result is known as de Finetti's theorem [22, 23, 25] and is fundamental to applied Bayesian inference. Thus the effort becomes focussed on the specifiation of the unit-level distribution of $(Y_i(0), Y_i(1))$ given the parameter θ and the prior (or marginal) distribution of θ, $\text{pr}(\theta)$.

Specifying the joint distribution of $(Y_i(0), Y_i(1))$ conditional on θ can still be a difficult task, and generally requires domain knowledge. However, in many cases, and particularly for randomized experiments, inferences are often robust to model specification. For ease of exposition, we will start with models in which conditional on θ, $Y_i(0)$ and $Y_i(1)$ are independent, and later (in Section 9.9, extend this specification to models where their mutual dependence is allowed.

For Example 9.1, we start with the simplifying assumption that the binary potential outcomes $Y_i(0)$ and $Y_i(1)$ are independently distributed Bernoulli random variables with parameters π_0 and π_1 respectively, so that, $\theta = (\pi_0, \pi_1)$. Then, the joint distribution of $Y_i(1)$ and $Y_i(0)$ conditional on parameter $\theta = (\pi_0, \pi_1)$ is:

$$\Pr(Y_i(0) = y_0, Y_i(1) = y_1 | \pi_0, \pi_1) = \pi_0^{y_0}(1-\pi_0)^{1-y_0}\pi_1^{y_1}(1-\pi_1)^{1-y_1}, \quad (9.4)$$

where $y_0, y_1 \in \{0,1\}$ and $0 \le \pi_0, \pi_1 \le 1$.

Next, we need to provide a *prior* distribution $\text{pr}(\theta)$ on the unknown vector parameter θ. Although this specification is not always straightforward, sometimes the practical results are not particularly sensitive to the choice of the prior distribution and sensitivity analyses can always be conducted to exmaine the impact of changing the prior distribution on the resultant analysis. In the current example, because the parameters π_0 and π_1 represent probabilities, it is natural in mathematical statistics to specifiy Beta distributions with parameters (α_0, β_0) and (α_1, β_1) for π_0 and π_1 respectively. Formally, we can write

$$\pi_0 \sim \text{Beta}(\alpha_0, \beta_0) \perp\!\!\!\perp \pi_1 \sim \text{Beta}(\alpha_1, \beta_1),$$

where the symbol $\perp\!\!\!\perp$ represents independence (Dawid, 1979 [21]). The respective probability density functions of π_0 and π_1 are then (see Appendix 10.2)

$$\begin{aligned} \text{pr}(\pi_0) &= \frac{1}{B(\alpha_0,\beta_0)}\pi_0^{\alpha_0-1}(1-\pi_0)^{\beta_0-1}, \ 0<\pi_0<1, \alpha_0>0, \beta_0>0, \\ \text{pr}(\pi_1) &= \frac{1}{B(\alpha_1,\beta_1)}\pi_1^{\alpha_1-1}(1-\pi_1)^{\beta_1-1}, \ 0<\pi_1<1, \alpha_1>0, \beta_1>0. \quad (9.5) \end{aligned}$$

where $\text{B}(\alpha_0, \beta_0)$ and $\text{B}(\alpha_1, \beta_1)$ are normalizing constants ensuring that the density functions integrate to one, and therefore satisfy $B(\alpha_1, \beta_1) = \int_0^1 \pi_1^{\alpha_1}(1-\pi_1)^{\beta-1}d\pi_1$ and $B(\alpha_0, \beta_0) = \int_0^1 \pi_0^{\alpha_0}(1-\pi_0)^{\beta-0}d\pi_0$.

The hyperparameters α_0, α_1, β_0 and β_1 represent the information we have about π_0 and π_1. For example, in the absence of much information or assumption, we can choose $\alpha_0 = \beta_0 = \alpha_1 = \beta_1 = 1$, so that both the prior distributions are Uniform over $[0,1]$. Now we calculate the joint probability distribution of the observed outcomes $\mathbf{Y}^{\text{obs}} = (y_1, \ldots, y_n)$ and the assignment vector $\mathbf{W}$, conditional on the parameter θ. Using the laws of conditional probability, this distribution can be written as

$$\text{pr}(\mathbf{Y}^{\text{obs}}, \mathbf{W}|\theta) = \text{pr}(\mathbf{Y}^{\text{obs}}|\mathbf{W}, \theta)\ \text{pr}(\mathbf{W}|\theta). \quad (9.6)$$

Because in a designed experiment, the distribution of the assignment vector $\mathbf{W}$ is known, the second term of (9.6) equals $\text{pr}(\mathbf{W})$, the known probability distribution of the assignment vector. Recall from (2.4) that for a completely randomized allocation of N units into two groups with N_0 and N_1 units, this probability distribution is

$$\Pr(\mathbf{W}=\mathbf{w}) = \begin{cases} \frac{1}{N!/(N_0!N_1!)} & \text{if } \sum_i w_i = N_1 \\ 0 & \text{otherwise.} \end{cases}.$$

Noting that $y_i = Y_i(w_i)$ if $W_i = w_i$, the first term of (9.6) can be written as

$$\begin{aligned}
&\mathrm{pr}(\mathbf{Y}^{\mathrm{obs}}|\mathbf{W},\theta) = \mathrm{pr}(Y_1(w_1),\ldots,Y_N(w_N)|\mathbf{W}=\mathbf{w},\theta)\\
&= \mathrm{pr}(Y_1(w_1),\ldots,Y_N(w_N)|\theta), \text{ because } Y_i(W_i) \perp\!\!\!\perp W_i \text{ for } i=1,\ldots,N\\
&= \prod_{i=1}^{N} \mathrm{pr}(Y_i(w_i)|\theta), \text{by the independence of potential outcomes across units}\\
&= \prod_{i:w_i=1} \mathrm{pr}(Y_i(1)|\theta) \prod_{i:w_i=0} \mathrm{pr}(Y_i(0)|\theta). \qquad (9.7)
\end{aligned}$$

Substituting (9.7) into (9.6), the joint probability distribution of observed outcomes and the assignment vector can be written as

$$\mathrm{pr}(\mathbf{Y}^{\mathrm{obs}},\mathbf{W}|\theta) \propto \mathrm{pr}(\mathbf{Y}^{\mathrm{obs}}|\mathbf{W},\theta) \propto \prod_{i:w_i=1} \mathrm{pr}(Y_i(1)|\theta) \prod_{i:w_i=0} \mathrm{pr}(Y_i(0)|\theta). \quad (9.8)$$

where $\propto$ denotes "proportional to". This probability function $\mathrm{pr}(\mathbf{Y}^{\mathrm{obs}},\mathbf{W}|\theta)$ can also be viewed as a function of θ conditional on the observed data and is commonly referred to as the *likelihood function* in standard textbooks on mathematical statistics, e.g., [42, 14]. Returning to example 9.1, substituting (9.4) into (9.8) yields

$$\begin{aligned}
\mathrm{pr}(\mathbf{Y}^{\mathrm{obs}},\mathbf{W}|\theta) &\propto \prod_{i:w_i=1} \pi_1^{w_i}(1-\pi_1)^{1-w_i} \prod_{i:w_i=0} \pi_0^{w_i}(1-\pi_0)^{1-w_i}\\
&= \pi_1^{n_{11}}(1-\pi_1)^{n_{10}}\pi_0^{n_{01}}(1-\pi_0)^{n_{00}} \qquad (9.9)\\
&= \pi_1^{10}(1-\pi_1)^{0}\pi_0^{7}(1-\pi_0)^{3}. \qquad (9.10)
\end{aligned}$$

where n_{11}, n_{10}, n_{01} and n_{00} are the summary counts in a 2×2 representation of the binary data shown in Table 2.5. The values of n_{11}, n_{10}, n_{01} and n_{00} for Example 9.1 as obtained from Table 9.1 are summarized in Table 9.2.

9.6 Posterior distribution of parameters and superpopulation inference

The next step is to draw inference for the unknown hyperparameter θ, conditional on the observed data. This step is an essential part of *Bayesian inference*

TABLE 9.2
2×2 table of observed binary outcomes from Table 9.1 and their marginal totals

	$y=1$	$y=0$	Total
$W=1$	$n_{11}=10$	$n_{10}=0$	$N_1=10$
$W=0$	$n_{01}=7$	$n_{00}=3$	$N_0=10$
Total	$n_{+1}=17$	$n_{+0}=3$	$N=20$

(see Appendix 11.4 and Gelman et al. 2013 [38]). From (11.4) of Appendix 11.4, for the prior distrubution $\text{pr}(\theta)$, the posterior distribution of θ given $\mathbf{Y}^{\text{obs}}, \mathbf{W}$ is

$$\text{pr}(\theta|\mathbf{Y}^{\text{obs}}, \mathbf{W}) \propto \text{pr}(\mathbf{Y}^{\text{obs}}, \mathbf{W}|\theta)\text{pr}(\theta), \tag{9.11}$$

Recall the superpopulation average treatment effect τ^{SP} defined in (9.2). Because both $\mathbb{E}[Y_i(1)]$ and $\mathbb{E}[Y_i(0)]$ depend on the parameter θ (for example, on π_0 and π_1 in Example 9.1), τ^{SP} is a function of θ. Therefore, the posterior distribution of τ^{SP} conditional on the observed data can be derived by using the posterior distribution of θ obtained from (9.11).

We now derive the posterior distribution of τ^{SP} and the inference on τ^{SP} for example 9.1. Because $Y_i(1)$ and $Y_i(0)$ are assumed to be independent Bernoulli random variables with parameters π_1 and π_0 respectively in (9.4), we have $\mathbb{E}[Y_i(1)] = \pi_1$ and $\mathbb{E}[Y_i(0)] = \pi_0$, and consequently, $\tau^{\text{SP}} = \pi_1 - \pi_0$. The posterior distribution of $\theta = (\pi_0, \pi_1)$ is obtained by substituting (9.5) without the proportionality constant and (9.9) on the right hand side of (9.11):

$$\begin{aligned}
&\text{pr}(\pi_0, \pi_1|\mathbf{Y}^{\text{obs}}, \mathbf{W}) \\
\propto\ & \pi_1^{n_{11}}(1-\pi_1)^{n_{10}}\pi_0^{n_{01}}(1-\pi_0)^{n_{00}}\pi_1^{\alpha_1-1}(1-\pi_1)^{\beta_1-1}\pi_0^{\alpha_0-1}(1-\pi_0)^{\beta_0-1} \\
\propto\ & \pi_1^{n_{11}+\alpha_1-1}(1-\pi_1)^{n_{10}+\beta_1-1}\pi_0^{n_{01}+\alpha_0-1}(1-\pi_0)^{n_{00}+\beta_0-1}.
\end{aligned} \tag{9.12}$$

The expression on the right hand side of (9.12) is the product of two probability density functions: first a $\text{Beta}(n_{11}+\alpha_1, n_{10}+\beta_1)$ random variable and second a $\text{Beta}(n_{01}+\alpha_0, n_{00}+\beta_0)$ random variable. Thus we can conclude that under the postulated model of independent Bernoulli potential outcomes, $Y_i(0)$ and $Y_i(1)$, governed by parameters π_0 and π_1, and Beta priors on the parameters π_0 and π_1, the posterior distributions of π_0 and π_1 are independent $\text{Beta}(n_{11}+\alpha_1, n_{10}+\beta_1)$ and $\text{Beta}(n_{01}+\alpha_0, n_{00}+\beta_0)$ random variables respectively.

Therefore, for the independent potential outcomes model with binary outcomes, the prior and posterior distributions of the parameters π_1 and π_0 are summarized in Table 9.3. Note that in both cases, π_1 and π_0 are independent Beta variables.

The posterior disribution of τ^{SP} is thus the distribution of the difference of independent $\text{Beta}(n_{11}+\alpha_1, n_{10}+\beta_1)$ and $\text{Beta}(n_{01}+\alpha_0, n_{00}+\beta_0)$ variables.

TABLE 9.3
Independent prior and posterior distributions of π_1 and π_0

Parameter	Prior distribution	Posterior distribution
π_1	Beta(α_1, β_1)	Beta($n_{11}+\alpha_1, n_{10}+\beta_1$)
π_0	Beta(α_0, β_0)	Beta($n_{01}+\alpha_0, n_{00}+\beta_0$)

Setting $\alpha_0 = \beta_0 = \alpha_1 = \beta_1 = 1$, i.e., assuming Uniform[0,1] prior distributions for π_0 and π_1, and substituting the summary data, $n_{11} = 10$, $n_{10} = 0$, $n_{01} = 7$ and $n_{00} = 3$ from Table 9.2 into the second column of (9.3), the posterior distribution of τ^{SP} in example 9.1 is

$$[\tau^{\text{SP}}|\mathbf{Y}^{\text{obs}}, \mathbf{W}] \sim [\pi_1|\mathbf{Y}^{\text{obs}}, \mathbf{W}] - [\pi_0|\mathbf{Y}^{\text{obs}}, \mathbf{W}],$$

where $\pi_1|\mathbf{Y}^{\text{obs}}, \mathbf{W}$ and $\pi_0|\mathbf{Y}^{\text{obs}}, \mathbf{W}$ are independent Beta$(11, 1)$ and Beta$(8, 4)$ variables. Figure 9.3 shows the prior and posterior densities of π_1, π_0 and τ^{SP} based on 10000 draws from the prior and posterior distributions using the following code (the first six draws are shown in the output).

```
------------------------------------------------------------
alpha0 = beta0 = alpha1 = beta1 = 1
n11 = 10
n10 = 0
n01 = 7
n00 = 3

set.seed(1120)
P0_prior <- rbeta(10000,alpha0,beta0)
P1_prior <- rbeta(10000,alpha1,beta1)
P0_post <- rbeta(10000,n01+alpha0, n00+beta0 )
P1_post <- rbeta(10000,n11+alpha1, n10+beta1 )

library(tidyverse)
tauSP <- P0 %>% mutate(P1) %>%
  mutate(tauSP_prior = P1_prior - P0_prior) %>%
  mutate(tauSP_post = P1_post - P0_post)

> head(tauSP)
# A tibble: 6 × 6
  P0_prior P0_post  P1_prior P1_post tauSP_prior tauSP_post
     <dbl>   <dbl>     <dbl>   <dbl>       <dbl>      <dbl>
1    0.352   0.822 0.547       0.993      0.196       0.171
2    0.827   0.549 0.898       0.917      0.0710      0.368
3    0.962   0.429 0.327       0.872     -0.635       0.443
4    0.443   0.553 0.851       0.895      0.407       0.343
5    0.718   0.613 0.661       0.933     -0.0572      0.321
6    0.897   0.583 0.0000769   0.884     -0.897       0.301
------------------------------------------------------------
```

Figure 9.3 shows how the knowledge about the parameters π_0 and π_1 changes from the state of almost no information (as indicated by the solid

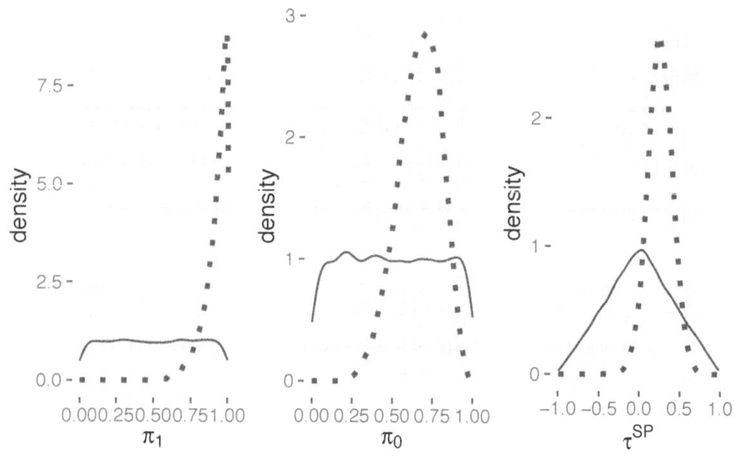

FIGURE 9.3
Prior (solid lines) and Posterior densities (dotted lines) of π_1, π_0 and τ^{SP} for Example 9.1 under the independent potential outcomes model (based on 10000 draws)

curves) before the experiment to states of higher concentrations around the modes, 0.67 and 0.92, after observing the outcomes. Further, the distribution of τ^{SP} shrinks from a distribution centered around 0 to one centered around 0.25 (see the R code below), indicating that the observed data suggests the presence of a superpopulation treatment effect.

To formalize the inference on the superpopulation average treatment effect, a point estimator of τ^{SP} is the expectation of its posterior distribution, i.e.

$$\widehat{\tau^{\text{SP}}} = \mathbb{E}[\tau^{\text{SP}}|\mathbf{Y}^{\text{obs}}, \mathbf{W}],$$

estimated as 0.25 from 10000 draws. A $100(1-\alpha)\%$ "posterior" or "credible" interval can be obtained as the interval between two values that cover $(1-\alpha)\%$ of the area under the posterior density curve. We can choose the lower α_1th and upper α_2th quantiles for $0 < \alpha_1, \alpha_2 < 1$ satisfying $\alpha_1 + \alpha_2 = 1 - \alpha$. For a symmetric distribution $\alpha_1 = \alpha_2 = \alpha/2$ would be a natural choice, but not necessarily for an asymmetric distribution like the posterior distribution of τ^{SP} shown in the right panel of Figure 9.3. Two such 95% intervals for τ^{SP} are derived for $\alpha_1 = \alpha_2 = 0.025$ and for $\alpha_1 = 0.049$ and $\alpha_2 = 0.001$ below.

```
----------------------------------------------------------
> mean(tauSP$tauSP_post)
[1] 0.2517442

> c(quantile(tauSP$tauSP_post,.049),quantile(tauSP$tauSP_post,.999))
       4.9%       99.9%
0.002823551 0.695342046
```

```
> # interval with alpha1= alpha2 = .025
> c(quantile(tauSP$tauSP_post,.025),quantile(tauSP$tauSP_post,.975))
       2.5%        97.5%
-0.04836514  0.55696566

> # interval with alpha1=.049 and alpha2=.001
> c(quantile(tauSP$tauSP_post,.049),quantile(tauSP$tauSP_post,.999))
       4.9%        99.9%
0.002823551 0.695342046
```

9.7 Model for imputing missing outcomes and finite population inference

We now return to the finite population setting and see how the model for potential outcomes can help to improve the naive strategies for imputation of missing potential outcomes presented earlier. To obtain an imputation model $\text{pr}(\mathbf{Y}^{\text{mis}}|\mathbf{Y}^{\text{obs}}, \mathbf{W})$, we can formulate the problem using two components. The first one is $\text{pr}(\mathbf{Y}^{\text{mis}}|\mathbf{Y}^{\text{obs}}, \mathbf{W}, \theta)$, the conditional distribution of the missing outcomes conditional on the observed outcomes, observed assignment *and* the parameter θ. The second component is the posterior distribution of θ, $\text{pr}(\theta|\mathbf{Y}^{\text{obs}}, \mathbf{W})$, derived in Section 9.6, that will be combined with $\text{pr}(\mathbf{Y}^{\text{mis}}|\mathbf{Y}^{\text{obs}}, \mathbf{W}, \theta)$ to "integrate out" or "average over" to obtain the imputation model:

$$\text{pr}(\mathbf{Y}^{\text{mis}}|\mathbf{Y}^{\text{obs}}, \mathbf{W}) = \int \text{pr}(\mathbf{Y}^{\text{mis}}|\mathbf{Y}^{\text{obs}}, \mathbf{W}, \theta)\ \text{pr}(\theta|\mathbf{Y}^{\text{obs}}, \mathbf{W})\ d\theta. \tag{9.13}$$

To obtain the first component of (9.13), i.e., $\text{pr}(\mathbf{Y}^{\text{mis}}|\mathbf{Y}^{\text{obs}}, \mathbf{W}, \theta)$, we note that

$$Y_i^{\text{mis}} = \begin{cases} Y_i(1) & \text{if} \quad W_i = 0 \\ Y_i(0) & \text{if} \quad W_i = 1. \end{cases}$$

Also, we can write the observed and missing potential outcomes for unit i as

$$Y_i^{\text{obs}} = W_iY_i(1) + (1 - W_i)Y_i(0), \tag{9.14}$$

$$Y_i^{\text{mis}} = W_iY_i(0) + (1 - W_i)Y_i(1), \tag{9.15}$$

Consequently,

$$\text{pr}(Y_i^{\text{mis}}|Y_i^{\text{obs}}, W_i, \theta) = \begin{cases} \text{pr}(Y_i(1)|Y_i(0), \theta) & \text{if} \quad W_i = 0 \\ \text{pr}(Y_i(0)|Y_i(1), \theta) & \text{if} \quad W_i = 1. \end{cases} \tag{9.16}$$

To obtain $\text{pr}(Y_i^{\text{mis}}|Y_i^{\text{obs}}, W_i, \theta)$ from (9.16), we need to derive the conditional distributions of $Y_i(1)|Y_i(0), \theta$ and $Y_i(0)|Y_i(1), \theta$. Such conditional distributions can be obtained from the joint model of $(Y_i(0), Y_i(1)|\theta)$. If the potential outcomes $Y_i(0)$ and $Y_i(1)$ are assumed independent, as in Section 8.3.1 and in the context of Example 9.1 so far, then (9.16) simply reduces to

$$\mathrm{pr}(Y_i^{\mathrm{mis}}|Y_i^{\mathrm{obs}}, W_i, \theta) = \begin{cases} \mathrm{pr}(Y_i(1)|\theta) & \text{if} \quad W_i = 0 \\ \mathrm{pr}(Y_i(0)|\theta) & \text{if} \quad W_i = 1. \end{cases} \tag{9.17}$$

Note that, to generate a draw of $\mathbf{Y}^{\mathrm{mis}}$ from $\mathrm{pr}(\mathbf{Y}^{\mathrm{mis}}|\mathbf{Y}^{\mathrm{obs}}, \theta)$ using (9.13), we neither need to obtain closed-form expressions from for $\mathrm{pr}(\theta|\mathbf{Y}^{\mathrm{obs}}, \mathbf{W})$ or $\mathrm{pr}(\mathbf{Y}^{\mathrm{mis}}|\mathbf{Y}^{\mathrm{obs}}, \mathbf{W}, \theta)$, nor need to evaluate the integral. Equation (9.13) provides us with the following simple two-step simulation strategy for generating draws of $\mathbf{Y}^{\mathrm{mis}}$ from $\mathrm{pr}(\mathbf{Y}^{\mathrm{mis}}|\mathbf{Y}^{\mathrm{obs}}, \mathbf{W})$:

Procedure 9.1 (Imputing missing outcomes) *Generating draws from* $pr(\mathbf{Y}^{mis}|\mathbf{Y}^{obs}, \mathbf{W})$ *involves the following two steps:*

1. *Generate a draw* $\tilde{\theta}$ *of* θ *from the posterior distribution* $pr(\theta|\mathbf{Y}^{obs}, \mathbf{W})$.

2. *For* $i = 1, \ldots, N$, *generate a draw of* Y_i^{mis} *from* $pr(Y_i^{mis}|Y_i^{obs}, W_i, \tilde{\theta})$ *using (9.16).*

An important point to notice is that, in step 2 of Procedure 9.1, we impute a complete set of missing data $Y_1^{\mathrm{mis}}, \ldots, Y_N^{\mathrm{mis}}$ on the basis of the same draw $\tilde{\theta}$ of parameter θ. Generating a new $\tilde{\theta}_i$ for each $i = 1, \ldots, N$ and using it to impute Y_i^{mis} would be incorrect.

Procedure 9.1 can be called a *phenomenological approach* (as in Rubin (1978) [80]) in the sense that it is driven by the observed phenomena, i.e., data, where the parameters are just hypothetical devices used to formalize the correct way to impute missing data.

Inference for finite population average treatment effect $\overline{\tau}$: Once the missing potential outcomes are imputed, we have the full set $(Y_i(0), Y_i(1))$ of potential outcomes for $i = 1, \ldots, N$, and can *compute* the finite population average treatment effect $\overline{\tau} = \left(\overline{Y}(1) - \overline{Y}(0)\right)$. Repeating such imputations multiple times, we can obtain a distribution of $\overline{\tau}$, and use it to draw inference, i.e., to construct point and interval estimates.

Returning to Example 9.1, recall that the joint distribution of $(Y_i(0), Y_i(1)|\pi_0, \pi_1)$ was given by (9.4), and the parameters π_0 and π_1 were assumed to have independent $\mathrm{Beta}(\alpha_0, \beta_0)$ and $\mathrm{Beta}(\alpha_1, \beta_1)$ prior distributions. In Section 9.6 we obtained the posterior distributions of π_0 and π_1 as independent Beta random variables with parameters specified in Table 9.3. Specifically, using the observed outcomes in Table 9.1, the posterior distributions of π_1 and π_0 were $\mathrm{Beta}(11, 1)$ and $\mathrm{Beta}(8, 4)$ respectively. Thus, following Procedure 9.1, to impute the missing outcomes in Table 9.1, we first draw $\widetilde{\pi}_1$ from $\mathrm{Beta}(11, 1)$ and $\widetilde{\pi}_0$ from $\mathrm{Beta}(8, 4)$. Next, for $i = 1, \ldots, 20$, impute

$$Y_i^{\mathrm{mis}} \sim \begin{cases} \mathrm{Bernoulli}(\widetilde{\pi}_1) & \text{if} \quad W_i = 0 \\ \mathrm{Bernoulli}(\widetilde{\pi}_0) & \text{if} \quad W_i = 1 \end{cases}$$

Below, we show a single imputation of the missing potential outcomes in Table 9.1 using the approach just described, implemented through a function Ypot_Imp_indep() that generates an imputed table of potential outcomes and thereby the value of the finite population estimand $\overline{\tau}$. The value of $\overline{\tau}$ obtained from the imputed matrix of potential outcomes is obtained as 0.25.

```
---------------------------------------------------------
# The following function takes a matrix like the one in Table 1 as input
# (i.e., with columns x1, x2, w, Y(0), Y(1))
# and returns (a) An imputed matrix of potential outcomes and (b) taubar.

Ypot_imp_indep <- function(Ypot){
pi0 = rbeta(1,8,4)
pi1 = rbeta(1,11,1)
for (i in 1:20) {
  if (W[i]==1) {
    ymis = rbinom(1,1,pi0)
    Ypot[i,4] = ymis
  } else {
    ymis = rbinom(1,1,pi1)
    Ypot[i,5] = ymis
  }
}
taubar = mean(Ypot[,5]) - mean(Ypot[,4])
ImpYpot <- Ypot[,4:5] # Imputed matrix of potential outcomes
return(list(ImpYpot,taubar))
} # end function

> set.seed(1)
> Ypot_imp_indep(Ypot)
[[1]]
   Y0 Y1
1   1  1
2   0  1
3   1  0
4   1  1
5   1  1
6   1  1
7   0  1
8   1  1
9   1  1
10  1  1
11  1  1
12  1  1
13  1  1
14  0  1
15  0  1
16  0  1
17  0  1
18  1  1
19  1  1
20  1  1

[[2]]
[1] 0.25
---------------------------------------------------------
```

The above imputation procedure was repeated 5000 times. Figure 9.4 shows the histogram of $\overline{\tau}$ obtained from these multiple imputations. The mean $\overline{\tau}$ and the lower and upper 0.025 quantiles of the distribution were obtained as 0.05 and 0.50, respectively.

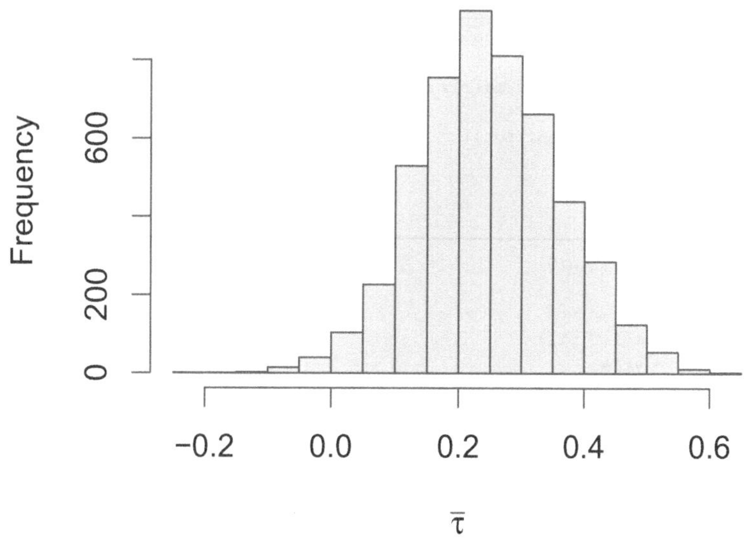

FIGURE 9.4
Histogram of $\overline{\tau}$ obtained from 5000 imputations of missing potential outcomes in Table 9.1

9.8 Independent normal potential outcomes model

Example 9.2 *In Chapter 2, we introduced a partial version of the experiment reported by Shadish et al. (2008) in [85] in which 235 undergraduate students from introductory psychology classes at a large mid-southern U.S. public university received either the treatment (vocabulary training) or control (math training) through a completely randomized treatment assignment. The number of students assigned to treatment versus control were 116 and 119, respectively. To demonstrate the model-based approach, we consider a subset of 20 students chosen randomly from this pool of 235, ensuring that 10 are treated and 10 are exposed to control. Here we assume that these 20 students constitute the finite population for our study, and for each of them, we have four covariates (Vpre, likelit, actcomp and collegegpa), the treatment* *assignment W and the outcome y (vocabulary test score) are shown in Table 9.4 for that 20 students.*

TABLE 9.4
20 units from partial Shadish experiment

Unit	Vpre	likelit	actcomp	collegegpa	W	y
1	27	5	22	3.00	0	6
2	24	7	23	3.20	0	9
3	15	6	22	2.60	1	19
4	23	7	22	2.37	1	18
5	10	7	15	1.68	1	17
6	21	6	20	2.29	1	16
7	24	7	26	3.19	0	10
8	20	9	19	1.62	0	6
9	26	8	24	3.84	0	9
10	24	3	24	3.25	0	8
11	17	5	17	2.39	0	6
12	18	6	18	1.18	1	17
13	19	7	19	3.14	1	12
14	19	7	20	3.25	0	9
15	19	5	22	2.81	0	12
16	26	9	24	3.68	1	20
17	21	7	20	2.78	1	19
18	21	7	16	2.13	1	8
19	20	5	20	1.96	1	15
20	21	5	22	3.08	0	6

We assume that $Y_i(0)$ and $Y_i(1)$ are independent normal variables with unknown means μ_0 and μ_1 and common known variances σ_0^2 and σ_1^2 respectively, i.e.,

$$Y_i(0)|\mu_0 \sim N(\mu_0, \sigma_0^2) \perp\!\!\!\perp Y_i(1)|\mu_1 \sim N(\mu_1, \sigma_1^2).$$

Independent normal priors are postulated for the means μ_0 and μ_1:

$$\mu_0|m_0, \tau_0^2 \sim \mathcal{N}(m_0, \tau_0^2) \perp\!\!\!\perp \mu_1|m_1, \tau_1^2 \sim \mathcal{N}(m_1, \tau_1^2)$$

where the "hyper-parameters" m_0, m_1, τ_0^2 and τ_1^2 have fixed values and are assumed known. Thus, the unknown parameter vector θ under this model is (μ_0, μ_1).

Then, following Appendix A of Chapter 8 of Imbens and Rubin (2015) [44], the posterior distributions of μ_0 and μ_1, given the observed data are independently distributed normal variables:

$$\begin{aligned} \mu_0|\mathbf{Y}^{\text{obs}}, \mathbf{W} &\sim \mathcal{N}\left(\frac{\overline{y}(0)\,(N_0/\sigma_0^2) + m_0\,(1/\tau_0^2)}{N_0/\sigma_0^2 + 1/\tau_0^2}, \frac{1}{N_0/\sigma_0^2 + 1/\tau_0^2}\right), \\ \mu_1|\mathbf{Y}^{\text{obs}}, \mathbf{W} &\sim \mathcal{N}\left(\frac{\overline{y}(1)\,(N_1/\sigma_1^2) + m_1\,(1/\tau_1^2)}{N_1/\sigma_1^2 + 1/\tau_1^2}, \frac{1}{N_1/\sigma_1^2 + 1/\tau_1^2}\right). \end{aligned} \quad (9.18)$$

The posterior means μ_0 and μ_1 shown in (9.18) are actually quite intuitive. The mean of μ_0 is a weighted combination of $\overline{y}(0)$, the mean of the observed potential outcomes in the control group, and the prior mean m_0, the weights being proportional to the inverse of the variances associated with the two estimates. The variance associated with $\overline{y}(0)$ is σ_0^2/N_0, whereas the variance of the prior distribution m_0 is τ_0^2. Similarly, the variance associated with $\overline{y}(1)$ is σ_1^2/N_1, whereas the variance of the prior distribution m_1 is τ_1^2. Thus, the posterior means of μ_0 and μ_1 shown in (9.18) can be written as the weighted sum

$$\mathbb{E}[\mu_0|\mathbf{Y}^{\text{obs}}, \mathbf{W}] = \delta_0\overline{y}(0) + (1 - \delta_0)m_0,$$
$$\mathbb{E}[\mu_1|\mathbf{Y}^{\text{obs}}, \mathbf{W}] = \delta_0\overline{y}(1) + (1 - \delta_1)m_1,$$

where

$$\delta_0 = \frac{N_0/\sigma_0^2}{N_0/\sigma_0^2 + 1/\tau_0^2}, \; \delta_1 = \frac{N_1/\sigma_1^2}{N_1/\sigma_1^2 + 1/\tau_1^2}. \tag{9.19}$$

Hence, the posterior distribution of the superpopulation estimand $\tau^{\text{SP}} = \mathbb{E}[Y_i(1)] - \mathbb{E}[Y_i(0)] = \mu_1 - \mu_0$ is normal with posterior mean

$$\mathbb{E}\left(\mu_1 - \mu_0|\mathbf{Y}^{\text{obs}}, \mathbf{W}\right) = \delta_1\overline{y}(1) + (1 - \delta_1)m_1 - \delta_0\overline{y}(0) - (1 - \delta_0)m_0, \tag{9.20}$$

where δ_0 and δ_1 are given by (9.19), and posterior variance

$$\text{Var}\left(\mu_1 - \mu_0|\mathbf{Y}^{\text{obs}}, \mathbf{W}\right) = \frac{1}{N_0/\sigma_0^2 + 1/\tau_0^2} + \frac{1}{N_1/\sigma_1^2 + 1/\tau_1^2}. \tag{9.21}$$

Let us now draw inference on τ^{SP} for Example 9.2, assuming that $\sigma_0^2 = \sigma_0^2 = 25$, $m_0 = m_1 = 0$ and $\tau_0^2 = \tau_1^2 = 100$. Substituting these values along with $\overline{y}(1) = 16.1$, $\overline{y}(0) = 8.1$ (calculated from Table 9.4 as shown in the code below) and $N_1 = N_0 = 10$ into (9.20) and (9.21), it follows that $\tau^{\text{SP}}|\mathbf{Y}^{\text{obs}}, \mathbf{W} \sim \mathcal{N}(7.80, 4.88)$. A 95% interval for τ^{SP} based on this posterior distribution is obtained as [3.48, 12.13].

```
-----------------------------------------------------------------
# The data in Table 8.4 is stored in data frame SRV (Shadish randomized verbal)
> library(tidyverse)
> head(SRV)
  Vpre likelit actcomp collegegpa W  y
1   27       5      22       3.00 0  6
2   24       7      23       3.20 0  9
3   15       6      22       2.60 1 19
4   23       7      22       2.37 1 18
5   10       7      15       1.68 1 17
6   21       6      20       2.29 1 16
> (y1bar = mean(SRV$y[SRV$W==1]))
[1] 16.1
> (y0bar = mean(SRV$y[SRV$W==0]))
[1] 8.1
> N1 = N0 = 10
> m1 = m0 = 0
```

```
> sigma0_sq = sigma1_sq = 25
> tau0_sq = tau1_sq = 100
> delta0 = (N0/sigma0_sq)/(N0/sigma0_sq + 1/tau0_sq)
> delta1 = (N1/sigma1_sq)/(N1/sigma1_sq + 1/tau1_sq)
> (mu1mean = delta1 * y1bar + (1-delta1)*m1)
[1] 15.70732
> (mu0mean = delta0 * y0bar + (1-delta0)*m0)
[1] 7.902439
> (mu1var = 1/(N1/sigma1_sq + 1/tau1_sq))
[1] 2.439024
> (mu0var = 1/(N0/sigma0_sq + 1/tau1_sq))
[1] 2.439024
> (mean_tauSP = mu1mean - mu0mean)
[1] 7.804878
> (var_tauSP = mu1var + mu0var)
[1] 4.878049
> c(qnorm(.025,mean_tauSP,sqrt(var_tauSP)),qnorm(.975,mean_tauSP,
  sqrt(var_tauSP)))
[1]  3.476042 12.133714
-----------------------------------------------------------------
```

To draw inference on the finite population average treatment effect $\overline{\tau}$, we need to impute the missing potential outcomes, for which we need to obtain the imputation model $\text{pr}(\mathbf{Y}^{\text{mis}}|\mathbf{Y}^{\text{obs}}, \mathbf{W})$ using (9.13). We have already obtained the second term within the integral in the right hand side of $\text{pr}(\mathbf{Y}^{\text{mis}}|\mathbf{Y}^{\text{obs}}, \mathbf{W})$, i.e., the posterior distribution $\text{pr}(\theta|\mathbf{Y}^{\text{obs}}, \mathbf{W})$, where $\theta = (\mu_0, \mu_1)$, in (9.18). To evaluate the first term $\text{pr}(\mathbf{Y}^{\text{mis}}|\mathbf{Y}^{\text{obs}}, \mathbf{W}, \mu_0, \mu_1)$, note that by the independence of $Y_i(1)$ and $Y_i(0)$ conditional on μ_0, μ_1, it follows that

$$(Y_i^{\text{mis}}|Y_i^{\text{obs}}, W_i, \mu_0, \mu_1) \sim \begin{cases} Y_i(1)|\mu_1 \sim N(\mu_1, \sigma_1^2) & \text{if} \quad W_i = 0 \\ Y_i(0)|\mu_0 \sim N(\mu_0, \sigma_0^2) & \text{if} \quad W_i = 1. \end{cases} \tag{9.22}$$

Following Procedure 9.1, we can first draw $\widetilde{\mu}_1$ and $\widetilde{\mu}_0$ from (9.18), and then generate the $N \times 1$ vector $\mathbf{Y}^{\text{mis}}$ from (9.22) by substituting $\widetilde{\mu}_0$ for μ_0 and $\widetilde{\mu}_1$ for μ_1.

To illustrate this procedure for Example 9.2, we first convert the pair of columns (y, W) in Table 9.4 to a 20×2 matrix of potential outcomes.

```
-----------------------------------------------------------------
# Split the column y in data frame SRV into two columns based on W
> Ypot = matrix(numeric(20*2),ncol=2)
> for (i in 1:20){
+   if (SRV$W[i]==1) {
+     Ypot[i,1] = NA
+     Ypot[i,2] = SRV$y[i]
+   } else {
+     Ypot[i,1] = SRV$y[i]
+     Ypot[i,2] = NA
+   }
+ }
# Print first six rows of the potential outcomes
> head(Ypot)
     [,1] [,2]
```

```
[1,]    6   NA
[2,]    9   NA
[3,]   NA   19
[4,]   NA   18
[5,]   NA   17
[6,]   NA   16
-------------------------------------------------------------------
```

The following R function imputes missing potential outcomes by first drawing $\widetilde{\mu}_1$ and $\widetilde{\mu}_0$ from (9.18), and then generating the $N \times 1$ vector $\mathbf{Y}^{\mathrm{mis}}$ from (9.22) by substituting $\widetilde{\mu}_0$ for μ_0 and $\widetilde{\mu}_1$ for μ_1. The function is run once with $\sigma_0^2 = \sigma_1^2 = 25$, $m_0 = m_1 = 0$ and $\tau_0^2 = \tau_1^2 = 100$ to generate a complete imputed matrix of all potential outcomes.

```
-------------------------------------------------------------------
# This function takes the following inputs:
# (1) An Nx2 matrix of potential outcomes with N0 and N1 observed outcomes
# for control and treatment, missing outcomes denoted by NA
# (2) Known variances of Y0 and Y1 denoted by sigma0_sq and sigma1_sq
      (set to default 25)
# (3) prior means m0 and m1 (set to default zero) of the normal means mu0
      and mu1
# (4) prior variances tau0_sq and tau1_sq (set to default zero) of the normal
      means mu0 and mu1

Ypot_imp_indep_normal <-
  function(Ypot,sigma0_sq=25,sigma1_sq=25,m0=0,m1=0,tau0_sq=100,tau1_sq=100){
  Y0 = Ypot[,1]
  Y1 = Ypot[,2]
  N = nrow(Ypot)
  W = is.na(Y0)
  N1 = sum(W)
  N0 = N - N1
  N = N0+N1
  y1bar = mean(Y1, na.rm=TRUE)
  y0bar = mean(Y0, na.rm=TRUE)
  delta0 = (N1/sigma0_sq)/(N1/sigma0_sq + 1/tau0_sq)
  delta1 = (N0/sigma1_sq)/(N0/sigma1_sq + 1/tau1_sq)
  mu1mean = delta1 * y1bar + (1-delta1)*m1 # posterior mean of mu1
  mu0mean = delta0 * y0bar + (1-delta0)*m0 # posterior mean of mu0
  mu1var = 1/(N1/sigma1_sq + 1/tau1_sq) # posterior var of mu1
  mu0var = 1/(N0/sigma0_sq + 1/tau0_sq) # posterior var of mu0

  mu0tilde = rnorm(1, mu0mean, mu0var) # draw from posterior of mu0
  mu1tilde = rnorm(1, mu1mean, mu1var) # draw from posterior of mu1
  for (i in 1:N) {
    if (W[i]==1) {
      ymis = rnorm(1,mu0tilde,sqrt(sigma0_sq))
      Ypot[i,1] = ymis
    } else {
      ymis = rnorm(1,mu1tilde,sqrt(sigma1_sq))
      Ypot[i,2] = ymis
    }
  }
  Ypot = round(Ypot)
  taubar = mean(Ypot[,2]) - mean(Ypot[,1])
```

```
  return(list(Ypot, taubar))
} # end function

>set.seed(20)
>Ypot_imp_indep_normal(Ypot)
# Here Ypot is the 20x2 matrix of potential outcomes derived from Table 8.4

[[1]]
      [,1] [,2]
 [1,]    6   23
 [2,]    9    8
 [3,]    9   19
 [4,]   14   18
 [5,]   -4   17
 [6,]    6   16
 [7,]   10   12
 [8,]    6   12
 [9,]    9   14
[10,]    8   14
[11,]    6   11
[12,]   17   17
[13,]    3   12
[14,]    9   12
[15,]   12   19
[16,]   11   20
[17,]   10   19
[18,]   13    8
[19,]   12   15
[20,]    6   14

[[2]]
[1] 6.4
-----------------------------------------------------------------
```

The imputed values of the missing potential outcomes and the corresponding imputed value of the finite population average treatment effect, $\overline{\tau} = 6.4$, are shown above as the output of the function. However, note that the imputed value for unit 5 is –4, which is an impossible value of the test score. The postulated normal model does not impose the constraint that all realistic imputed values must be positive. One way to reduce the probability of having negative values would be to reduce the values of the variance parameters σ_0^2 and σ_1^2, both of which were taken as 25 in this example. Another possibility, and the more preferred one, would be to specify a different model on the potential outcomes that would force them to be non-negative, e.g., a restricted normal model as in Rubin (1983).

```
-----------------------------------------------------------------
set.seed(555)
taubarvector <- NULL
for (i in 1:1000) {
  taubarvector = c(taubarvector, Ypot_imp_indep_normal(Ypot)[[2]])
}
> mean(taubarvector)
[1] 7.87665
```

```
> c(quantile(taubarvector,.025), quantile(taubarvector, .975))
    2.5%    97.5%
 3.85000 11.80125
>hist(taubarvector, main=NULL, xlab=expression(bar(tau)))
>abline(v=mean(taubarvector), lty=2, lwd=1.5)
------------------------------------------------------------
```

Imputing the missing potential outcomes 1000 times by repeated application of the imputation process (shown above) with the current choices of parameters, we obtain a distribution of the finite population estimand $\overline{\tau}$, shown in Figure 9.5. The mean of the distribution is 7.88, and a 95% interval that is symmetric around the mean is obtained as $[3.85, 11.80]$.

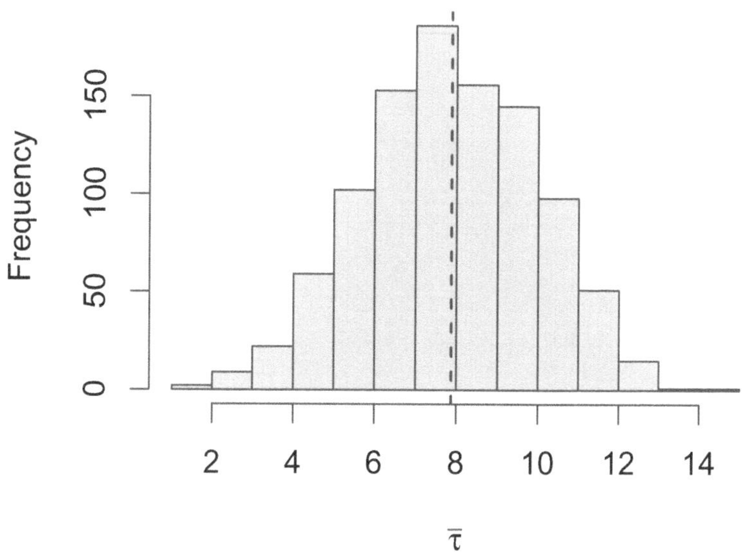

FIGURE 9.5
Histogram of $\overline{\tau}$ obtained from 1000 imputations of missing potential outcomes in Example 9.2

9.9 Dependent potential outcomes

As noted by many authors, an important decision in the model-based approach is the specification of the model of the joint distribution $\text{pr}(Y_i(0), Y_i(1)|\theta)$ of the unit-level potential outcomes. So far we have avoided this issue by assuming independence of $Y_i(0)$ and $Y_i(1)$ conditional on θ.

In this section, however, we consider a non trivial joint distribution of $(Y_i(0), Y_i(1))$. For a normal model, such a bivariate extension can be straightforward, although we believe that scientifically relevant further extensions are available. For binary outcomes as in Example 9.1, associations between potential outcomes can be based on the approach recommended by Ding and Dasgupta (2016) [27].

For a normal model, we can incorporate the dependence between the potential outcomes as follows:

$$\left(\begin{array}{c} Y_i(0) \\ Y_i(1) \end{array} \right) \Big| \theta \sim \mathcal{N}\left(\left(\begin{array}{c} \mu_0 \\ \mu_1 \end{array} \right), \left(\begin{array}{cc} \sigma_0^2 & \rho\sigma_0\sigma_1 \\ \rho\sigma_0\sigma_1 & \sigma_1^2 \end{array} \right) \right), \tag{9.23}$$

where as before, the variance parameters σ_0^2, σ_1^2 and the correlation coefficient ρ are considered known, so that $\theta = (\mu_0, \mu_1)$. The prior distribution for θ is assumed the same as in Section 9.8, i.e.,

$$\left(\begin{array}{c} \mu_0 \\ \mu_1 \end{array} \right) \Big| (m_0, m_1, \tau_0, \tau_1) \sim \mathcal{N}\left(\left(\begin{array}{c} m_0 \\ m_1 \end{array} \right), \left(\begin{array}{cc} \tau_0^2 & 0 \\ 0 & \tau_1^2 \end{array} \right) \right). \tag{9.24}$$

Under the model described by (9.23) and (9.24), the posterior distribution $\text{pr}(\theta|\mathbf{Y}^{\text{obs}}, \mathbf{W})$ is similar to the one in (9.18) of Section 9.8, that can be written as:

$$\begin{aligned} &\left(\begin{array}{c} \mu_0 \\ \mu_1 \end{array} \right) \Big| (\mathbf{Y}^{\text{obs}}, \mathbf{W}) \\ \sim\ &\mathcal{N}\left(\left(\begin{array}{c} \delta_0\overline{y}(0) + (1-\delta_0)m_0 \\ \delta_1\overline{y}(1) + (1-\delta_1)m_1 \end{array} \right), \left(\begin{array}{cc} \frac{1}{N_0/\sigma_0^2 + 1/\tau_0^2} & 0 \\ 0 & \frac{1}{N_1/\sigma_1^2 + 1/\tau_1^2} \end{array} \right) \right). \end{aligned}$$

Thus, under this model, inference for the superpopulation estimand τ^{SP} will remain unchanged in spite of introducing dependence between the potential outcomes. However, the imputation of missing values and the subsequent calculation of the finite population estimand $\overline{\tau}$ will change because now (9.16) will not reduce to (9.17). Because conditional on all parameters, $Y_i(0)$ and $Y_i(1)$ follow a bivariate normal distribution, it follows that the conditional distributions of one potential outcome given the other are also normal,

$$\begin{aligned} Y_i(1)|Y_i(0), \theta &\sim \mathcal{N}\left(\mu_1 + \rho\frac{\sigma_1}{\sigma_0}(Y_i(0) - \mu_0), (1-\rho^2)\sigma_1^2 \right) \\ Y_i(0)|Y_i(1), \theta &\sim \mathcal{N}\left(\mu_0 + \rho\frac{\sigma_0}{\sigma_1}(Y_i(1) - \mu_1), (1-\rho^2)\sigma_0^2 \right). \end{aligned}$$

Thus, in lieu of the imputation model (9.22), in this case (9.16) reduces to

$$\begin{aligned} &(Y_i^{\text{mis}}|Y_i^{\text{obs}}, W_i, \mu_0, \mu_1) \\ \sim &\left\{ \begin{array}{ll} Y_i(1)|Y_i(0), \mu_0, \mu_1 \sim \mathcal{N}\left(\mu_1 + \rho\frac{\sigma_1}{\sigma_0}(Y_i(0) - \mu_0), (1-\rho^2)\sigma_1^2 \right) & \text{if} \quad W_i = 0 \\ Y_i(0)|Y_i(1), \mu_0, \mu_1 \sim \mathcal{N}\left(\mu_0 + \rho\frac{\sigma_0}{\sigma_1}(Y_i(1) - \mu_1), (1-\rho^2)\sigma_0^2 \right) & \text{if} \quad W_i = 1. \end{array} \right. \end{aligned} \tag{9.25}$$

Following Procedure 9.1, we can first draw $\widetilde{\mu}_0$ and $\widetilde{\mu}_1$ from (9.18), and then generate the $N \times 1$ vector $\mathbf{Y}^{\text{mis}}$ from (9.25) by substituting $\mu_0 = \widetilde{\mu}_0$ and $\mu_1 = \widetilde{\mu}_1$. This change requires a minor adjustment in the R function Ypot_imp_indep_normal to create the new function Ypot_imp_dep_normal as shown below. The default values of $m_0, m_1, \sigma_0^2, \sigma_1^2, \rho, \tau_0^2$ and τ_1^2 are respectively set to 0, 0, 25, 25, 1, 100 and 100 respectively, but can be changed by the user.

```
------------------------------------------------------------------
# This function takes the following inputs:
# (1) An Nx2 matrix of potential outcomes with N0 and N1 observed outcomes
# for control and treatment, missing outcomes denoted by NA
# (2) Known variances of Y0 and Y1 denoted by sigma0_sq and sigma1_sq
      (set to default 25)
# (3) Known correlation rho (set to default 1)
# (4) prior means m0 and m1 (set to default zero) of the normal means mu0
      and mu1
# (5) prior variances tau0_sq and tau1_sq (set to default 100)
      of the normal means mu0 and mu1

Ypot_imp_dep_normal <-
  function(Ypot,sigma0_sq=25,sigma1_sq=25,rho=1, m0=0,m1=0,tau0_sq=100,tau1_
  sq=100){
  Y0 = Ypot[,1]
  Y1 = Ypot[,2]
  N = nrow(Ypot)
  W = is.na(Y0)
  N1 = sum(W)
  N0 = N - N1
  N = N0+N1
  y1bar = mean(Y1, na.rm=TRUE)
  y0bar = mean(Y0, na.rm=TRUE)
  delta0 = (N1/sigma0_sq)/(N1/sigma0_sq + 1/tau0_sq)
  delta1 = (N0/sigma1_sq)/(N0/sigma1_sq + 1/tau1_sq)
  mu1mean = delta1 * y1bar + (1-delta1)*m1 # posterior mean of mu1
  mu0mean = delta0 * y0bar + (1-delta0)*m0 # posterior mean of mu0
  mu1var = 1/(N1/sigma1_sq + 1/tau1_sq) # posterior var of mu1
  mu0var = 1/(N0/sigma0_sq + 1/tau0_sq) # posterior var of mu0

  mu0tilde = rnorm(1, mu0mean, mu0var) # draw from posterior of mu0
  mu1tilde = rnorm(1, mu1mean, mu1var) # draw from posterior of mu1

  sigma1 = sqrt(sigma1_sq)
  sigma0 = sqrt(sigma0_sq)
  for (i in 1:N) {
      if (W[i]==1) {
        mu01 = mu0tilde + rho*(sigma0/sigma1)*(Ypot[i,2]-mu1tilde)
        sd01 = sqrt((1-rho^2)*sigma0_sq)
        ymis = rnorm(1,mu01,sd01)
        Ypot[i,1] = ymis
      } else {
        mu10 = mu1tilde + rho*(sigma1/sigma0)*(Ypot[i,1]-mu0tilde)
        sd10 = sqrt((1-rho^2)*sigma1_sq)
        ymis = rnorm(1,mu10,sd10)
        Ypot[i,2] = ymis
      }
    }
```

```
  Ypot = round(Ypot)
  taubar = mean(Ypot[,2]) - mean(Ypot[,1])
  return(list(Ypot, taubar))
} # end function

>set.seed(20)
>Ypot_imp_dep_normal(Ypot)
# Here Ypot is the 20x2 matrix of potential outcomes derived from Table 8.4

[[1]]
      [,1] [,2]
 [1,]    6   10
 [2,]    9   13
 [3,]   15   19
 [4,]   14   18
 [5,]   13   17
 [6,]   12   16
 [7,]   10   14
 [8,]    6   10
 [9,]    9   13
[10,]    8   12
[11,]    6   10
[12,]   13   17
[13,]    8   12
[14,]    9   13
[15,]   12   16
[16,]   16   20
[17,]   15   19
[18,]    4    8
[19,]   11   15
[20,]    6   10

[[2]]
[1] 4
-----------------------------------------------------------------
```

The imputed values of the missing potential outcomes and the corresponding value of the finite population average treatment effect $\overline{\tau} = 4$ are shown above as the output of the function. The default values $m_0, m_1, \sigma_0^2, \sigma_1^2, \rho, \tau_0^2$ and τ_1^2 were used as inputs to the function. Note that this imputation leads to all the unit-level treatment effects τ_i, being equal to 4, i.e., the imputed potential outcomes are strictly additive. This result occured because substitution of $\sigma_0^2 = \sigma_1^2 = 25$ and $\rho = 1$ in (9.25) leads to

$$(Y_i^{\text{mis}} | Y_i^{\text{obs}}, W_i, \mu_0, \mu_1)$$
$$\sim \begin{cases} Y_i(1)|Y_i(0), \mu_0, \mu_1 \sim \mathcal{N}\left(\mu_1 + (Y_i(0) - \mu_0), 0\right) & \text{if} \quad W_i = 0 \\ Y_i(0)|Y_i(1), \mu_0, \mu_1 \sim \mathcal{N}\left(\mu_0 + (Y_i(1) - \mu_1), 0\right) & \text{if} \quad W_i = 1. \end{cases}$$

Because the above conditional imputation models are degenrate (i.e., have zero variance) for any pair $(\widetilde{\mu}_0, \widetilde{\mu}_1)$ drawn from the posterior distributions, the missing potential outcomes are imputed as:

$$Y_i^{\text{mis}} = \begin{cases} Y_i(0) + (\widetilde{\mu}_1 - \widetilde{\mu}_0) & \text{if} \quad W_i = 0 \\ Y_i(1) - (\widetilde{\mu}_1 - \widetilde{\mu}_0) & \text{if} \quad W_i = 1, \end{cases}$$

yielding constant unit-level treatment effects $\tau_i = Y_i(1) - Y_i(0) = \widetilde{\mu}_1 - \widetilde{\mu}_0$ for $i = 1, \ldots, N$.

Figure 9.6 shows a plot of the estimated value of $\overline{\tau}$ for example 9.2 and the corresponding 95% intervals for different values of ρ ranging from -1 to 1, under model (9.23) and (9.24), taking $m_0 = m_1 = 0$, $\sigma_0^2 = \sigma_1^2 = 25$ and $\tau_0^2 = \tau_1^2 = 100$ as values of the parameters and hyperparameters. The plot shows the intervals are the shortest when $\rho = -1$ and widest when $\rho = 1$. This phenomenon is consistent with the fact that the sampling variance of $\widehat{\overline{\tau}}$ is the smallest when the potential outcomes are perfectly negatively correlated and largest when they potential outcomes are perfectly positively correlated - a fact discussed in Chapter 6 of Imbens and Rubin (2015).

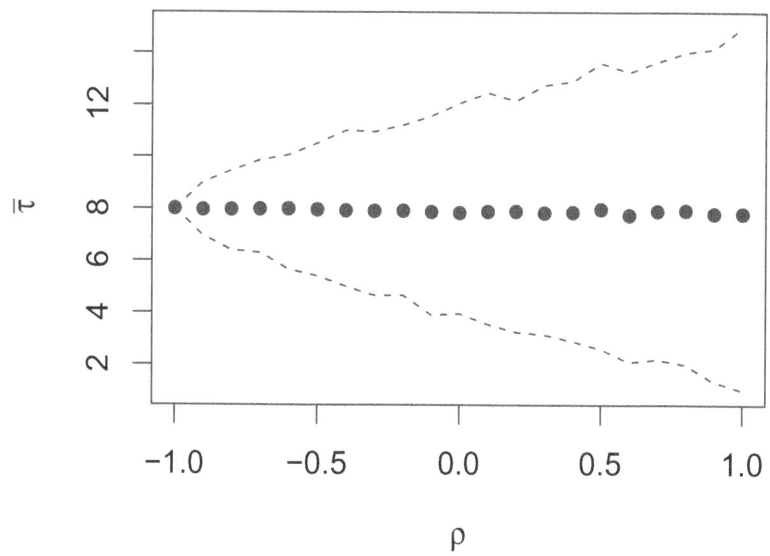

FIGURE 9.6
Plot of point and interval estimates of $\overline{\tau}$ versus the correlation parameter ρ for Example 9.2, assuming model (10.5) and (9.24) for the potential outcomes, taking for the values of the known parameters $\sigma_0^2 = \sigma_1^2 = 25$, and hyperparameters $m_0 = m_1 = 0$, and $\tau_0^2 = \tau_1^2 = 100$

9.10 2^K factorial experiments

The Bayesian method described so far in this Chapter can be extended to the case of 2^K factorial designs discussed in Chapters 6 and 7. Recall that a

2^K factorial experiment has $J = 2^K$ potential outcomes $Y_i(1), \ldots, Y_i(J)$, only one of which is observed, and the remaining $J - 1$ missing. The goal of the model-based approach is thus to impute the $J - 1$ missing potential outcomes for each unit based on the observed outcomes. We illustrate this procedure with the observed and missing potential outcomes from the hypothetical 2^2 experiment shown in Table 6.5.

Here, we can postulate a normal model on the potential outcomes. The simplest model is one that extends the model of Section 9.8, assuming that the four potential outcomes $Y_i(j)$, $j = 1, \ldots, 4$ for unit i are mutually independent with the distribution

$$Y_i(j)|\mu_j \sim N(\mu_j, \sigma_j^2),$$

where σ_j^2 is assumed known. Independent normal priors are postulated for the means μ_j:

$$\mu_j|m_j, \tau_j^2 \sim \mathcal{N}(m_j, \tau_j^2),$$

where the "hyper-parameters" m_j, and τ_j^2 and τ_1^2 are assumed known. Thus, the unknown parameter vector θ under this model is $(\mu_1, \ldots, \mu_4)$.

Then, following an argument analogous to the one in Section 9.8, the posterior distributions of μ_j, $j = 1, \ldots, 4$, given the observed data are independently distributed normal variables:

$$\mu_j|\mathbf{Y}^{\text{obs}}, \mathbf{W} \sim \mathcal{N}\left(\delta_j \overline{y}(j) + (1 - \delta_j)m_j, \frac{1}{N_j/\sigma_j^2 + 1/\tau_j^2}\right), \tag{9.26}$$

where

$$\delta_j = \frac{N_j/\sigma_j^2}{N_j/\sigma_j^2 + 1/\tau_j^2}.$$

Here, the superpopulation estimands or superpopulation-level factorial effects are contrasts of the μ_j's and their forms are analogous to the finite population factorial effects that are contrasts of $\overline{Y}(j)$'s in (6.14) - (6.16):

$$\tau^{\text{SP}}(F_1) = \frac{-\mu_1 - \mu_2 + \mu_3 + \mu_4}{2}, \tag{9.27}$$

$$\tau^{\text{SP}}(F_2) = \frac{-\mu_1 + \mu_2 - \mu_3 + \mu_4}{2}, \tag{9.28}$$

$$\tau^{\text{SP}}(F_1F_2) = \frac{+\mu_1 - \mu_2 - \mu_3 + \mu_4}{2}. \tag{9.29}$$

The posterior distributions of these superpopulation estimands can be obtained by substituting the posterior distributions of μ_j's from (9.26) into (9.27) - (9.29).

The following R code shows the derivation of the posterior means and variances of the μ_j's given $\mathbf{Y}^{\text{obs}}$ and $\mathbf{W}$ in Table 6.5 based on (9.26), and taking $\sigma_j^2 = 1, \tau_j^2 = 100$ for $j = 1, \ldots, 4$. Because this is a balanced design with $N_j = 4$ for $j = 1, \ldots, 4$, the δ_j's and the posterior variances of μ_j's are the same for each j.

```
# Enter data unit-wise
y=c(7.4,5.0,6.6,7.6,4.8,6.3,5.8,6.5,7.3,5.3,3.6,6.6,4.4,6.5,5.4,8.4,8.0,5.3,
    6.7,6.2)
N = length(y)

# Enter assignment vector
w=c(2,1,3,4,1,3,2,4,4,3,1,2,1,3,2,4,4,1,2,3)

## Generate vectors of ybar & s^2
ybarvector = varyvector = NULL
for (l in 1:4) {
  ybarvector = c(ybarvector,mean(y[w==l]))
  varyvector = c(varyvector,var(y[w==l]))
}

## Posterior distributions of mu_j
sigma_sq = 1
tau_sq = 100
delta = mu_mean = mu_var = m = numeric(4)
for (j in 1:4) {
delta[j] = (5/sigma_sq)/(5/sigma_sq + 1/tau_sq)
mu_mean[j] = delta[j] * ybarvector[j] + (1-delta[j])*m[j]
mu_var[j] = 1/(5/sigma_sq + 1/tau_sq)
}

> mu_mean
[1] 4.610778 6.367265 6.167665 7.544910
> mu_var
[1] 0.1996008 0.1996008 0.1996008 0.1996008
```

To draw posterior inferences on the superpopulation estimands $\tau^{\text{SP}}(F_1)$, $\tau^{\text{SP}}(F_2)$ and $\tau^{\text{SP}}(F_1F_2)$, we generate draws $\widetilde{\mu}_j$'s from the posterior distribution of $\mu_j|\mathbf{Y}^{\text{obs}}, \mathbf{W}$, and for each draw obtain $\tau^{\text{SP}}(F_1)$, $\tau^{\text{SP}}(F_2)$ and $\tau^{\text{SP}}(F_1F_2)$ using the same technique described in Section 6.3, i.e., first pre-multiplying the vector $(\widetilde{\mu}_1, \ldots, \widetilde{\mu}_4)^{\text{T}}$ by $(1/4)\mathbf{L}^T$, where $\mathbf{L}$ is the coefficient matrix, and then extracting the last three elements of the resultant vector and multiplying each by 2. The mean and lower and upper 2.5% quantiles obtained from 5000 simulations of the posterior distributions of the superpopulation factorial effects $\tau^{\text{SP}}(\cdot)$ are shown at the end of the R code chunk below.

```
mu_covmatrix = diag(mu_var) # covariance matrix (diagonal in this case)

library(AlgDesign)
D = gen.factorial(2,2) ## ME contrast columns in reverse order
D[, c(1:2)] = D[,c(2:1)] ## Reverse the order of the columns
colnames(D) = c("F1","F2")
L = model.matrix(~ F1 + F2 + F1*F2,D)

tauFS = NULL
library(MASS)
set.seed(888)
for (i in 1:5000) {
```

```
mu_post = mvrnorm(1,mu_mean, mu_covmatrix)
tau_post_adj = (t(L) %*% mu_post)/4
tauFS = rbind(tauFS, 2*tau_post_adj[2:4])
}

> apply(tauFS,2,mean)
[1]  1.3591476  1.5663546 -0.1878647
> apply(tauFS,2, quantile, probs=c(.025,.975))
           [,1]      [,2]       [,3]
2.5%  0.4767354 0.6922682 -1.0775794
97.5% 2.2603642 2.4498528  0.6624004
-------------------------------------------------------------------
```

Turning now to the finite population inference, if $W_i = j^*$, then Y_i^{mis} is a vector comprising elements $Y_i^{\text{mis}}(j) = Y_i(j)$ for $j = 1, \ldots, J$, $j \neq j^*$. The $J-1$ missing outcomes can be imputed using imputation model (9.30) that follows from (9.22):

$$(Y_i^{\text{mis}}(j)|Y_i^{\text{obs}}, W_i = j^*, \mu_1, \ldots, \mu_J) \sim Y_i(j)|\mu_j \sim N(\mu_j, \sigma_j^2), j \neq j^*. \quad (9.30)$$

In our current example, $W_1 = 2$, which means $\mathbf{Y}_1^{\text{mis}} = (Y_1(1), Y_1(3), Y_1(4))$, and $Y_1(j)$ for $j = 1, 3, 4$ can be imputed using (9.30) after drawing from the posterior distribution of μ_j's.

The advantage of using the Bayesian approach is that it allows us to easily incorporate prior information on factorial effects, often available when analyzing factorial experiments. Even in the absence of such information, assumptions like *effect hierarchy* (lower order effects are more likely to be important than higher order effects, and effects of the same order are equally likely to be important) and sometimes *effect heredity* (in order for an interaction to be significant, at least one of its parent factors should be significant) can be incorporated into the imputation model to distinguish active effects from inactive effects. Details of these principles can be found in Wu and Hamada (2009) [93]. A sequential approach for identifying active effects was proposed by Espinosa et al. (2015) [31].

9.11 Problems

9.1 For example 9.1, consider the same setting as in Section 9.5, i.e., assume independent Bernoulli models for $Y_i(0)$ and $Y_i(1)$ and independent Beta(α_0, β_0) and Beta(α_1, β_1) priors for the probabilities $\pi_0 = \Pr[Y_i(0) = 1]$ and $\pi_1 = \Pr[Y_i(1) = 1]$, where $\alpha_0, \beta_0, \alpha_1$ and β_1 are known hyperparameters. Using the data in Table 9.1, draw inferences (point and 95% interval estimates) for the superpopulation estimand, $\tau^{\text{SP}} = \pi_1 - \pi_0$, and the finite population estimand $\overline{\tau} = \overline{Y}(1) - \overline{Y}(0)$ for the following choices of $\alpha_0, \beta_0, \alpha_1$ and β_1.

(i) $(\alpha_0, \beta_0) = (2, 2)$, $(\alpha_1, \beta_1) = (2, 2)$.

(ii) $(\alpha_0, \beta_0) = (5, 5)$, $(\alpha_1, \beta_1) = (5, 5)$.

(iii) $(\alpha_0, \beta_0) = (2, 2)$, $(\alpha_1, \beta_1) = (3, 3)$.

(iv) $(\alpha_0, \beta_0) = (3, 2)$, $(\alpha_1, \beta_1) = (3, 2)$.

(v) $(\alpha_0, \beta_0) = (3, 2)$, $(\alpha_1, \beta_1) = (2, 3)$.

In each case, comment on the shapes of prior distributions. Compare the resulting inferences drawn on the basis of posterior distributions and discuss how sensitive they are to the choice of prior distributions. What would you expect to see with larger sample sizes?

9.2 Impute the missing outcomes in Table 9.1 using the following strategy: replace all missing outcomes in the treated group by 1 and impute the missing outcomes in the control group by generating Bernoulli random variables with probability of success $p_{01} = 0.75$ for females and $p_{02} = 0.67$ for males. Obtain the distribution of $\overline{\tau}$ obtained from 500 such imputations and provide a histogram analogous to the one in Figure 9.2

9.3 Consider the same situation as in Section 9.5 and in Problem 8.1. For binary outcomes, two popular causal estimands that are non-linear functions of the potential outcomes are the log risk ratio (LRR) or the log odds ratio (LOR) are defined as follows

$$\begin{aligned}
\text{LRR} &= \log \frac{\overline{Y}(1)}{\overline{Y}(0)} = \log \frac{P_1}{P_0}, \\
\text{LOR} &= \log \frac{\overline{Y}(1)/\left(1 - \overline{Y}(1)\right)}{\overline{Y}(0)/\left(1 - \overline{Y}(0)\right)} = \log \frac{P_1/(1-P_1)}{P_0/(1-P_0)},
\end{aligned}$$

where $\overline{Y}(0) = P_0$ and $\overline{Y}(1) = P_1$. The superpopulation counterparts of these estimands are:

$$\begin{aligned}
\text{LRR}^{\text{SP}} &= \log \frac{\mathbb{E}[Y_i(1)]}{\mathbb{E}[Y_i(0)]} = \log \frac{\pi_1}{\pi_0}. \\
\text{LOR}^{\text{SP}} &= \log \frac{\mathbb{E}[Y_i(1)]/\left(1 - \mathbb{E}[Y_i(1)]\right)}{\mathbb{E}[Y_i(0)]/\left(1 - \mathbb{E}[Y_i(0)]\right)} = \log \frac{\pi_1/(1-\pi_1)}{\pi_0/(1-\pi_0)},
\end{aligned}$$

where $\mathbb{E}[Y_i(0)] = \pi_0$ and $\mathbb{E}[Y_i(1)] = \pi_1$.

Using the data in Table 9.1, draw inferences (point and 95% interval estimates) on the above superpopulation and finite population estimands taking $\alpha_0 = \beta_0 = \alpha_1 = \beta_2 = 2$.

9.4 Analyze the data in Table 9.4 using independent normal models for the potential outcomes in the treated and control groups as in Section 9.8, but with unknown variance parameters σ_0^2 and σ_1^2. Assume the following model with independent (but common) inverse gamma prior distributions for σ_0^2 and σ_1^2:

$$\begin{aligned} Y_i(0)|\mu_0 \sim N(\mu_0, \sigma_0^2) \perp\!\!\!\perp Y_i(1)|\mu_1 \sim N(\mu_1, \sigma_1^2), \\ \mu_0|m_0, \tau_0^2 \sim \mathcal{N}(m_0, \tau_0^2) \perp\!\!\!\perp \mu_1|m_1, \tau_1^2 \sim \mathcal{N}(m_1, \tau_1^2) \\ \sigma_0^2 \sim \text{Inv-gamma}(a, b) \perp\!\!\!\perp \sigma_1^2 \sim \text{Inv-gamma}(a, b) \end{aligned}$$

where the "hyper-parameters" m_0, m_1, τ_0^2, τ_1^2, a and b have known fixed values.

Then, as before, the conditional posterior distributions of μ_0 and μ_1, given the observed data and variance parameters σ_0^2 and σ_1^2 are independently distributed normal variables:

$$\begin{aligned} \mu_0|\mathbf{Y}^{\text{obs}}, \mathbf{W}, \sigma_0^2 &\sim \mathcal{N}\left(\frac{\overline{y}(0)\left(N_0/\sigma_0^2\right) + m_0\left(1/\tau_0^2\right)}{N_0/\sigma_0^2 + 1/\tau_0^2}, \frac{1}{N_0/\sigma_0^2 + 1/\tau_0^2}\right), \\ \mu_1|\mathbf{Y}^{\text{obs}}, \mathbf{W}, \sigma_1^2 &\sim \mathcal{N}\left(\frac{\overline{y}(1)\left(N_1/\sigma_1^2\right) + m_1\left(1/\tau_1^2\right)}{N_1/\sigma_1^2 + 1/\tau_1^2}, \frac{1}{N_1/\sigma_1^2 + 1/\tau_1^2}\right). \end{aligned}$$

The posterior distribution of the variance parameters are:

$$\begin{aligned} \sigma_0^2|\mathbf{Y}^{\text{obs}}, \mathbf{W} &\sim \text{Inv-gamma}(a_0, b_0), \\ \sigma_1^2|\mathbf{Y}^{\text{obs}}, \mathbf{W} &\sim \text{Inv-gamma}(a_1, b_1), \end{aligned}$$

where

$$\begin{aligned} a_0 &= a + N_0/2, \\ a_1 &= a + N_1/2, \\ b_0 &= b + \frac{1}{2}\left[(N_0 - 1)s_0^2 + \frac{bN_0}{b + N_0 a\tau^2}\left(\overline{y}(0) - m_0\right)^2\right], \\ b_1 &= b + \frac{1}{2}\left[(N_1 - 1)s_1^2 + \frac{bN_1}{b + N_1 a\tau^2}\left(\overline{y}(1) - m_1\right)^2\right], \end{aligned}$$

with s_0^2 and s_1^2 being the sample variances obtained from the treated and control group observations respectively.

Using this model and its posterior distributions, the same choices for hyperparameters as in Section 9.8 along with $a = b = 1$, re-analyze the data in Table 9.4 and obtain 95% posterior intervals for the superpopulation estimand $\tau^{\text{SP}} = \mu_1 - \mu_0$ and the finite population estimand $\overline{\tau} = \overline{Y}(1) - \overline{Y}(0)$.

9.5 Using the independent potential outcomes model for a 2^3 factorial experiment

$$Y_i(j)|\mu_j \sim N(\mu_j, \sigma_j^2),$$

where σ_j^2 is assumed known, and independent normal priors are used for the means μ_j:

$$\mu_j | m_j, \tau_j^2 \sim \mathcal{N}(m_j, \tau_j^2),$$

where the "hyper-parameters" m_j and τ_j^2 are assumed known for $j = 1, \ldots, J(= 8)$,

(i) Write expressions for the superpopulation versions of the seven factorial effects in terms of the μ_j's.

(ii) Find 95% posterior intervals for the seven superpopulation estimands using the data in Table 7.5 on the basis of 10,000 simulations from the posterior distributions of μ_j's conditional on $\mathbf{Y}^{\text{obs}}$ and $\mathbf{W}$.

(iii) From the same simulations as in (ii), find 95% posterior intervals for the seven finite population estimands (contrasts of $\overline{Y}(j)$'s) by imputing the missing potential outcomes.

10

Including covariates in the design, model-based analysis and subsequent inference

10.1 A naive strategy for including covariates in the model and inference

As noted in Chapter 8 of Imbens and Rubin 2015 [44], "the presence of covariates in a completely randomized experiment does not fundamentally change the underlying method for imputing the missing potential outcomes in the model-based approach." Returning to example 9.1, we extend the naive strategy for imputing missing potential outcomes in Table 9.1 by taking covariates into consideration. Table 10.1 summarizes the outcomes in Table 9.1 by sex. Recall that our previous naive strategy used the last row of overall proportion of successes, imputing missing outcomes in the control and treated group by Bernoulli random variables with probability of success $p_0 = 0.7$ and $p_1 = 1.0$ respectively. Instead, we could impute missing outcomes in the control group by generating Bernoulli random variables with probability of success $p_{01} = 0.75$ for females and $p_{02} = 0.67$ for males. Obtaining the distribution of $\overline{\tau}$ obtained from repeated applications of this strategy is left as an exercise (Problem 8.2).

TABLE 10.1
Number and proportion of "successes" (not developing a peanut allergy) in the data of Table 9.1 stratified by sex. The last row shows the entries in the composite 2×2 table that summarizes the observed data using the same notation as in Table 2.5

Sex	Control ($w = 0$) Exposed	Control ($w = 0$) success ($y = 1$)	Control ($w = 0$) failure ($y = 0$)	Control ($w = 0$) prop	Treatment ($w = 1$) Exposed	Treatment ($w = 1$) success ($y = 1$)	Treatment ($w = 1$) failure ($y = 0$)	Treatment ($w = 1$) prop
Female	4	3	1	0.75	4	4	0	1
Male	6	4	2	0.67	6	6	0	1
Total	$N_0 = 10$	$n_{01} = 7$	$n_{00} = 3$	0.70	$N_1 = 10$	$n_{11} = 10$	$n_{10} = 0$	1

DOI: 10.1201/9781003048794-10

10.2 Inference for superpopulation and finite population estimands using models that involve covariates

We now extend the model-based approach developed and discussed in Sections 9.5 to 9.9 to include covariates. The first step of this extension is to extend the joint distribution $\mathrm{pr}(\mathbf{Y}(0), \mathbf{Y}(1)|\theta)$ to $\mathrm{pr}(\mathbf{Y}(0), \mathbf{Y}(1)|\mathbf{X}, \theta)$, where $\mathbf{X}$ is an $N \times p$-dimensional covariate matrix, i.e., there is a p-dimensional covariate vector $\mathbf{X}_i$ associated with unit i. As in Section 9.4, capitalizing on the exchangeability of units and applying de Finetti's theorem, we can write the joint distribution of $(\mathbf{Y}(1), \mathbf{Y}(0), \mathbf{X})$ conditional on a parameter θ as

$$\mathrm{pr}\left(\mathbf{Y}(0), \mathbf{Y}(1), \mathbf{X}|\theta\right) = \prod_{i=1}^{N} \mathrm{pr}\left(Y_i(0), Y_i(1), \mathbf{X}_i|\theta\right). \tag{10.1}$$

The joint distribution $\mathrm{pr}\left(Y_i(0), Y_i(1), \mathbf{X}_i|\theta\right)$ can be factored as:

$$\mathrm{pr}\left(Y_i(0), Y_i(1), \mathbf{X}_i|\theta\right) = \mathrm{pr}\left(Y_i(0), Y_i(1)|\mathbf{X}_i, \theta_{Y|X}\right) \mathrm{pr}\left(\mathbf{X}_i|\theta_X\right), \tag{10.2}$$

where $\theta_{Y|X}$ is the parameter (possibly vector) governing the conditional distribution of $Y_i(0), Y_i(1)$ given $\mathbf{X}_i$, and θ_X denotes the parameter (possibly vector) governing the distribution of $\mathbf{X}$.

Next, to specify prior distributions on the unknown parameter θ, often it is convenient to assume that its two components $\theta_{Y|X}$ and θ_X are independent, i.e.,

$$\mathrm{pr}(\theta_{Y|X}, \theta_X) = \mathrm{pr}(\theta_{Y|X})\mathrm{pr}(\theta_X).$$

In that case, we only need to postulate a model for $\mathrm{pr}\left(Y_i(0), Y_i(1)|\mathbf{X}_i, \theta_{Y|X}\right)$ because neither $\mathrm{pr}\left(\mathbf{X}_i|\theta_X\right)$ nor $\mathrm{pr}(\theta_X)$ will influence the inference for the finite population estimand. We can return to this issue later while obtaining the posterior distribution of superpopulation estimands. To simplify notations, we suppress the index $Y|X$ and write the required model as $\mathrm{pr}\left(Y_i(0), Y_i(1)|\mathbf{X}_i, \theta\right)$. As before, the superpopulation estimand τ^{SP} will be a function of θ.

Essentially the same approach as described in Section 9.4 can be used to (a) obtain the posterior distribution of θ, and thereof the super population estimand τ^{SP}, and (b) obtain an imputation model $\mathrm{pr}(\mathbf{Y}^{\mathrm{mis}}|\mathbf{Y}^{\mathrm{obs}}, \mathbf{W}, \mathbf{X})$ that can be used to impute the missing potential outcomes and thereby to draw inference on the finite population estimand $\overline{\tau}$. First, note that the likelihood function in (9.8) now becomes

$$\mathrm{pr}(\mathbf{Y}^{\mathrm{obs}}, \mathbf{W}, \mathbf{X}|\theta) \propto \prod_{i:w_i=1} \mathrm{pr}(Y_i(1)|\mathbf{X}_i, \theta) \prod_{i:w_i=0} \mathrm{pr}(Y_i(0)|\mathbf{X}_i, \theta). \tag{10.3}$$

This likelihood, in conjuction with the prior distribution for θ, leads to the following posterior distribution of θ, which is an extension of equation (9.11):

$$\mathrm{pr}(\theta|\mathbf{Y}^{\mathrm{obs}}, \mathbf{W}, \mathbf{X}) \propto \mathrm{pr}(\mathbf{Y}^{\mathrm{obs}}, \mathbf{W}, \mathbf{X}|\theta)\ \mathrm{pr}(\theta). \tag{10.4}$$

Because both $\mathbb{E}[Y_i(1)|\mathbf{X}=\mathbf{x}]$ and $\mathbb{E}[Y_i(1)|\mathbf{X}=\mathbf{x}]$ depend on the parameter θ, the posterior distribution of the conditional superpopulation estimand $\tau^{\text{SP}}(\mathbf{x})$ defined in (9.3) is found by using the posterior distribution obtained from (10.4). Furthermore, the unconditional superpopulation estimand τ^{SP} can be found as

$$\tau^{\text{SP}} = \mathbb{E}_{\theta_X}\left(\tau^{\text{SP}}(\mathbf{X})\right) = \int \tau^{\text{SP}}(\mathbf{x})\text{pr}(\mathbf{x}|\theta_X)d\mathbf{x}.$$

As an illustration, suppose the following model is postulated for the potential outcomes in a randomized experiment where there is one premeasured covariate X:

$$\left(\begin{array}{c} Y_i(0) \\ Y_i(1) \end{array}\right)\Bigg|\theta, X \sim \mathcal{N}\left(\left(\begin{array}{c} \alpha_0+\beta_0 X \\ \alpha_1+\beta_1 X \end{array}\right), \left(\begin{array}{cc} \sigma_0^2 & \rho\sigma_0\sigma_1 \\ \rho\sigma_0\sigma_1 & \sigma_1^2 \end{array}\right)\right), \tag{10.5}$$

where the variance parameters σ_0^2, σ_1^2 and the correlation coefficient ρ are considered known, so that $\theta = \theta_{Y|X} = (\alpha_0, \alpha_1, \beta_0, \beta_1)$ for which a suitable prior (typically normal) can be postulated.

For this model, the conditional superpopulation estimand

$$\tau^{\text{SP}}(\mathbf{x}) = \mathbb{E}\left[Y_i(1)|X=x\right] - \mathbb{E}\left[Y_i(0)|X=x\right] = (\alpha_1-\alpha_0)+(\beta_1-\beta_0)x.$$

If, in addition, it is assumed that $X \sim N(\mu_X, \sigma_X^2)$, then the unconditional superpopulation estimand is

$$\tau^{\text{SP}} = \mathbb{E}\left(\tau^{\text{SP}}(X)\right) = (\alpha_1-\alpha_0)+(\beta_1-\beta_0)\mu_X.$$

Next, to derive the imputation model, we first incorporate X in (9.13) to obtain

$$\text{pr}(\mathbf{Y}^{\text{mis}}|\mathbf{Y}^{\text{obs}}, \mathbf{W}, \mathbf{X}) = \int \text{pr}(\mathbf{Y}^{\text{mis}}|\mathbf{Y}^{\text{obs}}, \mathbf{W}, \mathbf{X}, \theta)\ \text{pr}(\theta|\mathbf{Y}^{\text{obs}}, \mathbf{X}, \mathbf{W})d\theta. \tag{10.6}$$

The second factor within the integral in (10.6) is the posterior distribution of θ in (10.4). To obtain the first factor, we note that for $i = 1, \ldots, N$, just as in (9.16),

$$\text{pr}(Y_i^{\text{mis}}|Y_i^{\text{obs}}, W_i, \mathbf{X}_i, \theta) = \begin{cases} \text{pr}(Y_i(1)|Y_i(0), \mathbf{X}_i, \theta) & \text{if} \quad W_i = 0 \\ \text{pr}(Y_i(0)|Y_i(1), \mathbf{X}_i, \theta) & \text{if} \quad W_i = 1. \end{cases} \tag{10.7}$$

The conditional distributions $\text{pr}(Y_i(1)|Y_i(0), \mathbf{X}_i, \theta)$ and $\text{pr}(Y_i(0)|Y_i(1), \mathbf{X}_i, \theta)$ in (10.7) can be derived from the specified distribution $\text{pr}(Y_i(0), Y_i(1)|\mathbf{X}_i, \theta)$.

Finally, Procedure 9.1 for imputation of missing outcomes can be updated in a straightforward manner to incorporate covariates as follows:

Procedure 10.1 (Imputing missing outcomes with covariates)
Generating draws from pr($\mathbf{Y}^{mis}|\mathbf{Y}^{obs}, \mathbf{W}, \mathbf{X}$) involves the following two steps:

1. *Generate a draw $\tilde{\theta}$ of θ from the posterior distribution pr($\theta|\mathbf{Y}^{obs}, \mathbf{W}, \mathbf{X}$) using (10.4).*

2. *For $i = 1, \ldots, N$, generate a draw of Y_i^{mis} from pr($Y_i^{mis}|Y_i^{obs}, W_i, \mathbf{X}_i, \tilde{\theta}$) using (10.7).*

10.3 Covariate-based model for normally distributed outcomes: an illustration

We now illustrate the approach outlined in Section 10.2 using Example 9.2 where we have four covariates X_1: Vpre, X_2: likelit, X_3: actcomp, X_4: collegegpa. Including a column of ones, $\mathbf{X}_i$ is a row vector of dimension 5 and we have a 20×5 covariate matrix, with the top six rows shown below (recall that the data frame in Table 9.4 was stored by the name SRV).

```
> X = as.matrix(SRV[,1:4])
> X = cbind(rep(1,20),X)
> head(X)
       Vpre likelit actcomp collegegpa
[1,] 1   27       5      22       3.00
[2,] 1   24       7      23       3.20
[3,] 1   15       6      22       2.60
[4,] 1   23       7      22       2.37
[5,] 1   10       7      15       1.68
[6,] 1   21       6      20       2.29
```

We postulate the following model (10.8) for the potential outcomes, conditional on the covariate row vector $\mathbf{X}_i$ and parameter $\theta = (\beta_0, \beta_1, \sigma_0^2, \sigma_1^2)$, where β_0 and β_1 are 5×1 column vectors and σ_0^2 and σ_1^2 are scalars. The parameter vectors β_0 and β_1 quantify the impact of the covariates on the outcomes in the control and treated group respectively. Unlike the previous sections, we now assume the variance parameters σ_0^2 and σ_1^2 to be unknown. Note that although here we have assumed the potential outcomes to be independent, such an assumption can be relaxed and the same procedure can be extended to incorporate dependence through a correlation parameter ρ as in Section 9.9. Under independence of the potential outcomes,

$$\begin{pmatrix} Y_i(0) \\ Y_i(1) \end{pmatrix} \Bigg| \mathbf{X}_i, \theta \sim \mathcal{N}\left(\begin{pmatrix} \mathbf{X}_i\beta_0 \\ \mathbf{X}_i\beta_1 \end{pmatrix}, \begin{pmatrix} \sigma_0^2 & 0 \\ 0 & \sigma_1^2 \end{pmatrix} \right). \tag{10.8}$$

To postulate a joint prior distribution on θ, we first factor the prior distributions of the parameters as

$$\mathrm{pr}(\beta_0, \beta_1, \sigma_0^2, \sigma_1^2) = \mathrm{pr}(\sigma_0^2 \sigma_1^2)\mathrm{pr}(\beta_0, \beta_1 | \sigma_0^2, \sigma_1^2),$$

and, for mathematical convenience, postulate independent inverse-gamma prior distributions on σ_0^2 and σ_1^2:

$$\sigma_0^2 \sim \text{Inv-gamma}(a, b) \perp\!\!\!\perp \sigma_1^2 \sim \text{Inv-gamma}(a, b), \tag{10.9}$$

where the hyperparameters a and b are assumed known.

Next, again for mathematical convenience, we postulate the following multivariate normal prior distribution on the vector parameters β_0 and β_1 conditional on σ_0^2 and σ_1^2:

$$\begin{pmatrix} \beta_0 \\ \beta_1 \end{pmatrix} \Bigg| (\sigma_0^2, \sigma_1^2) \sim \mathcal{N}\left(\begin{pmatrix} \mu_0 \\ \mu_1 \end{pmatrix}, \begin{pmatrix} \sigma_0^2\Omega & 0 \\ 0 & \sigma_1^2\Omega \end{pmatrix} \right). \tag{10.10}$$

where Ω is the known common 5×5 prior correlation matrix of β_0 and β_1. Alternatively, it is straightforward to extend (10.10) to incorporate two different prior covariance matrices Ω_0 and Ω_1 for β_0 and β_1, respectively. The vectors μ_0, μ_1 and correlation matrix Ω are assumed known.

To obtain the posterior distribution of the parameters, we split the observed data into two groups. Let $(\mathbf{X}_0, \mathbf{y}_0)$ denote the covariates and observed outcomes associated with units in the control group, and $(\mathbf{X}_1, \mathbf{y}_1)$ for those associated with the treated group. Then, following Gelman et al. (2013) [38], the joint posterior distribution of θ given the observed data $\mathbf{Y}^{\text{obs}}$, $\mathbf{X}$ and $\mathbf{W}$ can be obtained from the following set of equations. We drop the conditioning in the subsequent equations to reduce notational clutter but remember that all probability distributions in the following equations are conditional on $\mathbf{Y}^{\text{obs}}, \mathbf{X}, \mathbf{W}$,

$$\begin{aligned} & \operatorname{pr}(\beta_0, \beta_1, \sigma_0^2, \sigma_1^2) \propto \operatorname{pr}(\beta_0, \beta_1 | \sigma_0^2, \sigma_1^2) \operatorname{pr}(\sigma_0^2, \sigma_1^2) \\ \propto \; & \operatorname{pr}(\beta_0 | \sigma_0^2) \operatorname{pr}(\sigma_0^2) \operatorname{pr}(\beta_1 | \sigma_1^2) \operatorname{pr}(\sigma_1^2). \end{aligned} \tag{10.11}$$

The posterior distributions of β_0 and β_1 conditional on the variance parameters are:

$$\beta_0 | \sigma_0^2 \sim \mathcal{N}(\mu_{\beta_0}, \sigma_0^2 \Omega_{\beta_0}) \perp\!\!\!\perp \beta_1 | \sigma_1^2 \sim \mathcal{N}(\mu_{\beta_1}, \sigma_1^2 \Omega_{\beta_1}), \tag{10.12}$$

where

$$\Omega_{\beta_0} = \left(\mathbf{X}_0^{\mathrm{T}} \mathbf{X}_0 + \Omega^{-1}\right)^{-1} \tag{10.13}$$

$$\Omega_{\beta_1} = \left(\mathbf{X}_1^{\mathrm{T}} \mathbf{X}_1 + \Omega^{-1}\right)^{-1} \tag{10.14}$$

$$\mu_{\beta_0} = \Omega_{\beta_0} \left(\mathbf{X}_0^{\mathrm{T}} \mathbf{y}_0 + \Omega^{-1} \mu_0\right). \tag{10.15}$$

$$\mu_{\beta_1} = \Omega_{\beta_1} \left(\mathbf{X}_1^{\mathrm{T}} \mathbf{y}_1 + \Omega^{-1} \mu_1\right). \tag{10.16}$$

Finally, the posterior distributions of the variance parameters are:

$$\sigma_0^2 \sim \text{Inv-gamma}(a_{\sigma_0^2}, b_{\sigma_0^2}), \; \sigma_1^2 \sim \text{Inv-gamma}(a_{\sigma_1^2}, b_{\sigma_1^2}). \tag{10.17}$$

where

$$a_{\sigma_0^2} = a + N_0/2, \tag{10.18}$$

$$a_{\sigma_1^2} = a + N_1/2, \tag{10.19}$$

$$b_{\sigma_0^2} = b + \frac{1}{2}\left(\mathbf{y}_0^{\mathrm{T}} \mathbf{y}_0 + \mu_0^{\mathrm{T}} \Omega^{-1} \mu_0 - \mu_{\beta_0}^{\mathrm{T}} \Omega_{\beta_0}^{-1} \mu_{\beta_0}\right). \tag{10.20}$$

$$b_{\sigma_1^2} = b + \frac{1}{2}\left(\mathbf{y}_1^{\mathrm{T}} \mathbf{y}_1 + \mu_1^{\mathrm{T}} \Omega^{-1} \mu_1 - \mu_{\beta_1}^{\mathrm{T}} \Omega_{\beta_1}^{-1} \mu_{\beta_1}\right). \tag{10.21}$$

The function Bayes_normal() shown below generates one draw from the posterior distribution of β and σ^2 derived in (10.11)-(10.21) for an $N \times 1$ given vector of observed outcomes $\mathbf{y}$, an $N \times p$ covariate matrix $\mathbf{X}$ under the model specified in (10.8) - (10.10). Thus, this function can be used to draw from the posterior distributions for β and σ^2 separately for each (treated or control) group of observations, by substituting $y = \mathbf{y}_1$, $X = \mathbf{X}_1$, $\mu = \mu_1$ and $y = \mathbf{y}_0$, $X = \mathbf{X}_0$, $\mu = \mu_0$ as arguments for the treated and control groups respectively.

```
-------------------------------------------------------------------------
Bayes_normal <- function(y,X,a,b,mu,Omega) {
    n = length(y)
    p = ncol(X)

    # Calculating parameters of posterior distribution

    # Calculate Omega_beta and mu_beta

    Omega_beta = solve((t(X) %*% X)+ solve(Omega))
    mu_beta = Omega_beta %*% ((t(X) %*% y) + (solve(Omega)%*% mu))

    # parameters of sigma^2

    a_sigmasq = a + n/2
    b_sigmasq = b + (t(y) %*% y + t(mu) %*% solve(Omega) %*% mu - t(mu_beta)
    %*% solve(Omega_beta) %*% mu_beta)/2
    # Draw from sigma_sq

    new_sigmasq = 1/rgamma(1,a_sigmasq,b_sigmasq)

    # Draw from beta|sigma_sq

    var_beta = new_sigmasq * Omega_beta
    library(MASS)
    new_beta = mvrnorm(1,mu_beta, var_beta)

    return(list(new_beta, new_sigmasq))

  } # end function
-------------------------------------------------------------------------
```

We now use the function to generate the posterior distributions of the parameters β_0 and β_1 conditional on the observed data. To do this, note that these posterior distributionscan be written as:

$$\text{pr}(\beta_0|\mathbf{y}_0, \mathbf{X}_0) = \int \text{pr}(\beta_0|\mathbf{y}_0, \mathbf{X}_0, \sigma_0^2)\text{pr}(\sigma_0^2|\mathbf{y}_0, \mathbf{X}_0)d\sigma_0^2$$

$$\text{pr}(\beta_1|\mathbf{y}_1, \mathbf{X}_1) = \int \text{pr}(\beta_1|\mathbf{y}_1, \mathbf{X}_1, \sigma_1^2)\text{pr}(\sigma_1^2|\mathbf{y}_1, \mathbf{X}_1)d\sigma_1^2$$

Thus, we can draw from the posterior distributions of σ_0^2 and σ_1^2 shown in (10.17)-(10.21), and use the generated values to draw from the conditional distributions of $\beta_0|\sigma_0^2$ and $\beta_1|\sigma_1^2$ shown in (10.12). These calculations, using the function Bayes_normal(), are shown below. First, we split the data frame SRV into two sub-data frames for the treated and control groups, and set the hyperparameters $\mu_0 = \mu_1 = (0, 0, 0, 0, 0)$ and the matrix $\Omega = \mathbf{I}_5$, the identity matrix of dimension 5.

```
-------------------------------------------------------------------------
SRV1 = SRV[SRV$W==1,]
SRV0 = SRV[SRV$W==0,]
y1=SRV1$y
X1=cbind(rep(1,10),as.matrix(SRV1[,2:5]))
y0=SRV0$y
X0=cbind(rep(1,10),as.matrix(SRV0[,2:5]))
mu = c(rep(0,5))
Omega = as.matrix(diag(5))
-------------------------------------------------------------------------
```

We take $a = 2$ and $b = 25$ as the parameters of the inverse-gamma distribution for the variance parameters σ_0^2 and σ_1^2. This prior distribution has a mean of 25 (which is close to the observed mean) and effectively an infinite variance, and thus is a diffuse prior. Figure 10.1 shows a histogram obtained from 10,000 draws of an inverse-gamma distribution with paramaters $a = 2$ and $b = 25$.

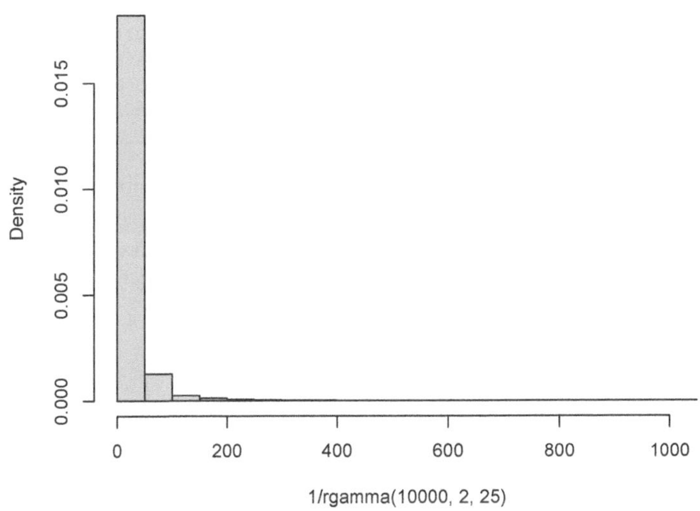

FIGURE 10.1
Histogram of 10000 random draws from inv-gamma(2,25)

```
-------------------------------------------------------------------------
# Posterior distributions of beta0, beta1
> set.seed(2222)
> hist(1/rgamma(10000,2,25),main=NULL, freq=FALSE)
-------------------------------------------------------------------------
```

Next, we generate 2000 draws from the posterior distributions of β_0 and β_1. Note that both $\beta_0 = (\beta_{00}, \beta_{01}, \beta_{02}, \beta_{03}, \beta_{04})$ and $\beta_1 = (\beta_{10}, \beta_{11}, \beta_{12}, \beta_{13}, \beta_{14})$ are five dimensional vectors, where β_{00} and β_{10} represent the intercepts for the two groups and the remaining parameters represent the coefficients of $X_1, \ldots, X_4$ in the models for the potential outcomes. Thus these 2000 draws result in two 2000×5 matrices denoted by beta_chain_treatment and beta_chain_control for the treated and control groups respectively.

```
------------------------------------------------------------------------
# Posterior distributions of beta0, beta1

set.seed(999)
beta_chain_treatment = beta_chain_control = NULL
sigmasq0_chain = sigmasq1_chain = NULL
for (i in 1:5000){
  out1 = Bayes_normal(y=y1,X=X1,a=2, b=25, mu = mu,Omega=Omega)
  out0 = Bayes_normal(y=y0,X=X0,a=2, b=25, mu = mu,Omega=Omega)
  newbeta1 = as.numeric(out1[[1]])
  newbeta0 = as.numeric(out0[[1]])
  beta_chain_treatment = rbind(beta_chain_treatment, t(newbeta1))
  beta_chain_control = rbind(beta_chain_control, t(newbeta0))
  newsigmasq1 = out1[[2]][1]
  newsigmasq0 = out0[[2]][1]
  sigmasq1_chain = c(sigmasq1_chain, newsigmasq1)
  sigmasq0_chain = c(sigmasq0_chain, newsigmasq0)
}
beta_chain_treatment <- as.data.frame(beta_chain_treatment[3001:5000,])
names(beta_chain_treatment) = c("beta10","beta11","beta12","beta13","beta14")
beta_chain_control <- as.data.frame(beta_chain_control[3001:5000,])
names(beta_chain_control) = c("beta00","beta01","beta02","beta03","beta04")
------------------------------------------------------------------------
```

Histograms for the coefficients of the four covariates for both the treated and control groups obtained using the package Hmisc are shown in Figures 10.2 and 10.3 respectively. These figures show that the coefficients for X_2 and X_4 for both the treated and control groups are aposteriori distributed around zero, suggesting that these two covariates X_2 (likelit) and X_4 (collegegpa) do not help predict the outcome.

```
------------------------------------------------------------------------
library(Hmisc)
hist.data.frame(beta_chain_treatment[,2:5])
hist.data.frame(beta_chain_control[,2:5])
------------------------------------------------------------------------
```

To draw inference on the superpopulation estimand

$$\tau^{\mathrm{SP}}(\mathbf{x}_{\mathrm{new}}) = \mathbb{E}[Y_i(1)|\mathbf{X} = \mathbf{x}_{\mathrm{new}}] - \mathbb{E}[Y_i(1)|\mathbf{X} = \mathbf{x}_{\mathrm{new}}] = \mathbf{x}_{\mathrm{new}}^{\mathrm{T}}\beta_1 - \mathbf{x}_{\mathrm{new}}^{\mathrm{T}}\beta_0,$$

at $\mathbf{X} = \mathbf{x}_{\mathrm{new}}$, we can use the function est_taux() shown below, that generates posterior draws from β_1 and β_0, again using the function Bayes_normal(). The function takes arguments $\mathbf{y}_0$, $\mathbf{y}_1$, $\mathbf{X}_0$, $\mathbf{X}_1$, the inverse-gamma parameters a and b, the mean parameters μ_0 and μ_1, the correlation matrix Ω, the covariate vector $\mathbf{x}_{\mathrm{new}}$ and ITER, the number of values of $\tau^{\mathrm{SP}}(\mathbf{x}_{\mathrm{new}})$ to be generated.

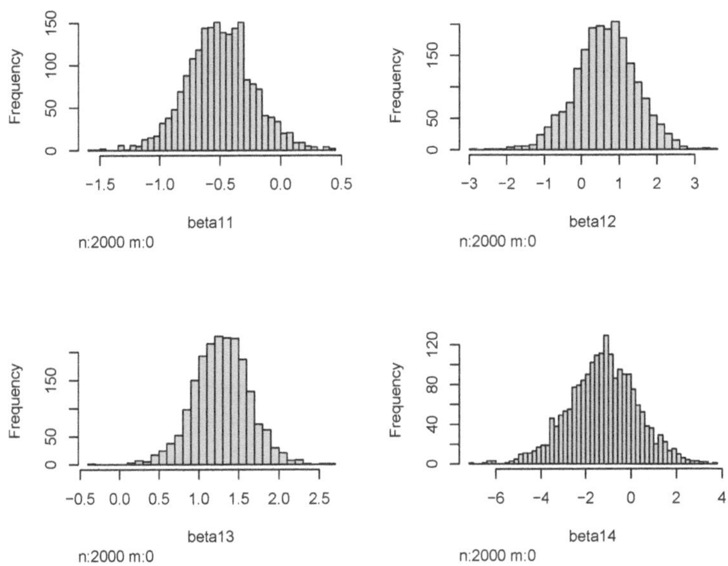

FIGURE 10.2
Histogram of 2000 posterior draws of model coefficients for the treated group

The function returns a vector of draws from the posterior distribution of $\tau^{\text{SP}}(\mathbf{x}_{\text{new}})$, the mean of these values, that can be considered an estimate of $\tau^{\text{SP}}(\mathbf{x}_{\text{new}})$ and a 95% posterior interval for $\tau^{\text{SP}}(\mathbf{x}_{\text{new}})$.

```
-----------------------------------------------------------------------------
est_taux = function(y0,y1,X1,X0,a,b,mu0,mu1,Omega,x_new) {
  tau_xnew_array = NULL
  for (i in 1:5000){
    out1 = Bayes_normal(y=y1,X=X1,a=a, b=b, mu = mu1,Omega=Omega)
    out0 = Bayes_normal(y=y0,X=X0,a=b, b=b, mu = mu0,Omega=Omega)
    newbeta1 = as.numeric(out1[[1]])
    newbeta0 = as.numeric(out0[[1]])
    tau_xnew = t(x_new)%*%newbeta1 - t(x_new)%*%newbeta0
    tau_xnew_array = c(tau_xnew_array, tau_xnew)
  }
  tau_xnew_hat = mean(tau_xnew_array)
  tau_xnew_CI = quantile(tau_xnew_array, probs=c(.025, .975))
  return(list(tau_xnew_array, tau_xnew_hat, tau_xnew_CI))
}
-----------------------------------------------------------------------------
```

Using the function est_taux with $\mathbf{x}_{\text{new}} = (1, 15, 7, 10, 3.50)^{\text{T}}$, the same mean vectors and covariance matrices for the prior distributions of the parameters, and $a = 2$, $b = 25$ as before, we obtain the posterior distribution of $\tau^{\text{SP}}(\mathbf{x}_{\text{new}})$ as shown in the left panel of Figure 10.4. The posterior mean

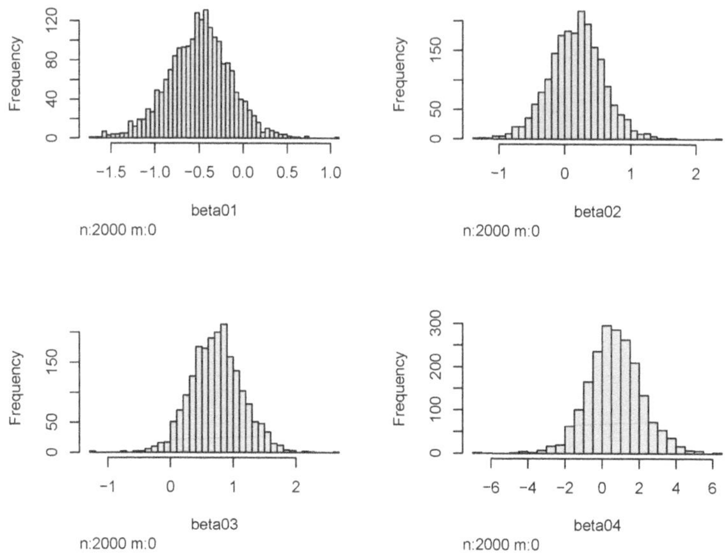

FIGURE 10.3
Histogram of 2000 posterior draws of model coefficients for the control group

is 1.55 and a 95% interval is $[-7.0, 10.4]$, suggesting that there is essentially no evidence that the average causal effect of the treatment for subjects in the population with $\mathbf{X} = \mathbf{x}_{\text{new}} = (1, 15, 7, 10, 3.50)^{\mathrm{T}}$ is different from zero. However, repeating this task with $\mathbf{x}_{\text{new}} = (1, 21, 5, 22, 3.08)^{\mathrm{T}}$ yields the histogram in the right panel of Figure 10.4. The posterior mean of 7.69 and a 95% interval is $[2.8, 12.6]$, suggesting with there is a treatment effect for subjects with $\mathbf{X} = \mathbf{x}_{\text{new}} = (1, 15, 7, 10, 3.50)^{\mathrm{T}}$. Note that the two covariate vectors mostly differ with respect to X_1 and X_3, the two covariates that were found to influence the outcome.

```
--------------------------------------------------------------------------
# Predicting tauSP(xnew) at (1,15,7,10,3.50)
set.seed(100)
output <- est_taux(y0,y1,X1,X0,a=2,b=25,mu0=mu, mu1=mu, Omega=Omega,
                   x_new = c(1,15,7,10,3.50))
> output[[2]]
[1] 1.549585
> output[[3]]
     2.5%     97.5%
-7.038542 10.416023
hist(output[[1]], main=NULL, xlab=expression(bar(tau)(x[new])), xlim=c(-20,20))
abline(v=output[[2]], lty=2, lwd=2)
abline(v=as.numeric(output[[3]]), lty=2, lwd=2)

# Predicting tauSP(xnew) at (1,21,5,22,3.08)
set.seed(200)
output <- est_taux(y0,y1,X1,X0,a=2,b=25,mu0=mu, mu1=mu, Omega=Omega,
```

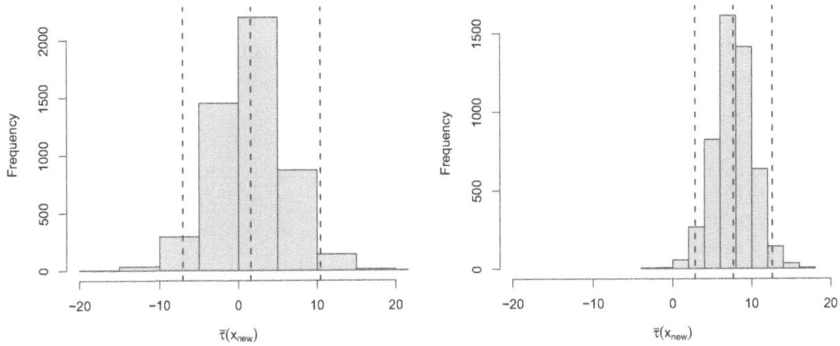

FIGURE 10.4
Histogram of the posterior distribution of $\tau^{\text{SP}}(\mathbf{x}_{\text{new}})$ for $\mathbf{x}_{\text{new}} = (1, 15, 7, 10, 3.50)^{\text{T}}$ (left) and $\mathbf{x}_{\text{new}} = (1, 21, 5, 22, 3.08)^{\text{T}}$ (right)

```
                    x_new = c(1,21,5,22,3.08))
[1] 7.692294
> output[[3]]
     2.5%     97.5%
 2.817786 12.586273
hist(output[[1]], main=NULL, xlab=expression(bar(tau)(x[new])), xlim=c(-20,20))
abline(v=output[[2]], lty=2, lwd=2)
abline(v=as.numeric(output[[3]]), lty=2, lwd=2)
-------------------------------------------------------------------------
```

Finally, to draw inference on the superpopulation estimand $\tau^{\text{SP}} = \mathbb{E}_{\theta_{Y|X}}[\tau^{\text{SP}}(X)]$, where $\mathbb{E}_{\theta_{Y|X}}$ denotes expectation with respect to the distribution of $\mathbf{X}$ governed by paramaters $\theta_{Y|X}$, we need to make assumptions about this distribution in the target superpopulation. In the current example, let us assume that $\mathbf{X}$ is multivariate normal with mean vector and covariance matrix equal to the sample mean vector and the sample covariance matrix of the covariates in Table 9.4, i.e., in the data frame SRV. This mean vector and the covariance matrices are shown below:

```
-------------------------------------------------------------------------
> (X_mu = apply(SRV[,2:5],2,mean))
     Vpre    likelit    actcomp collegegpa
   20.750      6.400     20.750      2.672
> (X_cov = cov(SRV[,2:5]))
                 Vpre     likelit    actcomp collegegpa
Vpre       16.4078947  0.47368421  8.3552632 1.76105263
likelit     0.4736842  2.14736842 -0.1578947 0.05021053
actcomp     8.3552632 -0.15789474  8.3026316 1.48684211
collegegpa  1.7610526  0.05021053  1.4868421 0.50309053
-------------------------------------------------------------------------
```

To conduct inference on τ^{SP}, we generate r draws $\mathbf{x}_1, \ldots, \mathbf{x}_r$ from the distribution of $\mathbf{X}$. For the ith draw, $i = 1, \ldots, r$, generate $\tau^{\text{SP}}(\mathbf{x}_{\text{new}})$ using the function est_taux() at $\mathbf{x}_{\text{new}} = (1, \mathbf{x}_i^{\text{T}})$, and finally obtain the point and interval estimates of the superpopulation causal effect as the mean and lower and upper quantiles of these r draws. Implementation of this procedure with $r = 100$ is shown below:

```
---------------------------------------------------------------------------
set.seed(1479)
tauSParray = NULL
for (i in 1:100) {
  library(MASS)
  x_new = c(1, mvrnorm(1,X_mu, X_cov))
  tauSP <- est_taux(y0,y1,X1=X1,X0=X0,a=2,b=25,mu0=mu, mu1=mu, Omega=Omega,
x_new = x_new)[[2]]
  tauSParray <- c(tauSParray, tauSP)
}
> (M = mean(tauSParray))
[1] 8.672671
> (Q = quantile(tauSParray, c(0.025,0.975)))
     2.5%     97.5%
 6.689789 10.887611
hist(tauSParray, main=NULL, xlab=expression(tau^SP))
abline(v=c(M,Q),lty=2, lwd=2)
---------------------------------------------------------------------------
```

The estimated superpopulation average causal effect is 8.67 with a 95% posterior interval [6.7, 10.9]. A histogram of the posterior distribution is shown in Figure 10.5. The result suggests that under the assumed model for the potential outcomes conditional on covariates and the model for the covariates, the treatment appears to have an average effect of 8.67 in the target superpopulation as defined by the assumed distribution of the covariates. Comparing this result with that obtained in Section 9.8, where a similar analysis of the same data was done without incorporating covariates into the model, yielding a substantially wider 95% posterior interval [3.5, 12.1] for τ^{SP}.

10.3.1 Inference for the finite population estimand

The function Ypot_imp_cov_normal() (shown below) imputes the missing potential outcomes for the model specified in (10.8)-(10.10). The function calls the Bayes_normal() function (introduced earlier) to generate draws $\tilde{\beta}_0$, $\tilde{\beta}_1$, $\tilde{\sigma}_0^2$ and $\tilde{\sigma}_1^2$ from the posterior distribution of parameters and imputes missing potential outcomes based on Procedure 10.1 using the following rule:

$$(Y_i^{\text{mis}} | Y_i^{\text{obs}}, W_i, \mathbf{X}_i, \tilde{\beta}_0, \tilde{\beta}_1, \tilde{\sigma}_0^2, \tilde{\sigma}_1^2) \sim \begin{cases} \mathcal{N}(\mathbf{X}_i\tilde{\beta}_0, \tilde{\sigma}_0^2) & \text{if} \quad W_i = 1 \\ \mathcal{N}(\mathbf{X}_i\tilde{\beta}_1, \tilde{\sigma}_1^2) & \text{if} \quad W_i = 0. \end{cases} \tag{10.22}$$

Among inputs to the function, the prior mean vectors μ_0 and μ_1 of β_0 and β_1 respectively are the same (specified as μ). Other inputs to the function include an $N \times 1$ vector of observed outcomes $\mathbf{y}$, an $N \times (p+1)$ matrix of covariates $\mathbf{X}$ (with the first column all ones), common inverse gamma parameters a, b for σ_0^2 and σ_1^2, and common correlation matrix Ω.

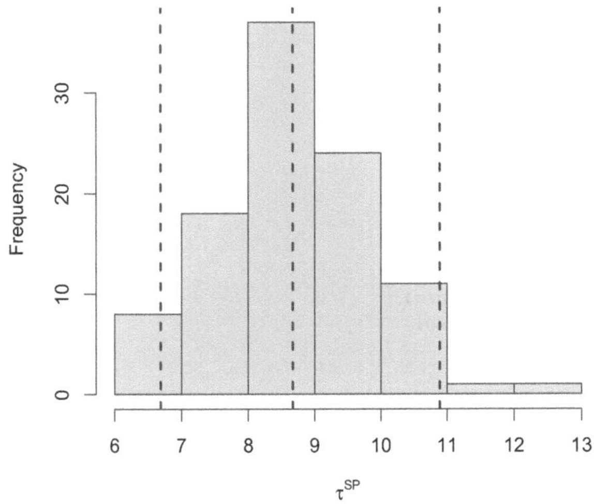

FIGURE 10.5
Histogram of the posterior distribution of τ^{SP} incorporating all four covariates

```
-----------------------------------------------------------------------
### function to impute missing outcomes with covariates (normal model)
## Inputs
## Nx1 vector y
## (p+1)x1 matrix X
## a, b (common inverse gamma parameters for variance)
## Omega: Common correlation matrix of beta0, beta1

# Outputs (as a list)
## One draw each from posterior beta0, beta1, sigma0, sigma1
## Imputed potential outcomes
## Estimated finite population ATE taubar

Ypot_imp_cov_normal <-
  function(y,W,X,a,b,mu0,mu1,Omega){
    Ypot = Nbytwo(y,W)
    Y0 = Ypot[,1]
    Y1 = Ypot[,2]
    N = length(y)
    N1 = sum(W)
    N0 = N - N1
    N = N0+N1
    y0 = Y0[W==0] # data for control group
    y1 = Y1[W==1] # data for treated group
    X0 = X[W==0,] # Xmatrix for control group
    X1 = X[W==1,] # X matrix for treated group

  # Draw from posteriors of beta and sigmasq separately for the two groups
  # using function Bayes_normal()
    newbeta0 = as.numeric(Bayes_normal(y0,X0,a,b,mu = mu0,Omega=Omega)[[1]])
```

```
    newsigmasq0 = as.numeric(Bayes_normal(y0,X0,a,b,mu = mu0,Omega=Omega)[[2]])
    newbeta1 = as.numeric(Bayes_normal(y1,X1,a,b,mu = mu1,Omega=Omega)[[1]])
    newsigmasq1 = as.numeric(Bayes_normal(y1,X1,a,b,mu = mu1,Omega=Omega)[[2]])

    for (i in 1:N) {
      if (W[i]==1) {
        m0 = X[i,] %*% newbeta0
        sd0 = sqrt(newsigmasq0)
        ymis = rnorm(1,m0,sd0)
        Ypot[i,1] = ymis
      } else {
        m1 = X[i,] %*% newbeta1
        sd1 = sqrt(newsigmasq1)
        ymis = rnorm(1,m1,sd1)
        Ypot[i,2] = ymis
      }
    }
    Ypot = round(Ypot)
    taubar = mean(Ypot[,2]) - mean(Ypot[,1])
    return(list(newbeta0, newbeta1, newsigmasq0, newsigmasq1, Ypot, taubar))
} # end function
------------------------------------------------------------------------
```

We impute the missing potential outcomes 1000 times and generate the distribution of $\overline{\tau}$ by repeated application of Procedure 10.1 with the imputation model (10.22) using the functions Ypot_imp_cov_normal().

```
------------------------------------------------------------------------
## Multiple imputations

set.seed(1101)
taubarvector <- NULL
for (i in 1:1000) {
newtaubar = Ypot_imp_cov_normal(y=y, W=W,X=X,a=2,b=25,mu0=mu,
mu1=mu,Omega=Omega)[[6]]
taubarvector = c(taubarvector,newtaubar)
}
> mean(taubarvector)
[1] 8.522
> c(quantile(taubarvector,.025), quantile(taubarvector, .975))
    2.5%    97.5%
 6.39875 10.85000
hist(taubarvector, main=NULL, xlab=expression(bar(tau)))
abline(v=mean(taubarvector), lty=2, lwd=1.5)
------------------------------------------------------------------------
```

The posterior distribution of the finite population estimand $\overline{\tau}$ is shown in Figure 10.6. The mean of the distribution is 8.52, and a symmetric 95% credible interval is $[6.40, 10.85]$. Recall that a 95% interval based on the model for independent normal potential outcomes without covariates was obtained as $[3.85, 11.80]$, which is wider than the interval obtained by incorporating covariates, suggesting that in this example, the covariates sharpened the inference, because at least two covariates (X_1 and X_3) helped predict the missing potential outcomes.

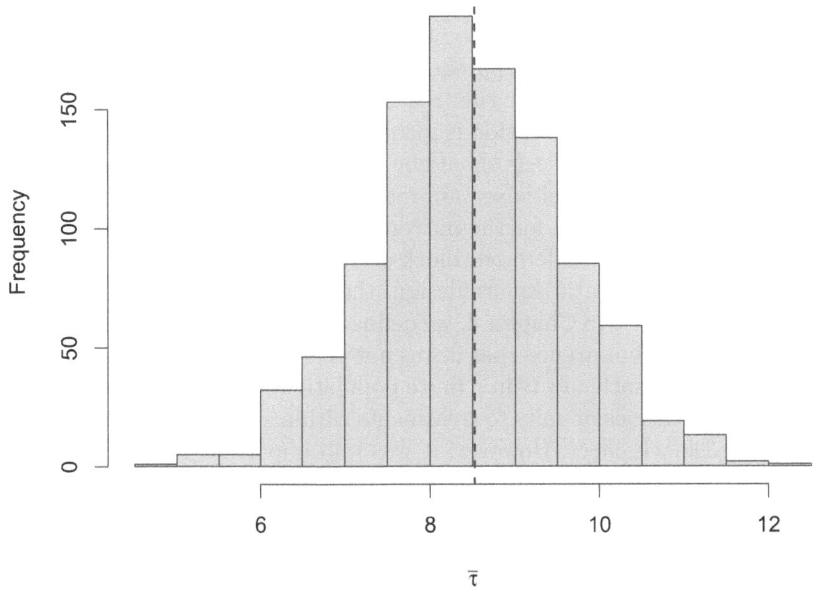

FIGURE 10.6
Histogram of $\overline{\tau}$ obtained from 1000 imputations of missing potential outcomes in Example 9.2 using Procedure 10.1 under models (10.8)–(10.10)

10.4 Role of design in Bayesian inference

In Section 9.5, we discussed how the randomized assignment mechanism played a key role in the developments of Chapters 8 and 9, and specifically in the derivation of the likelihood function. In (9.8), we established that

$$\mathrm{pr}(\mathbf{Y}^{\mathrm{obs}}, \mathbf{W}|\theta) \propto \mathrm{pr}(\mathbf{Y}^{\mathrm{obs}}|\mathbf{W}, \theta) \propto \prod_{i:w_i=1} \mathrm{pr}(Y_i(1)|\theta) \prod_{i:w_i=0} \mathrm{pr}(Y_i(0)|\theta),$$

a representation that was possible by the independence of the assignment variable $\mathbf{W}$ and the potential outcomes $(\mathbf{Y}(0), \mathbf{Y}(1))$. This independence allowed us to "remove" $\mathbf{W}$ from the subsequent inferential steps, except for its role in identifying which potential outcomes are missing.

This property (independence of assignment vector with potential outcomes) of an assignment mechanism is known as "ignorability" (Rubin 1976, 1978 [77, 79]), and is one of the most important conditions required to draw valid causal inference. Ignorability is automatically satisfied in randomized experiments, irrespective of whether we use a completely randomized design, or a randomized block design, or designs in which allocations are made on the basis of a pre-defined criterion of covariate imbalance - explored and expounded in Chapter 4. Unfortunately, this ubiquity of ignorability holding in randomized experiments often leads to the perception that as long as the design is randomized, the model-based approach to causal inference would essentially be the same irrespective of how the treatment allocations are made.

We believe that this perception is incorrect. The foundation of the model-based approach, *the model itself*, should be guided by the design. This principle is consistent with the model-based approach in classical experimental design literature, where the model for the outcomes incorporates features of the design. For example, in a randomized block design, the outcome model contains a "block effect". In a Latin square design, the model contains a "row effect" and a "column effect". In Chapter 3, we defined such block effects based on the potential outcomes but argued that it was not possible to draw randomization-based inference for such effects in a finite population setting, because different randomized allocations of units to treatments within a block would not change the estimated block effect. However, it was still important to recognize that the physical act of blocking would improve precision of treatment comparison within blocks, and lead to more efficient inference by eliminating noise due to variation of effects across heterogeneous blocks.

Following the same logic, the choice of a model in a model-based approach to causal inference should be closely tied to the choice of the design. Specifically, the way we incorporate covariates in the model should depend on the design.

10.5 Problems

10.1 A hypothetical LDL experiment was described in Chapter 3 where 12 units were divided into two blocks of size six each based on age, and two treatments (labeled 0 and 1) were assigned to these units using a balanced randomized block design. The covariate values (age) and observed outcomes for each unit are shown in Table 10.2

(i) Propose a suitable model for the potential outcomes consistent with the design that does not include the covariate age. Define the superpopulation estimand τ^{SP} in terms of model parameters.

(ii) Obtain the posterior distribution of τ^{SP} and the finite population ATE $\overline{\tau}$ under the model proposed in (i) and find their point and interval estimates.

(iii) Now propose a suitable model that includes the covariate age and re analyze the data to find the posterior distribution, point and interval estimates of the conditional average treatment effect $\mathbb{E}[Y(1) - Y(0) | X \geq 60]$ and $\overline{\tau}$.

TABLE 10.2
Observed and missing potential outcomes from hypothetical LDL experiment

Block (Age)	Unit (i)	Age (X_i)	$Y_i(0)$	$Y_i(1)$	W_i	y_i
	1	20	?	97	1	97
	2	25	99	?	0	99
	3	30	?	97	1	97
$1(\leq 45)$	4	35	97	?	0	97
	5	40	102	?	0	102
	6	45	?	99	1	99
	7	50	?	97	1	97
	8	55	?	98	1	98
	9	60	100	?	0	100
$2\ (> 45)$	10	65	99	?	0	99
	11	70	97	?	0	97
	12	75	?	101	1	101

10.2 Recall the hypothetical simplified stem cell experiment discussed in Chapter 4. The experimental units comprised 12 wells arranged in a rectangular 3×4 array, and two treatments (chemical modulators) labeled 0 and 1 were assigned to these 12 treatments using a rerandomized design with the following acceptance criterion:

$$|\overline{x}_1(1) - \overline{x}_1(0)| \leq 3/4, \quad |\overline{x}_2(1) - \overline{x}_2(0)| \leq 3/4,$$

where the covariates x_1 and x_2 represent the row number and column number of each cell. Propose a suitable model for the potential outcomes. Using the assignments and observed outcomes in Figure 4.5, obtain the posterior distribution, point and interval estimates of the finite population average treatment effect $\overline{\tau}$.

11

Appendix

11.1 Basic combinatorics

Combinatorics deals with the calculation of the number of ways in which a set of objects can be arranged, or chosen out of a collection of objects. For example, suppose we are interested in calculating the total number of ways in which two distinct alphabets can be chosen from four alphabets A, B, C and D. The first alphabet can be any one of the four, and thus can be chosen in four ways. Once the first alphabet is chosen as A or B or C or D, the second can be chosen from the remaining three. Note that for each choice of the first alphabet, there would be three choices of the second. Thus, the total number of choices are 4×3. However, these 12 choices will contain AB as well as BA, i.e., six pairs of alphabets, each occuring twice. Although AB and BA are viewed as different *arrangements*, they are essentially the same *choices*. Thus, the total number of choices of two out of four alphabets will be half of 12, i.e., 6.

The total number of ways in which N units can be allocated to J groups without any restriction on the number of units allocated to any group is J^N. This follows from the fact that the first unit can be allocated to any of the J groups. The second unit, the third unit, and proceeding this way, the N units can also be allocated to any of the J groups leading to $J \times J \times \cdots \times J = J^N$ total possibilities. This number is the total number of allocations possible under an independent Bernoulli treatment assignment of N units to J treatments.

The total number of arrangements of N objects chosen r at a time is known as the total number of *permutations* of N objects chosen r at a time, and is denoted as ${}^N P_r$. The total number of choices of N objects chosen r at a time is known as the total number of *combinations* of N objects chosen r at a time, and is denoted by ${}^N C_r$ or $\binom{N}{r}$. The following formulae are fundamental and useful:

$$ {}^n P_r = \frac{n!}{(n-r)!} = n(n-1)\ldots(n-r+1). \tag{11.1} $$

$$ {}^n C_r = \frac{n!}{k!(n-r)!} = \frac{n(n-1)\ldots(n-r+1)}{r!}. \tag{11.2} $$

DOI: 10.1201/9781003048794-11

Substitution of $r = N_0$ or $r = N_1$ in (11.2) gives the total number of allocations possible in a completely randomized experiments with N units, where N_0 units receive treatment 0 and the remaining $N - N_0 = N_1$ units receive treatment 1.

This formula can be generalized to find the number of ways to arrange N distinct units into J distinct groups such that exactly N_j units belong to group j. This number is

$$\binom{N}{N_1, N_2, \ldots, N_J} = \frac{N!}{N_1!N_2! \cdots N_J!} \tag{11.3}$$

where $N = N_1 + \cdots + N_J$ is the total number of units. This is the total number of allocations of N units to J treatments in a completely randomized design.

For a more detailed exposition on combinatorics and related topics, readers are referred to the following texts: Grimaldi (2003) [39] and Bóna (2016) [8].

11.2 Probability distributions and random variables

Useful references on this topic are Ross (2019) [74], Blitzstein and Hwang (2019) [7] and DeGroot and Schervish [24].

11.2.1 Classical definition of probability

If an experiment has N possible outcomes that are mutually exclusive (no two of them can occur simultaneously), exhaustive (no other outcomes beyond these outcomes are possible) and equally likely (none of these outcomes can be preferred over the other apriori), out of which n outcomes are favorable to an event A, then the probability of the event A is

$$\Pr(A) = \frac{n}{N}.$$

As an example, consider the event A of getting an even number in a single throw of a fair die with six sides. Out of six possible outcomes $\{1,2,3,4,5,6\}$ that are mutually exclusive, exhaustive and equally likely, three ($\{2,4,6\}$) are favorable to event A. Thus $\Pr(A) = 3/6 = 1/2$.

11.2.2 Random variables and probability distributions

A random variable is a numerical quantity whose value depends on the outcome of a random experiment. A discrete random variable can only take countable values (e.g., 0,1,2, ...), whereas a continuous variable can take any value within a given range.

Example 11.1 (Discrete random variable) *Let X be the number of heads when flipping a fair coin twice. The set of possible outcomes of this random experiment is $\{HH, HT, TH, TT\}$ where H denotes head and T denotes tail. The random variable X can take values 2,1 and 0 with respective probabilities 1/4, 1/2 and 1/4.*

The probability distribution of a discrete random variable X is described by a probability mass function (pmf) that gives $\Pr(X = x)$, i.e., the random variable X takes a specific value x. In example 11.1, the pmf of X is shown in Table 11.1. Any pmf should satisfy the following two properties: (i) $f(x) \geq 0$ for all x and (ii) $\sum_x f(x)dx = 1$, where the sum of over all possible values x of X.

TABLE 11.1
PMF of X in Example 11.1

x	0	1	2	Total
$\Pr(X = x)$	1/4	1/2	1/4	1

The probability distribution of a continuous random variable X is described by a probability density function (pdf) $f(x)$ that gives the probability that X lies in the interval $[a, b]$ when integrated over this interval, i.e.,

$$\Pr(a \leq X \leq b) = \int_a^b f(x)dx.$$

Any pdf should satisfy the following two properties: (i) $f(x) \geq 0$ for all x and (ii) $\int_{\mathcal{R}} f(x)dx = 1$, where $\mathcal{R}$ denotes the range of X.

Example 11.2 (PDF of a continuous random variable) *Waiting times of events, e.g., time before a household appliance fails, are often characterized by the exponential distribution defined as follows: Let X be a continuous random variable that can take only non-negative values, and has a pdf $f(x) = \lambda e^{-\lambda x}$ for $x \geq 0$. The constant $\lambda > 0$ is called the rate parameter.*

11.2.3 Expectation, variance and covariance

The expectation or mean of a discrete random variable X that takes values $x_1, \ldots, x_k$ with probabilities $p_1, \ldots, p_k$ respectively is

$$\mathbb{E}[X] = \sum_{i=1}^{k} x_i p_i.$$

In example 11.1, from Table 11.1 it follows that

$$\mathbb{E}[X] = 0 \times 1/4 + 1 \times 1/2 + 2 \times 1/4 = 1.$$

The expectation of a continuous random variable X with a pdf $f(x)$ is

$$\mathbb{E}[X] = \int_{\mathcal{R}} x f(x) dx,$$

where $\mathcal{R}$ denotes the range of X. In example 11.2,

$$\mathbb{E}[X] = \int_0^\infty x\lambda e^{-\lambda x} dx = 1/\lambda,$$

which means the expected or average life is $1/\lambda$.

The variance of a random variable is a measure of its dispersion and is defined as

$$\text{Var}[X] = \mathbb{E}\left[X - \mathbb{E}[X]\right]^2 = \mathbb{E}[X^2] - \mathbb{E}[X^2].$$

In example 11.1,

$$\mathbb{E}[X^2] = 0^2 \times 1/4 + 1^2 \times 1/2 + 2^2 \times 1/4 = 3/2.$$

Thus, $\text{Var}[X] = 3/2 - 1^2 = 1/2$.

In example 11.2,

$$\mathbb{E}[X^2] = \int_0^\infty x^2\lambda e^{-\lambda x} dx = 2/\lambda^2,$$

implying

$$\text{Var}[X] = 2/\lambda^2 - (1/\lambda)^2 = 1/\lambda^2.$$

The covariance between two random variables X and Y with a joint probability distribution function $f(x, y)$ is defined as

$$\text{Cov}(X, Y) = \mathbb{E}[X - \mathbb{E}(X)][Y - \mathbb{E}(Y)] = \mathbb{E}[XY] - \mathbb{E}[X]\mathbb{E}[Y].$$

where

$$\mathbb{E}[XY] = \begin{cases} \sum_x \sum_y xyf(x, y) & \text{if } X, Y \text{are discrete,} \\ \int\int xyf(x, y)dxdy & \text{if } X, Y \text{are continuous.} \end{cases}$$

Example 11.3 (Joint distribution and covariance) *The joint probability distribution $f(x, y) = Pr[X = x, Y = y]$ of two discrete random variables X and Y, each taking values 0 and 1, is shown in Table 11.2.*

In Example 11.3, $\mathbb{E}[X] = \mathbb{E}[X^2] = 0 \times 0.45 + 1 \times 0.55 = 0.55$ and $\mathbb{E}[Y] = \mathbb{E}[Y^2] = 0 \times 0.55 + 1 \times 0.45 = 0.45$. Thus, $\text{Var}[X] = 0.55 \times 0.55^2 = 0.55 \times 0.45 = 0.2475$ and $\text{Var}[Y] = 0.45 \times 0.55 = 0.2475$.

Next $\mathbb{E}[XY] = 0 \times 0 \times 0.25 + 0 \times 1 \times 0.20 + 1 \times 0 \times 0.30 + 1 \times 1 \times 0.25 = 0.25$.

Consequently, the covariance between X and Y is:

$$\text{Cov}(X, Y) = \mathbb{E}[XY] - \mathbb{E}[X]\mathbb{E}[Y] = 0.25 - 0.2475^2 = 0.1887.$$

TABLE 11.2
Joint probability distribution of (X, Y)

x	y		Total
	0	1	
0	0.25	0.20	0.45
1	0.30	0.25	0.55
Total	0.55	0.45	1

11.2.4 Some probability distributions used in this book

11.2.4.1 Bernoulli trial and Bernoulli distribution

A random experiment that has two possible outcomes, often labeled as 0 (failure) and 1 (success), is called a Bernoulli trial. A discrete random variable X is said to follow a Bernoulli distribution (notation: $X \sim \text{Bernoulli}(p)$) with parameter p if it can take two values - 0 and 1 - with probabilities p and $q = 1 - p$ respectively. The pmf of X is:

$$\Pr(X = x) = \begin{cases} p^x q^{1-x}, & x = 0, 1 \\ 0 & \text{otherwise.} \end{cases}$$

The number of heads in a single toss of a fair coin is a Bernoulli random variable with $p = 0.5$.

11.2.4.2 Binomial distribution

Let X be a random variable that denotes the number of ones or successes in n independent Bernoulli trials, each with the same probability of success p. Then X can take values $0, 1, \ldots, n$ and is said to follow as binomial distribution with parameters n and p. Notationally, $X \sim \text{Binomial}(n, p)$. The pmf of X is:

$$\Pr(X = x) = \begin{cases} \binom{n}{x} p^x (1-p)^{n-x}, & x = 0, 1, \ldots, n, \\ 0 & \text{otherwise.} \end{cases}$$

11.2.4.3 Beta distribution

A continuous random variable X with support over $(0, 1)$ is said to follow a Beta distribution with parameters α and β (notation: $X \sim \text{Beta}(\alpha, \beta)$) if it has probability density function:

$$f(x) = \begin{cases} \frac{1}{\text{Beta}(\alpha,\beta)} x^{\alpha-1}(1-x)^{\beta-1}, & 0 < x < 1, \alpha > 0, \beta > 0, \\ 0 & \text{otherwise.} \end{cases}$$

where $\text{Beta}(\alpha, \beta)$ is a normalizing constant (ensures that the density function integrates to one) called the Beta function, and is defined as:

$$\text{Beta}(\alpha, \beta) = \int_0^1 u^{\alpha-1}(1-u)^{\beta-1} du.$$

11.2.4.4 Gamma and inverse gamma distributions

A continuous non-negative random variable X is said to follow a Gamma distribution with parameters $\alpha > 0$ (shape) and $\theta > 0$ (scale) (notation: $X \sim \text{Gamma}(\alpha, \theta)$) if it has probability density function:

$$f(x) = \begin{cases} \frac{1}{\Gamma(\alpha)\theta^\alpha} x^{\alpha-1} e^{-x/\theta}, & 0 \leq x < \infty, \alpha > 0, \theta > 0, \\ 0 & \text{otherwise.} \end{cases}$$

where $\Gamma(\alpha)$ is a normalizing constant (ensures that the density function integrates to one) called the Gamma function, and is defined as:

$$\Gamma(\alpha) = \int_0^\infty u^{\alpha-1} e^{-u} du.$$

A continuous non-negative random variable X is said to follow an Inverse-gamma distribution (notation: $X \sim \text{IG}(\alpha, \theta)$) with parameters α and θ if its reciprocal $1/X$ follows a Gamma distribution with parameters α and θ.

11.2.4.5 Normal or Gaussian distribution

A random variable X has the normal or Gaussian distribution with parameters μ and σ^2 (notation: $X \sim \mathcal{N}(\mu, \sigma^2)$) if it has probability density function

$$f(x) = \frac{1}{\sqrt{2\pi}\sigma} e^{-(x-\mu)^2/2\sigma^2}, \quad -\infty < x < \infty.$$

The density function is bell shaped, symmetric, has expectation μ and variance σ^2.

11.2.4.6 Bivariate normal distribution

Random variables (X, Y) follow a bivariate normal distribution with parameters μ_X, μ_Y, σ_X^2, σ_Y^2, and ρ if their joint pdf is of the form:

$$f(x, y) = \frac{1}{2\pi\sigma_X\sigma_Y\sqrt{1-\rho^2}} e^{-\frac{1}{2(1-\rho^2}\left(\left(\frac{x-\mu_X}{\sigma_X^2}\right)^2 - 2\rho\left(\frac{x-\mu_X}{\sigma_X}\right)\left(\frac{y-\mu_Y}{\sigma_Y}\right) + \left(\frac{y-\mu_Y}{\sigma_Y^2}\right)^2\right)}.$$

The marginal distributions of X and Y are $\mathcal{N}(\mu_X, \sigma_X^2)$ and $\mathcal{N}(\mu_Y, \sigma_Y^2)$ respectively. Further, the conditional distribution of X given $Y = y$ is normal with mean $\mu_X + \rho(\sigma_X/\sigma_Y)(y - \mu_Y)$ and variance $\sigma_X^2(1 - \rho^2)$.

11.3 A Short Elementary Tutorial on Frequentist Statistical Inference

Frequentist statistical inference is a framework in which probability is interpreted as the long–run relative frequency of events. The goal is to use sample observations $Y_1, \ldots, Y_n$, often assumed to be independent and identical in distribution (iid) to learn about unknown population parameters θ such as a mean, variance, or proportion. Two central tasks in this framework are *point estimation* and *interval estimation*.

Point estimation: A **point estimator** is a summary of the observed sample data $Y_1, \ldots, Y_n$ that produces a single numerical value used to estimate an unknown parameter θ. An estimator of parameter θ is typically denoted by $\widehat{\theta}$. A specific value of the point estimator is called a *point estimate*.

Properties of estimators: Frequentist statistics evaluates estimators according to sampling behavior, i.e. how they behave over repeated samples. Following are some desired properties of good estimators $\widehat{\theta}$:

- **Unbiasedness:** $\hat{\theta}$ is unbiased for θ if $E(\hat{\theta}) = \theta$.
- **Consistency:** $\hat{\theta} \to \theta$ in probability as $n \to \infty$.
- **Efficiency:** among unbiased estimators, one with the smallest variance is preferred.

Example 11.4 *Let $Y_1, \ldots, Y_n$ be an iid random sample from a population with mean μ and variance σ^2. An unbiased estimator of μ is the sample mean:*

$$\widehat{\mu} = \overline{Y} = \frac{1}{n}\sum_{i=1}^{n} Y_i.$$

and an unbiased estimator of σ^2 is the sample variance

$$\widehat{\sigma^2} = s^2 = \frac{1}{n-1}\sum_{i=1}^{n}(Y_i - \overline{Y})^2.$$

Interval estimator: An **interval estimator** gives a range of plausible values for a parameter. A $(1-\alpha) \times 100\%$ **confidence interval** is a random interval constructed from observed data $Y_1, \ldots, Y_n$ that contains the true parameter with probability $1-\alpha$ under repeated sampling. In frequentist inference, a 95% confidence interval does not mean that there is a 95% probability the parameter lies in the interval. The parameter is fixed; the randomness is in the interval itself.

Example 11.5 *Suppose $X_1, \ldots, X_n$ are i.i.d. from a normal population, or n is large. A 95% confidence interval for μ is*

$$\bar{X} \pm t_{n-1,\,0.975}\frac{s}{\sqrt{n}},$$

where $\bar{X}$ is the sample mean, s is the sample standard deviation, and $t_{n-1,\,0.975}$ is the upper 0.975 quantile of the t distribution with $n-1$ degrees of freedom.

Readers are referred to Hogg and Craig (2019) [42] and Casella and Berger (2002) [15] for a formal exposition on frequentist statistical inference.

11.4 Bayesian inference

Unlike frequentist inference where the unknown parameter θ is viewed as a fixed quantity, in Bayesian inference θ is viewed as a random variable. Any early (before seeing the data) information available about the parameter is captured through a so-called "prior" distribution denoted by $\text{pr}(\theta)$. In a Bayesian setting, we denote the probability distribution function of random variable Y by $\text{pr}(y|\theta)$, to indicate that the model is conditional on the random variable θ. Let $\mathbf{y} = (y_1, \ldots, y_n)$ be the vector of n independent and identically distributed (iid) observations from $\text{pr}(y|\theta)$. Then $\mathbf{y}$ is the observed data from the model. The *likelihood function* $L(\theta|\mathbf{y})$ is the joint distribution $\text{pr}(\mathbf{y}|\theta)$ but regarded as a function of θ for fixed $\mathbf{y}$. By application of Bayes' theorem, the *posterior distribution* of θ given the data $\mathbf{y}$ is

$$\text{pr}(\theta|\mathbf{y}) = \frac{\text{pr}(\mathbf{y}|\theta)\ \text{pr}(\theta)}{\int \text{pr}(\mathbf{y}|\theta)\ \text{pr}(\theta) d\theta} \propto L(\theta|\mathbf{y})\text{pr}(\theta). \tag{11.4}$$

The denominator of (11.4) is known as the *normalizing constant.* Often, this normalizing constant is difficult to compute analytically, in which case writing

$$\text{pr}(\theta|\mathbf{y}) \propto \text{pr}(\mathbf{y}|\theta)\ \text{pr}(\theta). \tag{11.5}$$

typically suffices to make posterior inference. Sometimes, the exact form of the posterior distribution is directly recognizable from the right hand side of (11.5), and if not, the right hand side expression can be used to obtain draws of θ from the posterior distribution.

Inference about θ is made on the basis of the posterior distribution $\text{pr}(\theta|\mathbf{y})$. The observed data are expected to "shrink" the prior distribution, thereby leading to more precise inference about θ from the posterior distribution than from the prior distribution. Some of the issues related to Bayesian inference include: (i) how to choose prior distributions? (ii) how to obtain the posterior? (However, it may be noted that obtaining closed-form solution for posterior distribution is no longer a requirement due to the vast advancements of Markov chain Monte Carlo (MCMC) methods) , and (iii) how to use the posterior distribution to obtain point and interval estimators of θ?

11.4.1 Conjugate prior distributions

Loosely speaking, a prior distribution is conjugate to a certain statistical model if it yields the same form for the posterior distribution as for the prior distribution.

Example 11.6 (Beta-Binomial) *Let* $y \sim \text{Binomial}(n, \theta)$ *with* n *known. The probability model is* $pr(y|\theta) = \binom{n}{y}\theta^y(1-\theta)^{n-y}$. *Consider a* $\text{Beta}(\alpha, \beta)$

prior distribution on θ, that is, θ follows a Beta *distribution with parameters α and β and has probability density function:*

$$\pi(\theta) = \frac{1}{\mathrm{B}(\alpha, \beta)}\theta^{\alpha-1}(1-\theta)^{\beta-1}, \ 0 < \theta < 1, \alpha > 0, \beta > 0,$$

where $\mathrm{B}(\alpha, \beta)$ *is as defined in Section 11.2.4.3. Then, by (11.5), the posterior probability density function of θ is*

$$pr(\theta|y) \propto \mathrm{pr}(y|\theta)pr(\theta) \propto \theta^{y+\alpha-1}(1-\theta)^{n-y+\beta-1},$$

which is the unnormalized pdf of a $\mathrm{Beta}(y+\alpha, n-y+\beta)$ *distribution.*

Conjugate priors have certain advantages

(i) Posterior distributions are tractable and can be obtained in closed-form. This is, however, no longer a strong advantage due to the extraordinary development of MCMC methods.

(ii) Natural appeal: Prior and posterior distributions are alike, and this leads to a clear visualization of shrinkage.

(iii) Historical significance.

(iv) The mathematical theory for exponential family of distributions is elegant, leading to nice interpretations of the results.

11.4.2 Bayesian Point and Interval Estimation

Bayesian inference focuses on $\mathrm{pr}(\theta|y)$ instead of $\mathrm{pr}(y|\theta)$; that is, on the probability of parameters given data instead of the probability of data given parameters. Thus, a point estimator of θ is a single quantity that summarizes the center of the posterior distribution. The expectation or the mode of the posterior distribution are two popular and common choices of point estimators of θ.

A common Bayesian interval estimate is a *probability interval* (probability intervals are sometimes called by the short name "credible intervals" or the wordy name "Bayesian confidence intervals"; we prefer to simply call them "probability intervals"). This means that our interval includes θ with some fixed probability; however, this probability is over the posterior distribution of θ (holding y fixed), as opposed to the distribution of $Y|\theta$. Thus, although we still evaluate $P(\theta \in T(Y))$, this is now a direct probability statement about θ instead of observation Y. It should be noted that, although these intervals are constructed in a Bayesian way, they can still be evaluated over repeated data sets using frequentist criteria.

11.5 Simplified proof of Neyman's variance result (2.24)

First, we transform the binary assignment random variables $W_1, \ldots, W_N$ to centered random variables $D_1, \ldots, D_N$, where

$$D_i = W_i - \frac{N_1}{N} = \begin{cases} \frac{N_0}{N} & \text{if } W_i = 1, \\ -\frac{N_1}{N} & \text{if } W_i = 0. \end{cases}$$

This implies,

$$W_i = D_i + N_1/N, \quad 1 - W_i = N_0/N - D_i. \tag{11.6}$$

Then, from (2.5), for $i = 1, \ldots, N$

$$\mathbb{E}(D_i) = \mathbb{E}(W_i) - \frac{N_1}{N} = \frac{N_1}{N} - \frac{N_1}{N} = 0.$$

Also, from (2.6), for $i = 1, \ldots, N$

$$\text{Var}(D_i) = \text{Var}(W_i) = \frac{N_0 N_1}{N^2}. \tag{11.7}$$

For two distinct units i and i' $(i \neq i'$,

$$\mathbb{E}(W_i W_{i'}) = 1 \times \Pr(W_i = 1, W_{i'} = 1) = \frac{N_1(N_1 - 1)}{N(N-1)}. \tag{11.8}$$

Thus, for $i \neq i'$,

$$\begin{aligned} \mathbb{E}(D_i D_{i'}) &= \mathbb{E}\left(W_i - \frac{N_1}{N}\right)\left(W_{i'} - \frac{N_1}{N}\right) \\ &= \mathbb{E}\left(W_i W_{i'}\right) - \frac{N_1}{N}\mathbb{E}(W_{i'}) - \frac{N_1}{N}\mathbb{E}(W_i) + \frac{N_1^2}{N^2} \\ &= \mathbb{E}\left(W_i W_{i'}\right) - \frac{N_1}{N^2} - \frac{N_1}{N^2} + \frac{N_1}{N^2} \\ &= \frac{N_1(N_1-1)}{N(N-1)} - \frac{N_1}{N^2}, \text{ by (11.8)} \\ &= -\frac{N_0 N_1}{N^2(N-1)}. \end{aligned} \tag{11.9}$$

Now, the unbiased estimator of the average treatment effect $\overline{\tau}$ is:

$$
\begin{aligned}
\widehat{\overline{\tau}} &= \overline{y}(1) - \overline{y}(0) \\
&= \frac{1}{N_1}\sum_{i=1^N} W_i Y_i(1) - \frac{1}{N_0}\sum_{i=1}^{N}(1-W_i)Y_i(0), \\
&= \frac{1}{N}\left\{\frac{N}{N_1}\left(D_i + \frac{N_1}{N}\right)Y_i(1) - \frac{N}{N_0}\left(\frac{N_0}{N} - D_i\right)Y_i(0)\right\}, \text{ by (11.6)}, \\
&= \frac{1}{N}\sum_{i=1}^{N}\{Y_i(1) - Y_i(0)\} + \frac{1}{N}\sum_{i=1}^{N} D_i\left\{\frac{N}{N_i}Y_i(1) + \frac{N}{N_0}Y_i(0)\right\}, \\
&= \overline{\tau} + \frac{1}{N}\sum_{i=1}^{N} D_i Y_i^+, \qquad (11.10)
\end{aligned}
$$

where

$$
Y_i^+ = \frac{N}{N_1}Y_i(1) + \frac{N}{N_0}Y_i(0).
$$

Because $\mathbb{E}(D_i) = 0$ for $i = 1, \ldots, N$, from (11.10) it follows that $\mathbb{E}[\widehat{\overline{\tau}}] = \overline{\tau}$, i.e., $\widehat{\overline{\tau}}$ is an unbiased estimator of $\overline{\tau}$, and consequently,

$$
\begin{aligned}
&\mathrm{Var}[\widehat{\overline{\tau}}] \\
&= \frac{1}{N^2}\mathbb{E}\left[\sum_{i=1} D_i Y_i^+\right]^2 \\
&= \frac{1}{N^2}\mathbb{E}\left[\sum_{i=1}^{N} D_i^2 (Y_i^+)^2 + \sum\sum_{i\neq i'} D_i D_{i'} Y_i^+ Y_{i'}^+\right] \\
&= \frac{N_0 N_1}{N^4}\sum_{i=1}^{N}(Y_i^+)^2 - \frac{N_0 N_1}{N^4(N-1)}\sum\sum_{i\neq i'} Y_i^+ Y_{i'}^+ \text{ by (11.7), (11.9)} \\
&= \frac{N_0 N_1}{N^4}\sum_{i=1}^{N}(Y_i^+)^2 + \frac{N_0 N_1}{N^4(N-1)}\sum_{i=1}^{N}(Y_i^+)^2 - \frac{N_0 N_1}{N^4(N-1)}\sum_{i=1}^{N}\sum_{i'=1}^{N} Y_i^+ Y_{i'}^+ \\
&= \frac{N_0 N_1}{N^3(N-1)}\sum_{i=1}^{N}(Y_i^+)^2 - \frac{N_0 N_1}{N^4(N-1)}\sum_{i=1}^{N}\sum_{i'=1}^{N} Y_i^+ Y_{i'}^+ \\
&= \frac{N_0 N_1}{N^3(N-1)}\left(\sum_{i=1}^{N}(Y_i^+)^2 - \frac{1}{N}\sum_{i=1}^{N}\sum_{i'=1}^{N} Y_i^+ Y_{i'}^+\right) \\
&= \frac{N_0 N_1}{N^3(N-1)}\sum_{i=1}^{N}\left(Y_i^+ - \overline{Y}^+\right)^2, \qquad (11.11)
\end{aligned}
$$

where $\overline{Y}^+ = \frac{1}{N}\sum_{i=1}^{N} Y_i^+$.

The last step of (11.11) follows from the algebraic identity

$$\sum_{i=1}^{n}(x_i-\overline{x})^2=\sum_{i=1}^{n}x_i^2-n\overline{x}^2=\sum_{i=1}^{n}x_i^2-\frac{1}{n}\sum_{i=1}^{n}\sum_{i'=1}^{n}x_ix_{i'}.$$

Substituting $Y_i^+=\frac{N}{N_1}Y_i(1)+\frac{N}{N_0}Y_i(0)$ and $\overline{Y}^+=\frac{N}{N_1}\overline{Y}(1)+\frac{N}{N_0}\overline{Y}(0)$ in (11.11), we get

$$\begin{aligned}
&\mathrm{Var}[\widehat{\tau}]\\
=\ &\frac{N_0N_1}{N^3(N-1)}\sum_{i=1}^{N}\left(\frac{N}{N_1}\left(Y_i(1)-\overline{Y}(1)\right)-\frac{N}{N_0}\left(Y_i(0)-\overline{Y}(0)\right)\right)^2\\
=\ &\frac{N_0}{NN_1(N-1)}\sum_{i=1}^{N}\left(Y_i(1)-\overline{Y}(1)\right)^2\\
+\ &\frac{N_1}{NN_0(N-1)}\sum_{i=1}^{N}\left(Y_i(0)-\overline{Y}(0)\right)^2\\
+\ &\frac{2}{N(N-1)}\sum_{i=1}^{N}\left(Y_i(1)-\overline{Y}(1)\right)\left(Y_i(0)-\overline{Y}(0)\right)\\
=\ &\frac{N_0}{NN_1}S_1^2+\frac{N_1}{NN_0}S_0^2+\frac{2}{N}S_{10},
\end{aligned}\tag{11.12}$$

where S_1^2 and S_0^2 are the variances of potential outcomes under treatment and control respectively (with divisor $N-1$) as defined in (2.26) and (2.25), and S_{10} is the covariance term:

$$S_{10}=\frac{1}{N-1}\sum_{i=1}^{N}\left(Y_i(1)-\overline{Y}(1)\right)\left(Y_i(0)-\overline{Y}(0)\right).$$

From (2.27), S^2_{diff} , the variance of the N unit-level treatment effects $\tau_1,\ldots,\tau_N$ can be written as

$$S^2_{\text{diff}}=\frac{1}{N-1}\sum_{i=1}^{N}\left(\left(Y_i(1)-\overline{Y}(1)\right)-\left(Y_i(0)-\overline{Y}(0)\right)\right)^2=S_1^2+S_0^2-2S_{10},$$

which implies

$$2S_{10}=S_1^2+S_0^2-S^2_{\text{diff}}.\tag{11.13}$$

Finally, substitution of (11.13) into (11.12) yields

$$\begin{aligned}
\mathrm{Var}[\widehat{\tau}]&=\frac{N_0}{NN_1}S_1^2+\frac{N_1}{NN_0}S_0^2+\frac{1}{N}\left(S_1^2+S_0^2-S^2_{\text{diff}}\right)\\
&=\left(\frac{N_0+N_1}{NN_1}\right)S_1^2+\left(\frac{N_1+N_0}{NN_0}\right)S_0^2-\frac{1}{N}S^2_{\text{diff}}\\
&=\frac{S_1^2}{N1}+\frac{S_0^2}{N_0}-\frac{S^2_{\text{diff}}}{N},
\end{aligned}$$

thus proving (2.24).

Bibliography

[1] J. Angrist, D. Lang, and P. Oreopoulos. Incentives and services for college achievement: Evidence 155 from a randomized trial. *American Economic Journal: Applied Economics*, 1:136–163, 2009.

[2] A. T. Arnholt and B. Evans. *BSDA: Basic Statistics and Data Analysis*, 2017. R package version 1.2.0.

[3] A. Atkinson, A. Donev, and R. Tobias. *Optimum Experimental Designs, With SAS*. Oxford: Oxford University Press, 2007.

[4] S. Baert. Hiring Discrimination: An Overview of (Almost) All Correspondence Experiments Since 2005. In *Audit Studies: Behind the Scenes with Theory, Method, and Nuance*, pages 63–77. Springer International Publishing, 2018.

[5] R. A. Bailey. *Design of Comparative Experiments*. Cambridge: Cambridge University Press, 2008.

[6] M. Bertrand and S. Mullainathan. Are Emily and Greg more employable than Lakisha and Jamal? A field experiment on labor market discrimination. *The American Economic Review*, 94:991–1013, 2004.

[7] J. K. Blitzstein and J. Hwang. *Introduction to Probability*. Chapman and Hall/CRC, Boca Raton, 2 edition, 2019.

[8] M. Bóna. *A Walk Through Combinatorics*. World Scientific, Singapore, 4 edition, 2016.

[9] G. E. P. Box and J. S. Hunter. The 2^{k-p} fractional factorial designs (parts 1 and 2). *Technometrics*, 3:311–351, 449–458, 1961.

[10] G. E. P. Box, J. S. S. Hunter, and W. G. Hunter. *Statistics for Experimenters: Design, Innovation and Discovery (2nd ed)*. Hoboken, New Jersey: John Wiley & Sons, 2005.

[11] G. E. P. Box and R. D. Meyer. An analysis for unreplicated fractional factorials. *Technometrics*, 28:11–18, 1986.

[12] Z. Branson, T. Dasgupta, and D.B. Rubin. Improving Covariate Balance in 2^K factorial designs via Re-randomization. *The Annals of Applied Statistics*, 10:1958–1976, 2017.

[13] D. R. Brillinger, L. V. Jones, and J. W. Tukey. Report of the statistical task force for the weather modification advisory board. *The Management of Western Resources II: The Role of Statistics in Weather Resources Management*, 1978.

[14] G. Casella and R. L. Berger. *Statistical inference, 2nd ed.* Pacific Grove, CA, Thomson Learning, 2002.

[15] G. Casella and R. L. Berger. *Statistical Inference.* Duxbury Press, Pacific Grove, CA, 2 edition, 2002.

[16] S. Champely. *PairedData: Paired Data Analysis*, 2018. R package version 1.1.1.

[17] W. G. Cochran and G. M. Cox. *Experimental Designs.* John Wiley & Sons, Hoboken, NJ, 2nd edition, 1957.

[18] W.G. Cochran. *Sampling Techniques.* New York: Wiley, 1963.

[19] D. R. Cox. *Planning of Experiments.* London: Chapman & Hall, 1958.

[20] T. Dasgupta, N. Pillai, and D.R. Rubin. Causal Inference for 2^K factorial designs by using potential outcomes. *Journal of the Royal Statisical Society (Series B)*, 77(4):727–753, 2015.

[21] A. P. Dawid. Conditional independence in statistical theory. *Journal of the Royal Statistical Society. Series B)*, 41:1–31, 1979.

[22] B. de Finetti. Foresight: its logieal laws, its subjective sources. *Studies in Subjective Probability (H. E. Kyburg and H. E. Smokler, eds.)*, 1964.

[23] B. de Finetti. *Probability Induction and Statistics.* New York: Wiley, 1972.

[24] M. H. DeGroot and M. J. Schervish. *Probability and Statistics.* Pearson, Boston, 4 edition, 2012.

[25] P. Diaconis. Finite forms of de finetti's theorem on exchangeability. *Synthese*, 36:271–281, 1977.

[26] P. Ding. A paradox from randomization-based causal inference (with discussion). *Statistical Science*, 32(3):331–365, 2017.

[27] P. Ding and T. Dasgupta. A potential tale of two-by-two tables from completely randomized experiments. *Journal of the American Statistical Association*, 111(513):157–168, 2016.

[28] P. Ding and T. Dasgupta. A randomization-based perspective of analysis of variance. *Biometrika*, 105(1):45–56, 2018.

[29] G. D. Toit, G. Roberts, P. H. Sayre, H. T. Bahnson, S. Radulovic, A. F. Santos, H. A. Brough, D. Phippard, M. Basting, M. Feeney, V. Turcanu, M. L. Sever, M. G. Lorenzo, M. Plaut, and G. Lack. Randomized trial of peanut consumption in infants at risk for peanut allergy. *New England Journal of Medicine*, 372(9):803–813, 2015.

[30] O. J. Dunn. Multiple comparisons among means. *Journal of the American Statistical Association*, 56(293):52–64, 1961.

[31] V. Espinosa, T. Dasgupta, and D. B. Rubin. A bayesian perspective on the analysis of unreplicated factorial designs using potential outcomes. *Technometrics*, 58:62–73, 2015.

[32] R. A. Fisher. *Statistical methods for research workers.* Edinburgh: Oliver and Boyd, 1925.

[33] R. A. Fisher. *The Design of Experiments, 3rd ed.* New York: Hafner-Publishing, 1942.

[34] E. B. Foa, B. O. Rothbaum, D. R. Riggs, and T. B. Murdock. Treatment of Posttraumatic Stress Disoerder in Rape Victims: A Comparison Between Cognitive-Behavioral Procedures and Counseling. *Journal of Consulting and Clinical Psychology*, 59(5):715–723, 1991.

[35] R. G. (Jr.) Fryer and S. D. Levitt. The causes and consequences of distinctively black names. *The Quarterly Journal of Economics*, CXIX:767–805, 2004.

[36] G. L. Gadbury. Randomization inference and bias of standard errors. *The American Statistician*, 55:310–313, 2001.

[37] S. Michael Gaddis. *Audit Studies: Behind the Scenes with Theory, Method, and Nuance.* Springer International Publishing, Cham, 2018.

[38] A. Gelman, J.B. Carlin, H.S. Stern, and D.B. Rubin. *Bayesian Data Analysis (3rd ed.).* Chapman and Hall/CRC., 2013.

[39] R. P. Grimaldi. *Discrete and Combinatorial Mathematics: An Applied Introduction.* Pearson, Upper Saddle River, NJ, 5 edition, 2003.

[40] J. Hennessy, T. Dasgupta, L. Miratrix, C. W. Pattanayak, and P. Sarkar. A conditional randomization test to account for covariate imbalance in randomized experiments. *Journal of Causal Inference*, 4:61–80, 2015.

[41] K. Hinkelmann and O. Kempthorne. *Design and Analysis of Experiments, Volume 1: Introduction to Experimental Design 2nd Edition.* New York: Wiley, 2007.

[42] R. V. Hogg, Joseph W. McKean, and A. T. Craig. *Mathematical Statistics, eighth edition.* Boston : Pearson, 2019.

[43] K. Imai. Variance identification and efficiency analysis in randomized experiments under the matched-pair design. *Statistics in Medicine*, 27(24):4857–4873, 2008.

[44] G. W. Imbens and D. B. Rubin. *Causal Inference in Statistics and Social Sciences.* Cambridge: Cambridge University Press, 2015.

[45] Y. Jin and S. Ba. Toward optimal variance reduction in online controlled experiments. *Technometrics*, 65(2):231–242, 2023.

[46] O. Kempthorne. *The Design and Analysis of Experiments.* London: Chapman & Hall, 1952.

[47] O. Kempthorne. The randomization theory of experimental inference. *Journal of the American Statistical Association*, 50:946–967, 1955.

[48] R E Kirk. *Experimental Design: Procedures for the Behavioral Sciences.* Brooks/Cole, Monterey, CA, 1982.

[49] L. J. Kitchens. *Basic Statistics and Data Analysis.* Pacific Grove, CA: Brooks/Cole, 2003.

[50] W. H. Kruskal and W. A. Wallis. Use of ranks in one-criterion variance analysis. *Journal of the American Statistical Association*, 47(260):583—-621, 1952.

[51] X. Li and P. Ding. General forms of finite population central limit theorems with applications to causal inference. *Journal of the American Statistical Association*, 112:1759–1769, 2017.

[52] X. Li, P. Ding, and D. B. Rubin. Rerandomization in 2^k factorial experiments. *The Annals of Statistics*, 48:43–63, 2020.

[53] B. Libgober. Getting a lawyer while black: A field experiment. *Lewis & Clark L. Rev.*, 24:53–108, 2020.

[54] J. Lu. On randomization-based and regression-based inferences for 2^k factorial designs. *Statistics and Probability Letters*, 112:72–78, 2016.

[55] X. Luo, T. Dasgupta, M. Xie, and R. Y. Liu. Leveraging the Fisher Randomization Test using Confidence Distributions: Inference, Combination and Fusion Learning. *Journal of the Royal Statistical Society Series B: Statistical Methodology*, 83(4):777–797, 2021.

[56] R. M. Lyle, C. L. Melby, G. C. Hyner, J. W. Edmondson, J. Z. Miller, and M. H. Weinberger. Blood Pressure and Metabolic Effects of Calcium Supplementation in Normotensive White and Black Men. *The Journal of the American Medical Association*, 257(13):1772–1776, 2017.

[57] P. Macwilliams and Y.L. Lynfield. Shaving and hair growth. *Journal of Investigative Dermatology*, 55(3):170–172, 1970.

[58] P. C. Mahalanobis. On the generalised distance in statistics. *Proceedings of the National Institute of Sciences of India*, 2(1):49–55, 1936.

[59] M. D. McKay, R. J. Beckman, and W. J. Conover. A Comparison of Three Methods for Selecting Values of Input Variables in the Analysis of Output from a Computer Code. *Technometrics*, 21(2):239–245, 1979.

[60] D. C. Montgomery. *Design and analysis of experiments.* John Wiley & Sons, Inc., 2005.

[61] K. L. Morgan and D. B. Rubin. Rerandomization to improve covariate balance in experiments. *The Annals of Statistics*, 40:1263–1282, 2012.

[62] K. L. Morgan. Rerandomization to Improve Covariate Balance in Randomized Experiments. Doctoral dissertation, Harvard University, Cambridge, MA, 2013.

[63] R. Mukerjee and T. Dasgupta. Causal inference from possibly unbalanced split-plot designs: a randomization-based perspective. *Statistica Sinica*, 32:591–612, 2022.

[64] R. Mukerjee and C. F. J. Wu. *A Modern Theory of Factorial Designs.* New York: Springer, 2006.

[65] J. Neyman. On the Application of Probability Theory to Agricultural Experiments. Essay on Principles. Section 9. *Statistical Science*, 5(4):465–472, 1923/1990.

[66] J. Neyman. Statistical problems in agricultural experimentation (with discussion). *Supplement to the Journal of the Royal Statistical Society*, 2:107–180, 1935.

[67] N. Pashley, B. Libgober, and T. Dasgupta. Analysis and sample-size determination for 2^K audit experiments with binary response and application to identification of effect of racial discrimination on access to justice, https://arxiv.org/abs/2503.10591v1 2025.

[68] N. E. Pashley and M. A. C. Bind. Causal inference for multiple treatments using fractional factorial designs. *Canadian Journal of Statistics*, 51(2):444–468, 2023.

[69] E.J. Pitman. Significance tests which may be applied to samples from any populations: Iii. the analysis of variance test. *Biometrika*, 29:322–335, 1938.

[70] R. M. Pruzek and J. E. Helmreich. Enhancing dependent sample analyses with graphics. *Journal of Statistics Education*, 17(1), 2009.

[71] T. Dasgupta R. Mukerjee and D. B. Rubin. Using standard tools from finite population sampling to improve causal inference for complex experiments. *Journal of the American Statistical Association*, 113(522):868–881, 2018.

[72] A. Ravichandran, N. E. Pashley, B. Libgober, and T. Dasgupta. Optimal allocation of sample size for randomization-based inference from 2^K factorial designs. *Journal of Causal Inference*, 12(1):20230046, 2024.

[73] R. Rosenthal and D. B. Rubin. Ensemble-Adjusted p-values. *Psychological Bulletin*, 94:540–541, 1983.

[74] S. M. Ross. *A First Course in Probability.* Pearson, Boston, 10 edition, 2019.

[75] D. B. Rubin. Estimating Causal Effects of Treatments in Randomized and Nonrandomized Studies. *Journal of Educational Psychology*, 66(5):688 – 701, 1974.

[76] D. B. Rubin. Bayesian inference for causality: The importance of randomization. In *Proceedings of the Social Statistics Section of the American Statistical Association*, pages 233–239, 1975.

[77] D. B. Rubin. Inference and Missing Data. *Biometrika*, 63(3):581–592, 1976.

[78] D. B. Rubin. Assignment to Treatment Group on the Basis of a Covariate. *Journal of Educational Statistics*, 2:1–26, 1977.

[79] D. B. Rubin. Bayesian Inference for Causal Effects: The Role of Randomization. *The Annals of Statistics*, 6:34–58, 1978.

[80] D. B. Rubin. Multiple Imputations in Sample Surveys - A Phenomenological Bayesian Approach to Nonresponse. *Proceedings of the Survey Research Methods Section, American Statistical Association*, pages 10–28, 1978.

[81] D. B. Rubin. Randomization Analysis of Experimental Data: The Fisher Randomization Test (Comment). *Journal of the American Statistical Association*, 75:591–593, 1980.

[82] D. B. Rubin. Comment: The design and analysis of gold standard randomized experiments. *Journal of the American Statistical Association*, 103(484):1350–1353, 2008.

[83] D. B. Rubin. For objective causal inference, design trumps analysis. *Annals of Applied Statistics*, 2(3):808–840, 2008.

[84] D. B. Rubin. Reflections Stimulated by the Comments of Shadish (2010) and West and Thoemmes (2010). *Psychological Methods*, 15:38–40, 2010.

[85] W.R. Shadish, M.H. Clark, and P.M. Steiner. Can nonrandomized experiments yield accurate answers? a randomized experiment comparing random and nonrandom assignments. *Journal of the American Statistical Association*, 103:1334–1343, 2008.

[86] R. M. Shiffrin. Drawing causal inference from Big Data. *PNAS*, pages 7308–7309, 2016.

[87] S. S. Shrikhande. Designs for two-way elimination of heterogeneity. *Annals of Mathematical Statistics*, 22:235–247, 1951.

[88] C. A. B. Smith and H. O. Hartley. The construction of youden squares. *Journal of the Royal Statistical Society, Series B.*, 10:262–263, 1948.

[89] W. Snedecor and W. Cochran. *Statistical methods.* Iowa State University Press, Ames, IA, 1980.

[90] Student. The probable error of a mean. *Biometrika*, 6(1):1–25, 1908.

[91] B. Welch. On the z-test in randomized blocks and latin squares. *Biometrika*, 29:21–52, 1937.

[92] B. Wheeler. *AlgDesign: Algorithmic Experimental Design*, 2019. R package version 1.2.0.

[93] C. F. J. Wu and M. S. Hamada. *Experiments: Planning, Analysis, and Optimization.* New York:Wiley, 2009.

[94] W. J. Youden. Experimental designs to increase accuracy of green-house studies. *Cont. Boyce Thompson Inst.*, 11:219–228, 1940.

[95] W. J. Youden and H. P. Beale. A statistical study of the local lesion method for estimating tobacco mosaic virus. *In Contributions from Boyce Thompson Institute*, 6:437, 1934.

Index

Note: Page numbers followed by f, and t indicate figures and tables, respectively.

Experimental unit	Treatment 0	Treatment 1	Causal effects
1	$Y_1(0)$	$Y_1(1)$	$\tau_1 = Y_1(1) - Y_1(0)$
2	$Y_2(0)$	$Y_2(1)$	$\tau_2 = Y_2(1) - Y_2(0)$
...	...	...	...
N	$Y_N(0)$	$Y_N(1)$	$\tau_N = Y_N(1) - Y_N(0)$
Average	$\overline{Y}(0)$	$\overline{Y}(1)$	$\overline{Y}(1) - \overline{Y}(1)$
Median	$\operatorname{Med}_i[Y_i(0)]$	$\operatorname{Med}_i[Y_i(1)]$	$\operatorname{Med}_i(\tau_i)$
Variance	$\frac{\sum_{i=1}^N (Y_i(0)-\overline{Y}(0))^2}{N-1}$	$\frac{\sum_{i=1}^N (Y_i(1)-\overline{Y}(1))^2}{N-1}$	$\frac{\sum_{i=1}^N (\tau_i-\overline{\tau})^2}{N-1}$
Range	$\max_i[Y_i(0)] - \min_i[Y_i(1)]$	$\max_i[Y_i(0)] - \min_i[Y_i(1)]$	$\max_i[\tau_i] - \min_i[\tau_i]$

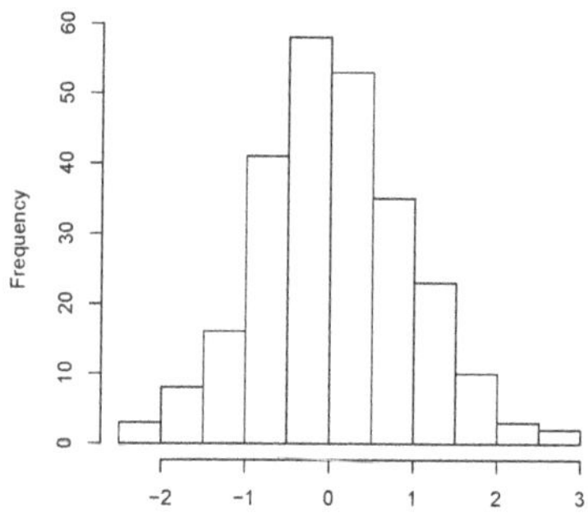

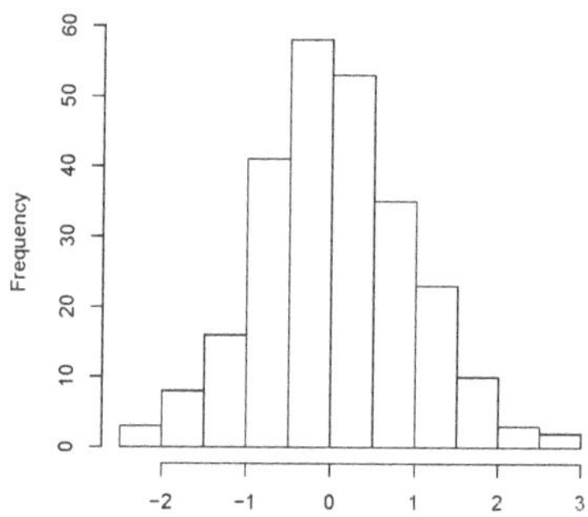

Block	Unit	Treatment 0	Treatment 1	Block average	Treatment effect Unit-level	Treatment effect Block-level
	1	$Y_1(0)$	$Y_1(1)$		τ_1	
1	...	...	...	$\overline{Y}_1(\cdot)$		$\overline{\tau}_1$
	M_1	$Y_{M_1}(0)$	$Y_{M_1}(1)$		τ_{M_1}	
...	...	...	...	...	...	...
...	...	...	...	...	...	...
...	...	...	...	...	...	...
	$N-M_B+1$	$Y_{N-M_B+1}(0)$	$Y_{N-M_B+1}(1)$		τ_{N-M_B+1}	
B	...	...	...	$\overline{Y}_B(\cdot)$		$\overline{\tau}_B$
	N	$Y_N(0)$	$Y_N(1)$		τ_N	
Treatment average		$\overline{Y}.(0)$	$\overline{Y}.(1)$	Grand avg $\overline{Y}.(\cdot)$	Simple avg $\overline{\tau}$	Wghtd avg $\overline{\tau}$

(ii) Calculate Δ for the two worst assignment vectors $(W_1, \ldots, W_{12})$ under the balanced CRD.

(iii) Write down two assignment vectors $(W_1, \ldots, W_{12})$, other than the worst ones, that are possible under the balanced CRD but not under a balanced RBD with 6 units per block and 3 units assigned at random to each treatment within each block, and calculate the Δ for each of these assignments.

(iv) What is the worst possible assignment vector under the RBD? Calculate Δ for this assignment vector. If there are multiple such "worst" vectors that are equally bad, write all of them down and calculate Δ for each.

3.4 Let $x_1, \ldots, x_N$ denote the pre-measure values of a covariate x for N experimental units. For a given assignment vector $(W_1, \ldots, W_N)$ generated by a completely randomized design with N_1 and N_0 units assigned to treatments labeled 1 and 0 respectively, let

$$\overline{x}(1) = \frac{1}{N_1} \sum_{i=1}^{N} W_i x_i$$

$$\overline{x}(0) = \frac{1}{N_0} \sum_{i=1}^{N} (1 - W_i) x_i$$

respectively denote the average covariate value in the two groups assigned to treatments 1 and 0. We define $\overline{x}(1) - \overline{x}(0)$ as the measure of balance for the assignment vector $(W_1, \ldots, W_N)$. Prove that $\mathbb{E}\left[\overline{x}(1) - \overline{x}(0)\right] = 0$.

3.5 What is the total number of possible allocations of the 235 subjects into the two treatment groups in the partial Shadish experiment with a hypothetical randomized treatment assignment with sex as the blocking factor, where the allocation of units into the treated and control groups is predetermined according to Table 3.2. What percentage of all possible allocations in a completely randomized design would be discarded in this hypothetical experiment?

3.6 For the partial Shadish experiment with hypothetical blocking by sex, Figure 3.4 suggests that the treatment effects for females are distributed much more tightly compared to that for females due. Provide an explanation for this phenomenon.

3.7 Assess the plausibility of the sharp null hypothesis of zero treatment effect for *all* of the N experimental units in the hypothetical RBD version of partial Shadish experiment using (a) the $T_{\text{avg-unegvar}}$ statistic defined in (2.15) and (b) The $T_{\text{wt-avg-unegvar}}$ defined in (3.14).

3.8 In the hypothetical RBD version of partial Shadish experiment, consider testing whether the average treatment effects in the two blocks (female and male) are the same.

   (a) State the conditions under which the estimator defined in (3.21) is an unbiased estimator of the true sampling variance $\mathrm{Var}\left(\widehat{\overline{\tau}}_1 - \widehat{\overline{\tau}}_2\right)$, where $\widehat{\overline{\tau}}_b$ denotes the estimator of the treatment effect in block b. Justify your answer with proper mathematical arguments.

   (b) Use the test statistic defined in (3.22) to determine whether the treatment effect in the partial Shadish experiment differs between female and male subjects.

3.9 In a matched-pair design with B pairs indexed by $1, \ldots, B$ and $N = 2B$ units denoted by $i = 1, \ldots, N$, let $Y_i(0)$ and $Y_i(1)$ denote the potential outcomes of the ith unit when exposed to treatments 0 and 1 respectively. Let W_i denote the binary treatment assignment variable defined in Problem 3.1.

Calculate the following:

   (i) $\mathbb{E}(W_i)$ and $\mathrm{Var}(W_i)$ for $i \in$ block b and $b \in \{1, \ldots, B\}$.

$$v_a = \frac{\Pr(\chi_{k+2}^2 \leq a)}{\Pr(\chi_k^2 \leq a)}$$

$$H_{0\mathrm{N}} : \overline{\tau}(\cdot) = 0,$$

$$\begin{pmatrix} Y_i(0) \\ Y_i(1) \end{pmatrix} \Bigg| \theta \sim \mathcal{N}\left(\begin{pmatrix} \mu_0 \\ \mu_1 \end{pmatrix}, \begin{pmatrix} \sigma_0^2 & \rho\sigma_0\sigma_1 \\ \rho\sigma_0\sigma_1 & \sigma_1^2 \end{pmatrix} \right), \tag{9.23}$$

$$\begin{pmatrix} \mu_0 \\ \mu_1 \end{pmatrix} \Bigg| (m_0, m_1, \tau_0, \tau_1) \sim \mathcal{N}\left(\begin{pmatrix} m_0 \\ m_1 \end{pmatrix}, \begin{pmatrix} \tau_0^2 & 0 \\ 0 & \tau_1^2 \end{pmatrix} \right). \tag{9.24}$$

$$\begin{aligned} &\begin{pmatrix} \mu_0 \\ \mu_1 \end{pmatrix} \Bigg| (\mathbf{Y}^{\text{obs}}, \mathbf{W}) \\ \sim\ &\mathcal{N}\left(\begin{pmatrix} \delta_0\overline{y}(0) + (1-\delta_0)m_0 \\ \delta_1\overline{y}(1) + (1-\delta_1)m_1 \end{pmatrix}, \begin{pmatrix} \frac{1}{N_0/\sigma_0^2+1/\tau_0^2} & 0 \\ 0 & \frac{1}{N_1/\sigma_1^2+1/\tau_1^2} \end{pmatrix} \right). \end{aligned}$$

Sex	Control ($w = 0$)				Treatment ($w = 1$)			
	Exposed	success ($y = 1$)	failure ($y = 0$)	prop	Exposed	success ($y = 1$)	failure ($y = 0$)	prop
Female	4	3	1	0.75	4	4	0	1
Male	6	4	2	0.67	6	6	0	1
Total	$N_0 = 10$	$n_{01} = 7$	$n_{00} = 3$	0.70	$N_1 = 10$	$n_{11} = 10$	$n_{10} = 0$	1

For Product Safety Concerns and Information please contact our EU representative GPSR@taylorandfrancis.com Taylor & Francis Verlag GmbH, Kaufingerstraße 24, 80331 München, Germany